physica status solidi c

www.physica-status-solidi.com

conferences and critical reviews

Editor-in-Chief

Martin Stutzmann, Garching

Regional Editors

Martin S. Brandt, Garching
Peter Deák, Budapest
José Roberto Leite, Saõ Paulo
John I. B. Wilson, Edinburgh

Managing Editor

Stefan Hildebrandt, Berlin

Proceedings

10th Conference on
Hopping and Related Phenomena (HRP10)

Miramare–Trieste, Italy
1–4 September 2003

Guest Editors

Harald Böttger and Thomas Damker

1 · 1 · 2004

WILEY-VCH

physica status solidi (c) – conferences and critical reviews

Editor-in-Chief:	Martin Stutzmann
Managing Editor:	Stefan Hildebrandt
Production Editors:	André Danelius, Heike Höpcke, Irina Juschak
Editorial Assistance:	Katharina Fröhlich, Margit Schütz
Editorial Office:	physica status solidi Bühringstr. 10, 13086 Berlin, Germany Telephone: +49 (0) 30/47 03 13 31, Fax +49 (0) 30/47 03 13 34 e-mail: pss@wiley-vch.de
Publishers:	WILEY-VCH Verlag GmbH & Co. KGaA
Postal Address:	Bühringstr. 10, 13086 Berlin, Germany
Publishing Director:	Alexander Grossmann
Ordering:	Subscription Service, WILEY-VCH Verlag GmbH & Co. KGaA Postfach 10 11 61, 69451 Weinheim, Germany Telephone +49 (0) 62 01/60 64 00, Fax +49 (0) 62 01/60 61 84 e-mail: subservice@wiley-vch.de or through a bookseller
Printing House:	Druckhaus Thomas Müntzer GmbH, Bad Langensalza, Germany Printed on chlorine- and acid free paper.

physica status solidi (c) – conferences and critical reviews is published twelve times per year by WILEY-VCH Verlag GmbH & Co. KGaA.

Annual subscription rates 2004 for pss (a) and (c) *or* pss (b) and pss (c):

		Institutional*	Personal
Europe	Euro	4244/3858	298
Switzerland	SFr	7269/6608	444
All other areas	US$	5406/4914	394

Annual subscription rates 2004 for pss (a), pss (b) and pss (c):

		Institutional*	Personal
Europe	Euro	8488/7716	596
Switzerland	SFr	14538/13216	888
All other areas	US$	10812/9828	788

* print **and** electronic/print only **or** electronic only
pss (a) and/or pss (b) subscriptions: Now including pss (c) – conferences and critical reviews

Postage and handling charges included. All WILEY-VCH prices are exclusive of VAT.
Prices are subject to change.

Single print issues may be ordered by ISBN at www.wiley-vch.de or through your local bookseller.

ISBN 3-527-40495-3

For our American customers:
physica status solidi (c) – conferences and critical reviews (ISSN 1610-1634) is published twelve times per year by WILEY-VCH Verlag GmbH & Co. KGaA, Boschstr. 12, 69469 Weinheim, Germany. Periodicals postage paid at Jamaica, NY 11431. Air freight and mailing in the USA by Publications Expediting Service Inc., 200 Meacham Ave., Elmont, NY 11003. US Postmaster: send address changes to: physica status solidi (c), c/o WILEY-VCH, 111 River Street, Hoboken, NJ 07030.

Contents

Visit our homepage on: http://www.physica-status-solidi.com
Full text on: http://www.interscience.wiley.com

This Table of Contents is organized according to the topics presented at the conference. Articles with page numbers marked (b) are reprinted from phys. stat. sol. (b) **241**, No. 1, 13–84 (2004). You may find papers with phys. stat. sol. (b) and phys. stat. sol. (c) citations in the two sections of this volume, separated by coloured sheets for easy orientation.
Note from the Publisher: This issue of physica status solidi (c) has been produced from publication-ready manuscript files, written by the authors using the provided Word or LaTeX templates.

Hopping

Optical properties

Low-dimensional systems

Organic systems and polymers

6   Contents

Charge carrier properties below and above the metal–insulator transition in conjugated polymers
– recent results

DOI: **The fastest way to find an article online** is the *Digital Object Identifier* (DOI).
Starting in Volume 198, issue 2 (January 2003), DOIs have been printed in the header of the first
page of every article in physica status solidi (b). On the WWW, one can find an article for example
with a DOI of 10.1002/pssa.200306608 at **http://dx.doi.org/**10.1002/pssa.200306608.

Please use the DOI of the article to link from your home page to the articles in Wiley Interscience.

The DOI is a result of a cross-publisher initiative to create a system for the persistent identification
of documents on digital networks. More information is available from **www.doi.org.**

Preface

The following papers were presented at the 10th Conference on Hopping and Related Phenomena (HRP10) held 1–4 September in Miramare–Trieste, Italy. The conference was organized by the International Centre for Theoretical Physics (ICTP). It was co-sponsored by the Deutsche Forschungsgemeinschaft.

The HRP10 is the 10th in a series of high-rank international conferences on transport and interactions in disordered systems, which took place in Trieste (1985), Bratislava (1987), Chapel Hill (1989), Marburg (1991), Glasgow (1993), Jerusalem (1995), Rackeve (1997), Murcia (1999), and Shefayim (2001).

The organizer of the first conference on hopping transport, at Trieste in 1985, was the late Paul N. Butcher, who at that time was one of the leading theoreticians in this field. Towards the end of the eighties, he successfully changed his interest to low-dimensional structures, which are now also an important topic of the HRP conferences. The HRP10 gave a suitable opportunity for commemoration of Paul N. Butcher and his work.

The HRP10 put some emphasis on the following topics: glassy relaxational behaviour of electronic systems, metal–insulator transition and glassy state in two-dimensional systems, noise, Coulomb glass, transport in biological systems and hopping transport.

The list of invited speakers included: N. P. Armitage, O. Bleibaum, V. Dobrosavljevic, A. L. Efros, V. Filinov, E. Frey, R. Haug, H. Kamimura, P. Kleinert, K. Kohary, S. Kravchenko, D. Kumar, B. Narozhny, Z. Ovadyahu, D. Popovic, V. M. Pudalov, M. Reznikov, A. K. Savchenko, W. Wu, A. G. Zabrodskii. In addition to these 20 invited talks the program encompassed 16 contributed talks and 40 posters.

Not every talk and poster given at the HRP10 is present with a paper in these proceedings. However, the reader should be able to come up with a general picture of the HRP10, which illustrates that transport and interactions in disordered systems (TIDS) is still an active field of research.

The next conference in this series will be held in the Netherlands in 2005, then under the new name TIDS-11.

Harald Böttger
Conference Chairman
November 2003

The Conference Proceedings are dedicated to the memory of
Professor Paul N. Butcher (1929–1999)

Committees

Organizer

Harald Böttger (Magdeburg)

International Advisory Committee

Zvi Ovadyahu (Jerusalem)
Michael Pepper (Cambridge)
Michael Pollak (Riverside)
Myriam Sarachik (New York)
Boris Shklovskii (Minneapolis)
Peter Thomas (Marburg)
Igor Zvyagin (Moscow)

Sponsors

Deutsche Forschungsgemeinschaft
The Abdus Salam International Centre for Theoretical Physics (ICTP), Trieste

physica status solidi **b**

www.physica-status-solidi.com

basic research

The following pages have been reprinted from
phys. stat. sol. (b) **241**, No. 1, 13–84 (2004) as part of the
Proceedings of the 10th Conference on
Hopping and Related Phenomena (HRP10),
held in Miramare–Trieste, Italy, 1–4 September 2003.

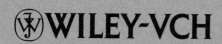

WILEY-VCH

phys. stat. sol. (b) **241**, No. 1, 13–19 (2004) / **DOI** 10.1002/pssb.200303606

Relaxation of non-equilibrium charge carriers in the hopping regime

O. Bleibaum[*,1], **E. Haba**[1], **H. Böttger**[1], and **V. V. Bryksin**[2]

[1] Institut für Theoretische Physik, Otto-von-Guericke Universität Magdeburg, Germany
[2] A. F. Ioffe Physico-Technical Institute, St. Petersburg, Russia

Received 1 September 2003, accepted 16 September 2003
Published online 15 December 2003

PACS 72.20Ee, 72.80.Ng, 72.80.Sk

We discuss a kinetic model for glass-like relaxation processes in a strongly localized electron system. In our hopping model we assume that the charge carriers can exchange maximally an amount ω of energy in one hop. The investigation of the system shows that the energy scale ω discriminates between two different relaxation modes. If the thermal energy is large compared to ω the relaxation is percolation like. In this regime the relaxation is determined by the same paths as the close-to-equilibrium conductivity in a transport problem. If the thermal energy is small compared to ω then the relaxation becomes Miller-Abraham like. In this relaxation mode the charge carriers always jump to the next closest site. The differences between these transport regimes manifest themselves in the typical time scales governing the relaxation and in the structure of the transport coefficients. To discuss them we focus on energy relaxation processes.

1 Introduction Investigations on glass transitions and glassy relaxation processes are of much current interest. At present particular attention is paid to investigations on glassy relaxation processes of non-equilibrium charge carriers in electron glasses, in which transport proceeds by hopping between localized states [1–4]. In contrast to structural glasses or to spin glasses such systems can be investigated experimentally very easily, since the resistance can be measured much easier and with higher precision than the viscosity or the magnetic susceptibility. Surprisingly, experiments have shown that the character of the relaxation at low temperatures is qualitatively distinct from that at high temperatures (see, e.g., Refs. [1, 2]). For high temperatures the relaxation of non-equilibrium charge carriers is fast on a macroscopic time scale. At low temperatures the relaxation is extremely slow and strongly dispersive. Relaxation currents lasting for many hours could be observed after initial excitation. According to experiments the relaxation rate depends on the degree of disorder, on electron concentration and on temperature in this regime. A rise of the temperature reflects itself in an increase of the energy relaxation rate. While arguments based on the observed dependence of the relaxation rate on the electron concentration indicate that a mixture of disorder and interaction is necessary in order to account for the situation [1], there are also experimental indications, as the absence of any impact of external sources for screening on the relaxation rate, which show that interaction effects are not too important [2].

So far there is no compelling theory which can account for the strange behavior of such systems. The concepts of the close to equilibrium transport theory seem to fail. It is therefore important to realize why the transport in the non-equilibrium situation is different from that close to equilibrium. In

---

[*] Corresponding author: e-mail: olaf.bleibaum@physik.uni-magdeburg.de, Phone: +49 391 67 12474, Fax: +49 391 67 11217

the literature two mechanism leading to glassy relaxation are discussed. The first mechanism is based on the notion that the system posses a complex energy landscape with many local energy minima. Glass-like transitions, which are based on this picture, have in particular been studied in spin systems on the mean field level [5]. These investigations have shown that below a dynamical transition temperature such a complex energy landscape can indeed lead to vitrification and to the absence of thermalization over many time scales. The sharp transition is, however, not observed in real materials, also not in the electron glasses referred to above. Instead a smooth crossover is found. In the second mechanism the transition results from a kinetic constraint. A model of this type is, e.g, the Kob–Andersen model [6], which has recently been discussed in Ref. [7]. In such systems a sharp transition is only observed if activated hopping processes are ignored.

The fact that the concepts of the equilibrium transport theory fail to describe the situation means that the processes which determine the equilibrium conductivity are not the most relevant ones in the non-equilibrium situation. To develop some understanding for the question how the concepts of the equilibrium transport theory can break down in the non-equilibrium situation we investigate a simple model of non-interacting charge carriers, in which a smooth change of the character of the relaxation can be observed if the temperature is changed. In this model the change of the character of the relaxation results from a kinetic constraint: every charge carrier can only exchange a small amount of energy with the phonon system in one hop.

2 The model   For our investigation we use the simple rate equation

$$\frac{\mathrm{d}\rho_m}{\mathrm{d}t} = \sum_n \left[\rho_n W_{nm} - \rho_m W_{mn} \right], \tag{1}$$

where ρ_m is the probability to find a charge carrier at site m and W_{nm} is the transition probability for a hop from the site n to the site m. In Eq. (1) it is assumed that the charge carriers are far from the Fermi energy and that the number of charge carriers is small so that Fermi-correlation is negligible. Our main assumption is that the charge carriers can exchange at most an amount ω of energy in one hop. Accordingly, the transition probability for a hop is given by

$$W_{nm} = \theta(\omega - |\epsilon_{nm}|) \, \nu \exp\left(-2\alpha|\boldsymbol{R}_{nm}| + \frac{\beta}{2}(\epsilon_{nm} - |\epsilon_{nm}|) \right). \tag{2}$$

Here $\epsilon_{nm} = \epsilon_n - \epsilon_m$, where ϵ_n is the site energy of the site n with position vector $\boldsymbol{R}_n$, $\boldsymbol{R}_{nm} = \boldsymbol{R}_n - \boldsymbol{R}_m$, α^{-1} is the localization radius, ν is the attempt-to-escape frequency, and β is the inverse temperature. Both the position vectors and the site energies are random quantities. We assume that the sites are distributed uniformly in space and that the site energies are random quantities distributed according to a distribution function proportional to the density of states. For simplicity we focus on a constant density of states with $N\omega$ sites per volume in a strip of width ω in energy space.

Our assumption, that in every hop the charge carriers can only exchange a small amount of energy with the phonon system, manifests itself in the step function in Eq. (2). Such a step function arises from the integration over all possible intermediate phonon states in the calculation of the transition probability with the Golden rule. If only one-phonon processes are taken into account and the electron phonon coupling constant in this integration is treated as a constant then ω is equal to the Debye-energy. In an Anderson insulator, however, the assumption that the electron phonon coupling constant is independent of the energy transferred in one jump, that is independent of the phonon momentum, does not hold. This coupling constant is proportional to the overlap integral between the localized electron wave-functions and the phonon wave-function and to the square of the magnitude of the Fourier-transformed Coulomb potential, which provides the electromagnetic coupling between the electron and the phonon wave. The overlap integral tends rapidly to zero if the phonon wavelength becomes small compared to the localization length. The Fourier transformed Coulomb potential is usually treated as a constant. However, this approximation assumes that the electrons can move fast so that they can screen out the electric field produced by the phonon wave immediately. In an Anderson insulator, however, there are no electrons which could move fast enough to screen out the electric field of the phonon wave for most phonon frequencies. Therefore, the Fourier-transformed Coulomb potential is of very long range and consequently already drops to zero for tiny momentum transfer. The energy ω in our approximation is the characteristic phonon energy at which the electron phonon coupling constant starts decaying. Accordingly, jumps with large energy transfer are ignored in this approximation since the preexponential factor for the transition probability for such jumps is small. Since this restriction decreases the number of accessible sites it competes with the tendency to minimize the hopping length as strong as possible. Therefore, we expect that this restriction is important in a hopping regime where the electrons are localized, but the localization is not too strong.

In the theory of hopping transport Eq. (1) has mainly been used for the investigation of relaxation and transport processes of localized charge carriers in band tails of amorphous semiconductors (see, e.g., [8]). However, in these investigations the step function in front of the right hand side of Eq. (2) was replaced by unity. It is important to realize that the presence of this step function increases the degree of complexity of the problem and renders the traditional methods for the investigation of relaxation processes in hopping systems far from equilibrium inapplicable. In fact, these methods are based on the notion that there is a separation of time scales in such a way that the evolution of a particle packet at time t is determined by jumps with hopping rates of the order of $1/t$ [8]. This notion breaks down if the hopping length does not increase rapidly enough from hop to hop, as it is the case in the presence of our kinetic constraint, or if the density of states depends only weakly on energy, as in the materials in the experiments referred to above. Consequently, the investigation of Eq. (1) in the presence of the kinetic constraints requires a different description than that used for relaxation phenomena in a strongly energy dependent density of states with large amount of energy transfer.

In investigating Eq. (1) we focus on energy relaxation processes. We are particularly interested in the moments of the energy distribution function $F(\epsilon_0, \epsilon \mid t)$. The energy distribution function is the probability to find an electron on a site with energy ϵ at time t if it was on a site with energy ϵ_0 at $t = 0$. The first moment of the distribution function,

$$v(t) = \frac{\mathrm{d}}{\mathrm{d}t} \int_{-\infty}^{\infty} \mathrm{d}\epsilon \, \epsilon F(\epsilon_0, \epsilon \mid t), \tag{3}$$

is the energy relaxation rate. From the structure of the transport equation it follows that the energy distribution function satisfies the equation [9]

$$sF(\epsilon', \epsilon \mid s) = \delta(\epsilon' - \epsilon) + \int \mathrm{d}\epsilon_1 \left[F(\epsilon', \epsilon_1 \mid s) \, \tilde{W}(\epsilon_1, \epsilon \mid s) \, N(\epsilon) - F(\epsilon', \epsilon \mid s) \, \tilde{W}(\epsilon', \epsilon_1 \mid s) \, N(\epsilon_1) \right]. \tag{4}$$

Here s the Laplace-frequency, which corresponds to a Laplace transformation with respect to time, $N(\epsilon)$ is the density of states and $\tilde{W}(\epsilon', \epsilon \mid s)$ is the effective probability to jump from a site with energy ϵ' to any site with energy ϵ. While Eq. (4) itself is exact the effective transition probabilities, which contain all impact of the disorder, can only be calculated approximately. To do so we use effective medium methods and the CTRW-method. The results of the analytical calculations are supported further by Monte-Carlo calculations.

3 Relaxation at high temperatures

In order to investigate the physical content of the integral Eq. (4) further we focus on the situation that the energy scale ω is the smallest energy scale in the problem. In this case we can use the quasi-elastic approximation to simplify this equation. In the quasi-elastic approximation the equation for the calculation of the energy distribution function takes the form

$$sF(\epsilon', \epsilon \mid s) = \delta(\epsilon' - \epsilon) + \beta^{-1} \frac{\mathrm{d}}{\mathrm{d}\epsilon} \left[N(\epsilon) \, v(\epsilon, s) \left(\frac{\mathrm{d}}{\mathrm{d}\epsilon} \frac{F(\epsilon', \epsilon \mid s)}{N(\epsilon)} + \beta \frac{F(\epsilon', \epsilon \mid s)}{N(\epsilon)} \right) \right] \tag{5}$$

in the limit $kT \gg \omega$. In this equation $v(\epsilon, s)$ is the spectral energy relaxation rate. The energy relaxation rate defined by Eq. (3) is related to the spectral energy relaxation rate by the integral

$$v(t) = \frac{1}{2\pi i} \int_c \mathrm{d}s \, \mathrm{e}^{st} \int_{-\infty}^{\infty} \mathrm{d}\epsilon \, F(\epsilon_0, \epsilon \mid s) \, v(\epsilon, s), \tag{6}$$

where the contour c is chosen in such way that all of the non-analyticities of the integrand are on its left side. To calculate the spectral energy relaxation rate we use the effective medium theory by Gouchanour, Andersen and Fayer [10]. Doing so, we find that for $s = 0$ the energy relaxation rate is given by

$$v(\epsilon, 0) = \frac{\beta\omega^2}{3} \nu \exp(-\rho_c(\epsilon)), \tag{7}$$

where

$$\rho_c(\epsilon) = \frac{2\alpha}{(2\omega N(\epsilon))^{1/d}} \left(\frac{d}{S_d}\right)^{1/d}. \tag{8}$$

Here d is the spatial dimension of the system and S_d is the solid angle in d dimensions. Thus, up to pre-exponential factors the equation for the calculation of the energy relaxation rate for $s = 0$ has the same structure as the equation for the calculation of the conductivity close to equilibrium, which can be derived under the assumption that the density of states is bounded from below. Accordingly, the energy relaxation is percolation like at high temperatures. For small s we find that the energy relaxation rate satisfies the equation

$$\frac{v(\epsilon, s)}{v(\epsilon, 0)} \ln \frac{v(\epsilon, s)}{v(\epsilon, 0)} = \frac{s}{\omega_0(\epsilon)}, \tag{9}$$

where

$$\omega_0(\epsilon) = \frac{2\,d\nu}{\rho_c(\epsilon)} \exp(-\rho_c(\epsilon)). \tag{10}$$

This equation also agrees with that which would be obtained for the dispersion of the conductivity close to equilibrium. Therefore, at high temperatures also the time scales, which govern the relaxation, are the same as those for conduction processes close to equilibrium. If the density of states is constant a particle packet, in which all particles have the same energy at $t = 0$, and which starts sinking down at $t = 0$, reaches a stationary velocity in energy space for $t > t_{\text{perc}}$, where $t_{\text{perc}} = e/\omega_0$ is the percolation time (Note that, for a constant density of states both $\omega_0(\epsilon)$ and $\rho_c(\epsilon)$ are independent of ϵ). For $t \gg t_{\text{perc}}$

$$v(t) = v_0\left(1 + \sqrt{\frac{e}{2\pi}} \frac{1}{(\omega_0 t)^{3/2}} \exp(-\omega_0 t/e)\right). \tag{11}$$

and for $t \ll t_{\text{perc}}$

$$v(t) = v_0 \frac{\sqrt{2\pi}}{e\omega_0 t} \frac{1}{(\ln(e/(\omega_0 t)))^2}. \tag{12}$$

Here $v_0 = v(s = 0)$. The same dependencies would be obtained for the diffusion of a particle in an electric field. In this case Eq. (11) would describe the regime of normal diffusion and Eq. (12) the anomalous diffusion regime.

The fact that the relaxation in the energy space is dispersive sets the situation in a disordered hopping system apart from that in an ordered system. This fact manifests itself also in the structure of the energy distribution function. At high temperatures the equation for the energy distribution function contains both the first and the second derivative with respect to energy, as it can be seen from Eq. (5). In an ordered system the first derivative is connected with the energy relaxation rate. The second derivative determines the shape of the distribution function. In this case a particle packet moves without distortion if the second derivative is ignored. In a disordered hopping system, however, the situation is different since in such a system the transport coefficients depend on the Laplace frequency s already at very low frequencies (see Eq. (9)). The dispersion provides the system with a second mechanism to generate the width of a particle packet. At low temperatures, that is for

$kT \ll \omega \sqrt{\rho_c(\epsilon)}$, the dispersive contribution to the width of the particle packet exceeds the thermal contribution, which results from the second derivative with respect to energy. Therefore, the second derivative in Eq. (5) can be ignored at low temperatures. Accordingly, this equation takes the form

$$sF(\epsilon', \epsilon \mid s) = \delta(\epsilon' - \epsilon) + \frac{d}{d\epsilon}[v(\epsilon, s) F(\epsilon', \epsilon \mid s)]. \tag{13}$$

In this case the width of the particle packet is governed entirely by the dispersion. For the constant density of states Eq. (13) yields

$$F(\epsilon_0, \epsilon \mid t) = \frac{1}{\sqrt{4\pi kT v_0 t}} \exp\left(-\frac{(\epsilon_0 - \epsilon - v_0 t)^2}{4kT v_0 t}\right) \tag{14}$$

for $t \gg t_{\text{perc}}$.

4 Relaxation at zero temperature If we decrease the temperature then the condition $\beta\omega \ll 1$ becomes violated. Despite this fact a quasi-elastic expansion of the energy distribution function can be established in the limit of zero temperature. It turns out that in this limit the equation for the calculation of the energy distribution function has again the structure of Eq. (13). If we use the effective medium approximation to calculate the spectral energy relaxation rate we find that in the limit of zero temperature [9]

$$v(\epsilon, 0) = \frac{\omega v}{2} \exp(-\rho_c(\epsilon)). \tag{15}$$

The quantity $\rho_c(\epsilon)$ in Eq. (15) differs from the quantity $\rho_c(\epsilon)$ at high temperatures, as defined by Eq. (8), only in that ω is replaced by $\omega/2$, since the width of the strip of accessible sites has decreased by a factor of $1/2$. For small frequencies the dispersion of the spectral energy relaxation rate is again given by Eqs. (9) and (10). Accordingly, the effective medium approximation predicts that the relaxation is qualitatively the same as at high temperatures. While at the first glance this result seems to be plausible its consequence, that the relaxation at zero temperature is determined by the same percolation paths which are appropriate for close-to-equilibrium transport problems, is surprising. Although most of the sites have an accessible neighbor within the average site spacing there is also a substantial part of sites without neighbor within the average site spacing. We call such sites distribution holes. Since the transition probability is an exponentially small quantity with respect to the hopping length the charge carriers prefer jumping to the closest site. Doing so, it does not matter whether this site is a distribution hole or not. If the temperature is zero they can not return. Accordingly, they can only leave the distribution hole by an extraordinary hard jump, that is a jump over a length which is large compared to the average site spacing. Therefore, at large times the relaxation should be governed by such jumps, and not by jumps of the order of the average site spacing, as suggested by Eq. (15). Thus, the effective medium theory proves to be inapplicable to the relaxation problem at low temperatures.

To calculate the spectral energy relaxation rate we therefore extend the CTRW-method of Ref. [11] to the presence of inelastic processes. Since this method is based on an approximation scheme which assumes that the particle never returns to its initial site the energy relaxation rate can be calculated exactly with this method for systems in which the bonds connecting successive sites are considered as independent random variables. In our calculation we average over the positions of the sites and their site energies. From the physical point of view there should be no difference between site averaging and bond averaging. Therefore, we expect that our results are valid although we average over the sites. If we compare the results with those of the Monte-Carlo simulations of three-dimensional systems of Ref. [12] we find excellent quantitative agreement. Our investigations show that at zero temperature the spectral energy relaxation rate is given by

$$v(\epsilon, s) = \frac{\omega}{2} \frac{1}{f(\epsilon, s)}, \tag{16}$$

where

$$f(\epsilon, s) = \int\limits_0^\infty dt \exp\left(-st - \frac{d}{\rho_c^d(\epsilon)} \int\limits_0^\infty dx\, x^{d-1}(1 - \exp\left(-vt \exp\left(-x\right)\right))\right).$$ (17)

In this equation $\rho_c(\epsilon)$ is the quantity defined in Eq. (8) with ω replaced by $\omega/2$. For $s = 0$ the spectral energy relaxation rate

$$v(\epsilon, 0) \propto \frac{\omega v}{2} \exp\left(-(d-1)\left(\rho_c(\epsilon)/d\right)^{d/(d-1)}\right).$$ (18)

This result differs strongly from Eq. (15). It shows that the relaxation at zero temperature is far from percolation like. In the case of the equilibrium conductivity a result with the same exponential dependence on ρ_c was first obtained by Miller and Abrahams. Therefore, we refer to this transport regime as Miller–Abraham transport regime. While for the close-to-equilibrium conductivity the Miller–Abraham contribution to the current has proven to be insignificant the relaxation at zero temperature is governed by this contribution due to the reasons discussed above.

The differences between the high temperature relaxation regime and the Miller–Abraham relaxation regime manifest themselves also in the characteristic time scale. For the constant density of states the time scale t_{perc} is replaced by

$$t_{MA} = \frac{1}{v} \exp\left((d-1)\left(\rho_c/d\right)^{d/(d-1)}\right).$$ (19)

For $t \ll t_{MA}$ we obtain

$$v(t) = \frac{\omega}{2}\frac{1}{t} \frac{\exp\left(\dfrac{\ln^d\left(v't\right)}{\rho_c^d}\right) d \ln^{d-1}(v't)}{\rho_c^d},$$ (20)

where $v' = v \exp\left(\gamma\right)$ (γ-Euler constant) and for $t \gg t_{MA}$

$$v(t) = v(s=0)\left(1 + \frac{2v(s=0)}{\omega v'}\frac{\rho_c^d v't}{d \ln^{d-1}(v't)} \exp\left(-\frac{\ln^d(v't)}{\rho_c^d}\right)\right).$$ (21)

If we investigate the dispersion of the transport coefficients further we notice that in contrast to the situation at high temperatures, in which the dispersion truly depends only on the parameter s/ω_0, in the Miller–Abrahams transport regime the dispersion truly depends on both on s and on ρ_c. This fact manifests also in the higher moments of the energy distribution function. If we calculate the higher moments of the energy distribution as an expansion with respect to large t then every moment needs a new scale. While for the constant density of states the result of the large t-expansion for the the first moment (Eq. (22)) becomes already valid if $t \gg t_{MA}$ the result for the large t-expansion of the second moment, the mean squared deviation

$$\sigma^2(t) \propto \omega^2 v't \exp\left((d-1)\left(\frac{\rho_c^d}{d}\right)^{d/(d-1)}\left(2^{d/(d-1)} - 3\right)\right),$$ (22)

applies only if $t \gg t_{MA}^{(2)}$, where

$$vt_{MA}^{(2)} = \left(vt_{MA}\right)^{2^{d/(d-1)}-1}.$$ (23)

Note that, the result (22) depends strongly on the dimension of the system. While in three dimensions $\sigma^2(t)/(v't)$ is exponentially small with respect to the parameter $\rho_c^{d/(d-1)}$, it is exponentially large in two dimensions. The reason for this strong difference is not entirely clear. It might reflect that in two-dimensional systems it is much harder to avoid the distribution hole than in three-dimensional systems.

5 Suitable observation conditions   If we accept the kinetic constraint then the question arises under which conditions the Miller–Abrahams relaxation can be observed. Since the presence of the Miller–Abrahams regime relies on the distribution holes the charge carriers have to have a good chance to fall into such holes. To this end either the concentration of distribution holes has to be sufficiently large or the charge carriers have to be far enough from the Fermi energy, so that they are able to perform a large enough number of hops. The concentration of the distribution holes is determined by Poisons distribution. The probability, that there is no site in a sphere with radius R in a d-dimensional system, is $\exp\left(-\kappa_d\right)$, where

$$\kappa_d = \left(\frac{2\alpha R}{\rho_c}\right)^d.$$

(24)

According to Eq. (18) the characteristic hopping length in the Miller–Abrahams regime is

$$R_{MA} = \frac{(d-1)}{2\alpha}\left(\frac{\rho_c}{d}\right)^{d/(d-1)}.$$

(25)

Thus, $\kappa_3 = 3.65\rho_c^{3/2}$ and $\kappa_2 = \frac{1}{16}\rho_c^2$. Given that in a typical experimental situation ρ_c is rarely larger than 5 we see that Miller–Abraham like relaxation processes should in particular be important in two-dimensional systems.

However, also not all two-dimensional systems provide equally well suitable conditions for the observation of the differences between the Miller–Abraham relaxation and the percolation like relaxation. In systems with a strong energy dependence of the density of states the structure of the particle packed is strongly affected by the fact that the number of accessible sites decreases strongly with decreasing energy, if the density of states decreases with decreasing energy. In an exponential density of states, e.g, the functional dependence of the energy relaxation rate on time for large times in the Miller–Abraham like relaxation regime agrees with that in the percolation-like relaxation regime up to numbers which only occur in logarithms. Due to this fact it becomes difficult to distinguish truly between both transport regimes. Therefore, we expect that two-dimensional systems with a weakly energy dependent density of states provide the most favorable conditions for the observation of Miller–Abraham like relaxation processes.

References

[1] Z. Ovadyahu and M. Pollak, Phys. Rev. Lett. **79**, 459 (1997).
[2] M. Pollak and Z. Ovadyahu, J. Phys. I (France) **7**, 1595 (1997).
[3] E. Bielejec and Wenhao Wu, Phys. Rev. Lett. **87**, 256601 (2001).
[4] G. Martinez-Arizala, D. E. Grupp, C. Christiansen, A. M. Mack, N. Marcovic, Y. Seguchi, and A. M. Goldman, Phys. Rev. Lett. **78**, 1130 (1997).
[5] J. P. Bouchad, L. F. Cugliandolo, J. Kurchan, and M. Mezard, in: Spin Glasses and Random Fields, edited by P. Young (World Scientific, Singapore, 1998).
[6] W. Kob and H. C. Andersen, Phys. Rev. E **48**, 4364 (1993).
[7] S. Franz, R. Mulet, and G. Parisi, Phys. Rev. E **65**, 21506 (2002).
[8] D. Monroe, in: Hopping Transport in Solids, edited by M. Pollak and B. I. Shklovskii (North-Holland, Amsterdam, 1991).
[9] O. Bleibaum, H. Böttger, V. V. Bryksin, and A. N. Samukhin, Phys. Rev. B **62**, 13440 (2000).
[10] C. R. Gouchanour, H. C. Andersen, and M. D. Fayer, J. Chem. Phys. **70**, 4254 (1979).
[11] H. Scher and M. Lax, Phys. Rev. B **7**, 4491 (1973).
[12] E. Haba, O. Bleibaum, H. Böttger, and V. V. Bryksin, Phys. Rev. B in print.

phys. stat. sol. (b) **241**, No. 1, 20–25 (2004) / **DOI** 10.1002/pssb.200303616

Long-time relaxation of conductivity in hopping regime

A. L. Efros[*,1], **D. N. Tsigankov**[1], **E. Pazy**[2], and **B. D. Laikhtman**[3]

[1] Department of Physics, University of Utah, Salt Lake City, Utah 84112
[2] Department of Physics, Ben-Gurion University of the Negev, Beer-Sheva 84105, Israel
[3] Department of Physics, Hebrew University, Jerusalem 91904, Israel

Received 1 September 2003, revised 17 September 2003, accepted 2 October 2003
Published online 15 December 2003

PACS 72.20.Ee, 72.80.Ng, 78.30.Ly

Using numerical simulations we studied the long-time relaxation of the hopping conductivity. We perturbed the system through insertion of electrons and monitored the conductivity as a function of time. Even though employing numerical simulations one can only follow the system for very short time scales we have shown that during such available times we can reach an apparent saturation of conductivity and energy. In order to investigate the long-time relaxation of the system we studied the difference between the saturated values of the conductivities obtained by the short-time relaxation from initial excited states with different electron distribution. We have related these two typical time scales to relaxation in one pseudoground state and to very slow transitions between pseudoground states. By employing two different two-dimensional models with electron–electron interactions we were able to show the effect of disorder on the relaxation of conductivity. In the strong-disorder case the universality of the Coulomb gap, which is responsible for the universal Efros–Shklovskii law for the conductivity, suppresses the long-time relaxation of conductivity since the universality strongly decreases the dispersion of conductivities of the pseudoground states. In the second model with a weak external disorder we found a difference between saturated values of conductivity in agreement with the experimental data.

1 Introduction Theoretical predictions regarding the existence of glassy properties in electronic systems were first made by Davies, Lee, and Rice [1]. This so called Electron (or Coulomb) glass is a result of the competition between disorder and the long-range Coulomb interaction. Experimental evidence for such properties has been observed in Anderson-localized crystalline and amorphous-indium-oxide [2] as well as in ultra-thin films of metals near the superconductor–insulator transition [3].

The complexity of the problem, i.e., long-range interactions, randomness and non-equilibrium effects, suggest that the best way to address it theoretically is through Monte-Carlo simulations. The first simulations were already performed 25 years ago by Baranovskii et al. [4]. In these simulations the authors identified states which have very close total energies but substantially different sets of the occupation numbers n_i. On the other hand the transitions between those states should be of many-electron nature and they are expected to be slow. These states have been called "pseudoground states" (PGS's). We believe that, in the same way as in real glasses, the glassy properties in this electron system are due to a slow relaxation through these states toward the ground state.

The properties of the PGS's have been studied recently, mostly by computational methods [5–7]. But can such computer simulations help us study relaxation processes? The true manifestation of a glassy system is in its dynamics, the most interesting effects are observed when the system is per-

---
[*] Corresponding author: e-mail: efros@physics.utah.edu, Phone: +01 801 585 5018, Fax: +01 801 581 4801

turbed and its relaxation path is examined. In the experiments performed by the group of Ovadyahu [2, 8], the glassy properties were investigated by inserting or extracting electrons from it. Then conductivity as a function of time and perturbation parameters were monitored exhibiting such interesting phenomena as memory and aging effects [8] as well as inverse–Arrhenius temperature dependence of relaxation [9].

At first glance simulation of the conductivity relaxation seems a hopeless task. Performing a simple time estimate shows that, due to the long-range of the Coulomb interaction and to the fact that in a real sample many transitions take place simultaneously while a computer processor is limited to performing them one at a time, the system's dynamics can only be simulated on very short time scales. A few additional processors cannot help much. Therefore, even using the most sophisticated program, we may simulate no more than 40 µs of the real time in a system of the size of 100×100 lattice sites. In order to realize that numerical simulations are still able to give us very important insights regarding the glassy behavior one has to question the conclusions reached by Baranovskii et al. [4], which found that the PGS's have the same universal Coulomb gap and have concluded that the existence of PGS's is not important for the variable range hopping (VRH) conductivity because transitions between them are very slow. In comparing this conclusion with experiments by the group of Ovadyahu [2, 8] one observes that the difference in the conductivities of the PGS's can be as large as 10–12%. Similar phenomena were observed by other groups [3, 10]. One therefore realizes there are two time scales defining the behavior of the conductivity: the first is a very short time scale corresponding to the average value of the conductivity, the second a very long-time scale defined by the long-time relaxation of the conductivity. We relate these two scales to the following physical picture, in which the relatively short time scale is a consequence of the relaxation of the system within one PGS, well described by our simulations, whereas the long-time relaxation of the conductivity, is related to transitions between different PGS's.

In this paper we aim at understanding the origin of the long-time relaxation of conductivity, observed in the experimental papers cited above. To do so we performed Monte-Carlo simulations employing the lattice model and the random site model for two dimensional systems. In our simulation we perturb the system by adding some extra electrons and then trace the relaxation of the conductivity with time. We have shown that during such available times we can reach an apparent saturation of conductivity. We interpret this "saturation" as a saturation within one PGS. Thus, it is rather the transition from the fast relaxation within one PGS to a very slow relaxation to PGS's with lower energy, that we cannot observe.

To overcome this shortcoming we came up with a new idea. We create different PGS's by relaxation from states with different initial distributions of electrons. Then we study the difference between the saturated values of the conductivities of the different PGS's. If these conductivities are different, one should expect that a long-time relaxation exists and the total change of the conductivity should be of the order of this difference. For the lattice model we obtain no such effects. The conductivities of different PGS are the same in the limits of our accuracy (about 1–2%). Since PGS's with different energies were observed by Baranovskii et al. [4] as well as in other papers, we arrive at the conclusion that the effect of long-time relaxation of the conductivity is absent due to the universality of the Coulomb gap for the lattice model, in the temperature range which we consider.

We also performed simulations for the random site model. In this model all disorder comes from the random position of sites and an "external" disorder is absent. Our results for these simulations exhibit a difference in the conductivities of the PGS's whose value is sufficiently large so as to explain the experimental results of Ovadyahu's group. Note that experiments are performed on thin films and it is not quite clear which dimensonality they have. Nevertheless we consider here two-dimensional models only.

2 Relaxation after addition of extra electrons

To study the relaxation of the energy and conductivity of the system we perturbed it by the addition of a few percents of extra electrons, mimicking the procedure employed in the experiments of Ovadyahu's group. However the time we are able to ob-

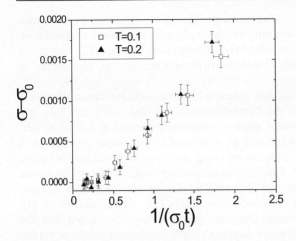

Fig. 1 The time dependence of the conductivity of the system is shown at different temperatures as a function of $1/\sigma_0 t$. At time $t = 0$ the filling factor ν has been increased from 0.5 to 0.52, such that the total number of extra electrons equals $N_{ex} = 0.02L^2/2$. The value of σ_0 is taken to be the saturated value of the conductivity at the longest time measured. The localization radius $a = 1$, the system size is 100×100.

serve the relaxation of the system is very short. We use the lattice model with the Hamiltonian

$$H = \sum_i \phi_i n_i + \frac{1}{2} \sum_{i \neq j} \frac{(n_i - \nu)(n_j - \nu)}{r_{ij}}, \tag{1}$$

where $n_i = 0, 1$ are occupation numbers and the filling factor $\nu = 1/2$. The "external" disorder is described by random energies ϕ_i homogeneously distributed within the interval $[A, -A]$ The extra electrons are added randomly on empty sites and the background is adjusted such that the system remains neutral.

The time dependence of the conductivity is shown on Fig. 1 for two different temperatures. Our algorithm for computation of the conductivity is described in details in the Appendix to paper [11]. One can see that for both temperatures the conductivity decreases with time finally reaching a kind of saturation, which may also be interpreted as a transition to a substantially slower rate of relaxation. We consider this 'saturation' as the end of the short time relaxation in our finite system. The value of the conductivity at the largest time we considered, is denoted by σ_0. One can see that as a function of $1/\sigma_0 t$ the results for both temperatures coincide.

The time corresponding to the saturation $t = 4/\sigma_0$ is very short. For $T = 0.1$ and $a = 1$ it is $\approx 0.8 ns$. The saturation itself might be a manifestation of the size effect. However, even if we assume an infinite system and extrapolate the law $\sigma - \sigma_0 \sim (\sigma_0 t)^{-1}$ to much larger times, we find that in the microsecond range all relaxation which can be observed is over.

Thus, the saturation of the conductivity averaged over different initial distributions of extra electrons, presented in this section, happens at very short times. It is a Maxwell-type relaxation of the extra charge. We have checked that the relaxation of the Coulomb gap inside the energy interval that is responsible for the VRH occurs even faster than the relaxations of the energy and conductivity. These results do not support the idea that long-time relaxation is due to the slow formation of the Coulomb gap [12].

However experimental data [2, 8] show long-time relaxation occurring on time scales of the order of seconds and hours. These time scales are 9–11 orders of magnitude longer than those which we are able to simulate. Thus, if a long-time relaxation exists in this system, the change of the averaged value of the conductivity is negligible during the physical time scales we have simulated. This explains the apparent saturation of the conductivity, we observe. If the long-time relaxation results from the transitions between different PGS's, the conductivity we have observed should be considered as the conductivity within a single PGS. The encouraging result obtained in this section is that the saturation of the conductivity of a single PGS can be achieved during the time scale available for our computation. Based on this result another approach to the problem of long-time relaxation, is proposed in the next section.

3 Conductivity of different pseudoground states The idea behind our new approach is rather simple. We want to compare the conductivity of the system in the different PGS's. If the values of the obtained conductivities are different for different PGS's then the long-time relaxation of the conductivity can be attributed to the slow transition between those PGS's. Even though we are unable to track these transitions with our computers the effect of long-time relaxation can be explained and the magnitude of this effect can be found if we know the differences in the conductivities of different PGS's. In the experiments of the group of Ovadyahu [8] such differences are of the order of 10%.

To observe the conductivity of different PGS's, we measured the conductivity of the same sample with the different initial distributions of electrons during the longest time we are able to simulate. The sample is characterized by the total set of random energies ϕ_i. Starting with different initial distribution of electrons the system relaxes to the different PGS's. If the saturated values of conductivities, measured as in previous section, is different, one should expect that the system will have a long-time relaxation due to transitions between different pseudoground states.

The results for the lattice model simulations performed at the lowest temperature are presented in Fig. 2A. The time evolution of the conductivity averaged over the time of measurement is shown for different initial distributions of electrons for the same sample. As one can see there is no appreciable difference in the values of the conductivities within our accuracy which is about 1–2%. We obtained the same result for all higher temperatures even with greater accuracy.

We therefore conclude that there is no apparent difference in the conductivities of different PGS's for this model down to the lowest temperatures we are able to simulate. We think that the reason for this is the following: It has been shown [13] that the VRH conductivity of the system is provided by the single–electron excitations. The properties of these excitations is defined by the structure of the Coulomb gap. In the lattice model at large A the DS in the Coulomb gap has a universal form [14]. It is independent of the properties of the system and is the same for all PGS's. For our temperature range, the case $A = 1$ can be considered as a large disorder. Thus all PGS's have the same conductivities and no long-time relaxation can be observed. We realize that the lattice model with large disorder cannot account for the effect of the long-time relaxation of the conductivity which is observed experimentally. Thus, to observe the difference in the conductivities of different PGS's one should take a system with smaller disorder, where the Coulomb gap is not universal. Unfortunately, it is difficult to find such a regime in the lattice model, due to the Wigner crystallization. Therefore we switch to the random site model without any external disorder. Such a model can be applicable for example to a system consisting of sites with negative Hubbard energy. The Hamiltonian of this model

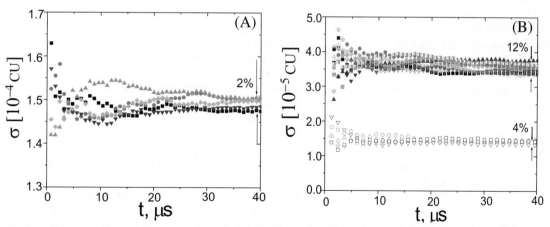

Fig. 2 (online colour at: www.interscience.wiley.com) Time evolution of the conductivity averaged over the time of measurement shown A) for different initial distributions of electrons in the lattice model for the same set of ϕ_i where $A = 1$, B) for the model with the random site distribution (filled symbols). Open symbols show the same evolution for a "compound" system with $A = 1$. The values of parameters: $T = 0.04$, localization radius $a = 1$, the system size $L = 70$ and the filling factor $\nu = 1/2$.

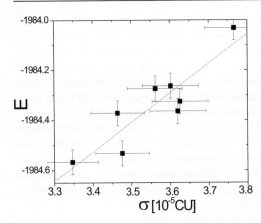

Fig. 3 (online colour at: www.interscience.wiley.com) The correlation between the total energy of the system and the conductivity is shown for 8 different initial distributions of electrons in the random site model. Both the total energy and the conductivity are averaged over the time 40 μs. The straight line is given as a guide for the eye. The values of the parameters used are the same as were used in Fig. 2B.

has the form

$$H = \frac{1}{2}\sum_{i \neq j} \frac{(n_i - 1/2)\,(n_j - 1/2)}{r_{ij}}\,,\qquad(2)$$

where r_{ij} is the distance between two random positions i, j on the plane. It differs from the lattice model in two important respects. The first: it is formulated on sites i, j, which have random positions on the plane. The second: any random energies ϕ_i, that are not correlated with the interaction, are absent. Thus, the random positions of the sites is the only source of disorder. Therefore it is interesting to study the conductivity of different PGS's in this model.

The simulation results for the time evolution of the conductivities of different PGS's for the model with the random spatial site distribution are shown in Fig. 2B by filled symbols at the same temperature as for the lattice model. In this case the conductivity value of each individual PGS saturates within 1% during the time of the simulation. However, the saturated values of the conductivities differ by 12% for the different initial distributions of electrons in the same sample. This is by an order of magnitude greater than the simulation accuracy. The study of the conductivities in the random site model shows that the slow transitions between PGS's may result in the long-time relaxation of the conductivity observed in the experiment. An important question now is whether the obtained result is due to the randomness in the site distribution or to the absence of the disorder which is not correlated with the interaction.

In order to answer this question we consider a "compound" model. Namely, we have added to the Hamiltonian of the random site model Eq. (2) the term $\sum \phi_i n_i$, where ϕ_i are random energies within the interval $[-A, A]$. One can see from Fig. 2B that the difference in the conductivities of different PGS's disappears at $A = 1$ within the simulation accuracy. Thus the effect exists at small "external" disorder and strong interaction.

Another important question is whether the values of the conductivity of the different PGS's are correlated with the energies of these PGS's. The total energy of the system in a given PGS is averaged over time and plotted versus the value of the conductivity for the same PGS's in Fig. 3. The data shows that the energy dependence of the conductivity is close to a linear behavior. PGS's with higher average total energy have also a larger value for the conductivity which is reasonable, since in the states with lower energy the electrons are in positions where they are more tightly bound and therefore their conductivity is lower. The same behavior is apparent in the experimental results of Ovadyahu [8]. The linear dependence is also reasonable because the energy difference between PGS's is small and conductivity as a function of energy may be expanded into the Taylor series, taking the first term. Similar to Anderson et al. [15] one can show that possibility of Taylor expansion leads to the relaxation laws $\delta E \sim -\log t$ and $\delta \sigma \sim -\log t$.

4 Conclusions Using a Monte-Carlo simulation we were able to follow the relaxation of the conductivity for short time scales which were large enough to observe apparent saturation. We were also able to probe the behavior of the system on very large time scales. We relate the short time scale to relaxation of the system within one PGS and the long-time scale to transitions between PGS's. We considered two different two-dimensional models with electron-electron interaction: the lattice model and the random site model, corresponding to "strong" and "weak" effective disorder. For the random site model, effective "weak" disorder, i.e., with random distances between sites and no external disorder we have found the difference of the conductivities to be within 10–12% which is large enough to explain the experimental data. We have shown also that the density of states in this model is not universal and that hopping conductivity does not obey the $T^{-1/2}$ law. We think that universality of the Coulomb gap, that manifests itself in the $T^{-1/2}$ law for the VRH, suppresses the long-time relaxation because in this case since the conductivities of different pseudoground states are very close to each other which originates from a large disorder as compared to an interaction strength. It is possible that the long-time relaxation might be observed in the lattice model with $A = 1$ but at lower temperatures than those which we are able to simulate, because at low temperatures the universality of the Coulomb gap might be violated [14]. With increasing A this temperature range should become lower.

References

[1] J. H. Davies, P. A. Lee, and T. M. Rice, Phys. Rev. Lett. **49**, 758 (1982).
[2] M. Ben-Chorin, Z. Ovadiahu, and M. Pollak, Phys. Rev. B **48**, 15025 (1993).
[3] G. Martinez-Arizala, D. E. Grupp, C. Christyiansen, A. M. Mack, N. Markovic, Y. Seguchi, and A. M. Goldman, Phys. Rev. Lett. **78**, 1130 (1997).
[4] S. D. Baranovskii, A. L. Efros, B. L. Gelmont, and B. I. Shklovskii, J. Phys. C **12**, 1023 (1979).
[5] Sh. Kogan, Phys. Rev. B **57**, 9736 (1998).
[6] A. Perez-Garrido, M. Ortuno, and A. Diaz-Sanches, phys. stat. sol. B **205**, 31 (1998).
[7] D. Menashe, O. Biham, B. D. Laikhtman, and A. L. Efros, Phys. Rev. B **64**, 115209 (2001).
[8] A. Vakhin, Z. Ovadyahu, and M. Pollak, Phys. Rev. Lett. **84**, 3402 (2000).
[9] A. Vakhin, Z. Ovadyahu, and M. Pollak, Europhys. Lett. **42**, 307 (1998).
[10] D. Monroe, A. C. Gossard, J. H. English, B. Golding, W. H. Haemmerle, and M. A. Kastner, Phys. Rev. Lett. **59**, 1148 (1987).
[11] D. N. Tsigankov, E. Pazy, B. D. Laikhtman, and A. L. Efros, Phys. Rev. B **68**, 184205 (2003).
[12] C. C. Yu, Phys. Rev. Lett. **82**, 4074 (1999).
[13] D. N. Tsigankov and A. L. Efros, Phys. Rev. Lett. **88**, 176602 (2002).
[14] F. G. Pikus and A. L. Efros, Phys. Rev. Lett. **73**, 3014 (1994).
[15] P. W. Anderson, B. I. Halperin, and C. M. Varma, Philos. Mag. **25**, 1 (1972).

phys. stat. sol. (b) **241**, No. 1, 26–32 (2004) / **DOI** 10.1002/pssb.200303640

Shot noise in mesoscopic transport through localised states

A. K. Savchenko[*,1], **S. S. Safonov**[1], **S. H. Roshko**[1], **D. A. Bagrets**[2], **O. N. Jouravlev**[2], **Y. V. Nazarov**[2], **E. H. Linfield**[3], and **D. A. Ritchie**[3]

[1] School of Physics, University of Exeter, Stocker Road, Exeter, EX4 4QL, United Kingdom
[2] Department of Applied Physics, Delft University of Technology, Lorentzweg 1, 2628 CJ Delft, The Netherlands
[3] Cavendish Laboratory, University of Cambridge, Madingley Road, Cambridge, CB3 0HE, United Kingdom

Received 2 September 2003, revised 10 October 2003, accepted 10 October 2003
Published online 15 December 2003

PACS 72.20.Ee, 72.70.+m, 73.20.Hb, 73.40.Gk

We show that shot noise can be used for studies of hopping and resonant tunnelling between localised electron states. In hopping via several states, shot noise is seen to be suppressed compared with its classical Poisson value $S_I = 2eI$ (I is the average current) and the suppression depends on the distribution of the barriers between the localised states. In resonant tunnelling through a single impurity an enhancement of shot noise is observed. It has been established, both theoretically and experimentally, that a considerable increase of noise occurs due to Coulomb interaction between two resonant tunnelling channels.

1 Introduction In the past few years much attention has been drawn to the properties of shot noise in mesoscopic structures [1]. So far experimental studies of shot noise have been primarily concentrated on ballistic and diffusive systems [2, 3], with only a few exceptions [4, 5] where shot noise in electron tunnelling or hopping between localised states was investigated.

In short, mesoscopic barriers, where the condition $L/L_c < 1$ is realised (L_c is the correlation length which represents the typical distance between the dominant hops in the hopping network [6]), we have studied shot noise in hopping. A suppression of shot noise in this regime was detected in [5]: $S_I = F2eI$ where $F < 1$ is the Fano factor. We have found that the result depends significantly on the geometry of the sample: the suppression of shot noise for a wide sample (2D geometry) is found to be much stronger than for a narrow (1D geometry) sample for the same sample length [7]. We explain this by the reconstruction of the hopping network in 2D [8, 9]: in a short and wide sample the current is carried by a set of most conductive parallel hopping chains.

With further decreasing the sample length and temperature, resonant tunnelling through a single localised state is seen. In this situation the suppression factor F has been predicted to be $\left(\Gamma_L^2 + \Gamma_R^2\right)/\left(\Gamma_L + \Gamma_R\right)^2$ [10], where $\Gamma_{L,R}$ are the leak rates from the state to the left and right contacts. In our study of shot noise in resonant tunnelling we have observed not only the suppression of noise, but also its significant enhancement [11]. We have proved that this effect is caused by the Coulomb interaction between two parallel resonant tunnelling channels, when two localised states carry the current in a correlated way. Experimental observation of increased noise in this case is confirmed by theoretical calculations.

---

[*] Corresponding author: e-mail: A.K.Savchenko@ex.ac.uk, Phone: +44 1392 264109, Fax: +44 1392 264111

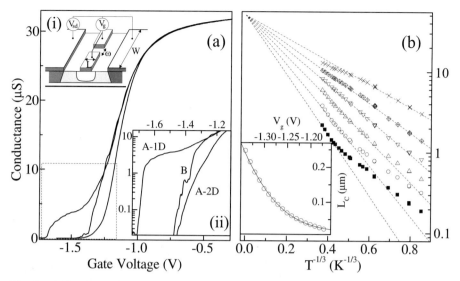

Fig. 1 a) Conductance as a function of the gate voltage for 2D and 1D configurations (samples A-2D and A-1D, B, respectively) at $T = 4.2$ K. Insets: (i) Cross-section of the transistor structure with two contacts and a split gate between them. (ii) Enlarged $G(V_g)$ near the pinch-off shown by the dashed box on the main figure. b) Temperature dependence of the conductance of the sample A-2D for different gate voltages: $V_g = -1.18$ V (crosses), -1.22 V (diamonds), -1.25 V (down triangles), -1.28 V (up triangles), -1.31 V (circles), and -1.34 V (squares). Solid lines are the fit by T-dependence of VRH. Inset: Correlation length for the hopping network as a function of gate voltage.

The experiment has been carried out on a n-GaAs MESFET consisting of a GaAs layer of 0.15 μm thickness (donor concentration $N_d = 10^{17}$ cm^{-3}). On the top of the GaAs layer Au gates are deposited with dimensions $L = 0.4$, $W = 4$ μm, and $L = 0.2$, $W = 20$ μm, Fig. 1a (i). By applying a negative gate voltage, V_g, a lateral potential barrier is formed between the ohmic contacts (source and drain). Some gates contain a split of width $\omega = 0.3$, 0.4 μm with the aim to define a one-dimensional hopping channel.

2 Suppression of shot noise in 1D and 2D hopping We have studied the sample with two gates on the same structure ($L = 0.4$ μm and $W = 4$ μm) with splits of different width: $\omega = 0.3$ μm (sample A) and $\omega = 0.4$ μm (sample B). The sample with the narrow split shows different behaviour from one cooldown to another: two-dimensional (referred to as A-2D) and one-dimensional (A-1D), Fig. 1a. This difference in $G(V_g)$ for the 1D and 2D configurations of the same sample can be caused by randomly trapped charge near the 1D constriction. When the 1D channel is blocked by strong fluctuations, electron transport is only possible under the continuous parts of the gate and the sample is effectively 2D. In the 1D configuration electron transport through the split is seen in Fig. 1a (ii) as a characteristic bend in the conductance. The second sample (B) has a $G(V_g)$ that does not look similar to either of the two curves of sample A.

The resistance of a macroscopic hopping network is determined by the most resistive (dominant) hops separated by the distance of about L_c. To determine the characteristic length L_c in the structure at different gate voltages, the T-dependence of the conductance of sample A in its 2D configuration has been measured, Fig. 1b. In the gate voltage region from -1.34 to -1.18 V good agreement with the T-dependence of variable range hopping (VRH) $G = G_0 \exp(-\xi)$ is found at $T > 4$ K (where $\xi = (T_0/T)^{1/3}$, $T_0 = \beta/(k_B a^2 g)$, g is the density of states at the Fermi level, $\beta = 13.8$ and a is the localisation radius). The parameter ξ at $T = 4.2$ K ranges from 2 to 6 as the negative gate voltage is increased, reflecting the decrease of the density of states as the Fermi level goes down. The correlation length $L_c \simeq a\xi^{2.33}/2$ is presented in Fig. 1b (inset).

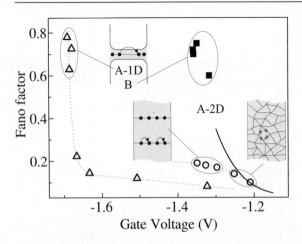

Fig. 2 Fano factor as a function of gate voltage for different samples: A-2D (circles), A-1D (triangles) and B (squares), with a schematic representation of hopping paths in each case. Small crosses show dominant hops. The solid line shows the $(L/L_c)^{-1}$ dependence and dotted lines are guides to the eye.

In order to measure shot noise at $T = 4.2$ K at a fixed V_g different currents I_{sd} are put through the sample, and voltage fluctuations are measured by two low-noise amplifiers. The cross-correlated spectrum of the two signals, $S_I(I_{sd})$, is detected by a spectrum analyser – this technique removes uncorrelated noise of the amplifiers and leads. The shot noise power is determined from the flat part of the excess noise spectrum above 30 kHz, where one can neglect the contribution of $1/f$ noise [7].

In order to determine the Fano factor we used the following fit for excess noise: $S_I(I_{sd}) = F2eI_{sd}$ $\coth\left(\dfrac{FeV_{sd}}{2k_BT}\right) - 4k_BTG_S$, where G_S is the ohmic conductance. This expression describes the evolution of noise from thermal noise for $eV_{sd} \ll k_BT/F$ into shot noise ($eV_{sd} \gg k_BT/F$) with the Fano factor $F = 1/N$, where N is the number of the dominant barriers along the chain [12, 13].

The Fano factor as a function of gate voltage for different structures is shown in Fig. 2. In the 2D case the Fano factor slowly increases from 0.1 to 0.2 with increasing negative gate voltage. According to the $F = 1/N$ model this change corresponds to a decrease in the number of dominant hops from 10 to 5. The value L/L_c from Fig. 1b (insert) is also plotted in Fig. 2, where in the range V_g from -1.34 to -1.18 V we find agreement between the Fano factor and the number of the dominant hops in the VRH network. With depleting the conducting channel the correlation length increases and approaches the length of the sample, while the Fano factor shows a saturation around $F \sim 0.2$. The difference between $(L/L_c)^{-1}$ and F confirms that electron transport at these gate voltages is dominated by chains of most conductive hops [9, 8]. High conductance of these 'optimal' chains is provided by the close position of localised states in them and the fact that all the barriers between the states are equal.

In sample A-1D the 1D channel is only formed at $V_g < -1.3$ V, Fig. 1a. In the range of V_g from -1.32 to -1.63 V, the Fano factor increases from 0.07 to 0.15 which corresponds to the number of dominant hops $N \simeq 1/F = 7$. With further increasing negative gate voltage ($V_g < -1.65$ V) the Fano factor in Fig. 2 rapidly increases to 0.8 as the distribution of the resistances of random hops in the 1D channel is exponentially broad. In this case a single hop dominates the whole conductance of the 1D channel, and the Fano factor is close to 1.

The conductance of the second sample (B) has not shown a clear definition of the 1D channel in Fig. 1a. For this sample shot noise in the range of V_g from -1.36 V to -1.31 V has shown an increase of the Fano factor from 0.6 to 0.8, Fig. 2. This large value compared with $F \sim 0.2$ expected for 2D hopping implies that, similar to A-1D, hopping in this sample occurs mainly through the 1D split and is dominated by one or two hard hops. The differences in $G(V_g)$ of samples A-1D, A-2D and B emphasise the importance of the random fluctuation potential in the formation of 1D channels [15]. If the random potential of impurities near the split is large, the conducting channel in the split cannot be formed (as in the case of sample A-2D). Sample B has a larger width of the split, and in several cooldowns the 1D channel has not been 'blocked' by the random potential. In the hopping regime (at $V_g < -1.31$ V) shot noise shows that the sample conduction is entirely determined by the split region.

It is interesting to note that for all studied samples, including the one with length 0.2 μm, the Poisson value of the Fano factor, $F = 1$, has never been observed. In Fig. 2 the maximum Fano factor fluctuates near $F \sim 0.7$. Although similar average values of F have been predicted for electron tunnelling through one impurity randomly positioned along the barrier ($F = 0.75$ [10]) and for tunnelling through two impurities ($F \sim 0.71$ [14]), this fact deserves further attention.

3 Enhancement of shot noise in resonant tunnelling via interacting localised states

Shot noise in the case of resonant tunnelling through a single impurity has been studied on a 2D sample with gate length 0.2 μm and width 20 μm. One can see in Fig. 3a that as the temperature is lowered the background conduction due to hopping decreases and the amplitude of the peaks increases, which is a typical feature of resonant tunnelling through an impurity [16]. The box in Fig. 3b indicates the range of V_g where shot noise has been studied at $T = 1.85$ K and 4.2 K. In Fig. 3c (inset) an example of the noise spectrum is shown at a gate voltage near the resonant tunnelling peak in Fig. 3b. The power spectral density of shot noise was determined at frequencies above 40 kHz where the contribution of $1/f^\gamma$ noise (shown in Fig. 3c (inset) by a solid line with $\gamma = 1.6$) can be totally neglected.

Fig. 3c shows the dependence of the shot noise power on V_{sd} at two temperatures. At small biases ($V_{sd} < 3$ mV) a pronounced peak in noise is observed, with an unexpectedly large Fano factor $F > 1$. At large biases ($V_{sd} > 3$ mV), shot noise decreases to a conventional sub-Poisson value, $F \sim 0.6$. The figure shows the dependence $S_I(V_{sd})$ with different F plotted using the phenomenological expression for shot noise for resonant tunnelling through a single impurity (cf. Eq. (62) in [1] and Eq. (11) in [10]): $S_I = F2eI_{sd} \coth\left(\dfrac{eV_{sd}}{2k_BT}\right) - F4k_BTG_S$. We have established that this increase of shot noise appears only in a specific range of V_g [11]. It will be shown below that the region V_{sd}–V_g with enhanced shot noise corresponds to the resonant current carried by two interacting impurities.

Consider two spatially close impurity levels, R and M, separated in the energy scale by $\Delta\epsilon$. If impurity M gets charged, the level R is shifted upwards by the Coulomb energy $U \sim e^2/4\epsilon_0\epsilon r$, where r is the separation between the impurities, Fig. 4a (diagram 1). Thus, depending on the occupation of M impurity R can be in two states: $R1$ or $R2$. If V_{sd} is small enough, state $R2$ is above the Fermi level

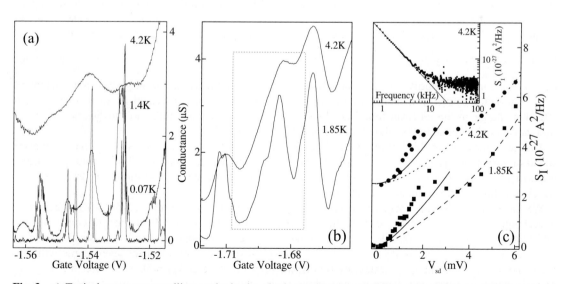

Fig. 3 a) Typical resonant tunnelling peaks in the ohmic conductance at different temperatures. b) Conductance peaks in the region of V_g where the current noise has been measured. c) Shot noise power as a function of V_{sd}: at $V_g = -1.6945$ V for $T = 1.85$ K and $V_g = -1.696$ V for $T = 4.2$ K. Lines show the dependences $S_I(V_{sd})$ expected for resonant tunnelling through a single impurity, with $F = 1$ (solid), $F = 0.63$ (dashed), and $F = 0.52$ (dotted). Inset: Excess noise spectrum at $V_g = -1.696$ V and $V_{sd} = 1.5$ mV.

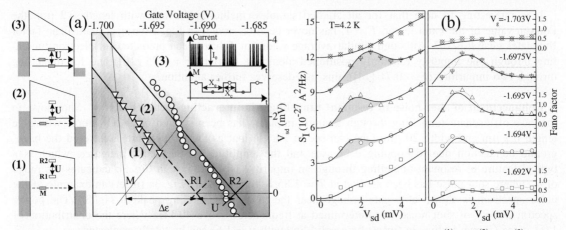

Fig. 4 a) Left panel: Energy diagrams of the two impurities for different positive V_{sd}: $V_{sd}^{(1)} < V_{sd}^{(2)} < V_{sd}^{(3)}$. Inset: Schematic representation of the modulation of the current through impurity R by changing the occupancy of modulator M. Main part: Grey-scale plot of the differential conductance as a function of V_g and V_{sd} at $T = 1.85$ K (darker regions correspond to higher differential conductance, background hopping contribution is subtracted). Lines show the positions of the conductance peaks of impurity R and modulator M obtained from the fitting of the noise data in (b). b) Shot noise and the corresponding Fano factor as a functions of source–drain bias at different gate voltages. Solid lines show the results of the numerical calculation.

in the left contact and electrons are transferred via $R1$ with the rates $\Gamma_{L,R}$ only when M is empty, Fig. 4a (diagram 2).

It can be shown that for such correlated transport via two impurities, a significant increase of noise can be seen if the leak rates of the two impurities are very different. For illustration, let us assume that, independent of the occupancy of R, impurity M (modulator) gets charged with rate X_c and empties with rate X_e. If $X_{e,c} \ll \Gamma_{L,R}$, the contribution of M to the total current through the two impurities is negligible. As a result, the current through impurity R jumps randomly between two values: zero and I_0, dependent on the occupancy of M, Fig. 4a (inset). If the bias is further increased, the upper state $R2$ is shifted down and the modulation of the current via impurity R vanishes, Fig. 4a (diagram 3). In this modulation regime, the corresponding Fano factor can be written as

$$F \simeq \frac{\Gamma_L^2 + \Gamma_R^2}{(\Gamma_L + \Gamma_R)^2} + 2 \frac{\Gamma_L \Gamma_R}{\Gamma_L + \Gamma_R} \frac{X_c}{(X_e + X_c)^2}.$$ The first term describes the conventional (suppressed)

Fano factor for one-impurity resonant tunnelling [10], whereas the second term gives an enhancement of F. The latter comes from bunching of current pulses, which can be treated as an increase of the effective charge in the electron transfers. Physically, this situation is similar to the increase of flux-flow noise in superconductors when vortices move in bundles [17].

The generalisation of this simple model for any relation between X and Γ is based on the master equation formalism [10, 18]. As a result of these calculations, the current through two interacting impurities and the Fano factor are obtained as functions of the energy positions of the impurities which are shifted with changing V_{sd} and V_g.

We have shown that the increase of shot noise occurs exactly in the region of V_g–V_{sd} where two interacting impurities carry the current in a correlated way. Fig. 4a presents the grey scale of the differential conductance versus V_g and V_{sd}. When a source-drain bias is applied, a single resonant impurity gives rise to two peaks in $dI/dV(V_g)$, which occur when the resonant level aligns with the Fermi levels $\mu_{L,R}$. On the grey scale they lie on two lines crossing at $V_{sd} = 0$. Therefore, for impurity M the central area between the thin lines of the cross at point M in Fig. 4a corresponds to its level being between μ_L and μ_R, that is to the situation when its occupancy changes with time. On the left of the central region the impurity is empty and on the right it is filled.

Experimentally, the cross of M is not seen as the modulator gives a negligible contribution to the current – it is plotted in accordance with the analysis below. However, at small V_{sd} a cross-like feature is clearly seen near point $R2$ – the exact positions of the maxima of the conductance peaks of this line are indicated by circles. With increasing V_{sd}, a new parallel line $R1$ appears at $V_g \approx -1.694$ V and $V_{sd} \approx 1$ mV, shifted to the left by $\Delta V_g \approx 4$ mV. This happens when the line $R2$ enters the central area of cross M – the maxima of the conductance peaks of the new line are shown by triangles. The lines $R1$ and $R2$ reflect the two states of impurity R shifted due to the Coulomb interaction with impurity M. The modulation of the current should then occur in region (2) of the central area of the cross, Fig. 4a, which corresponds to diagram (2). In region (3) there is no modulation as both states $R1$ and $R2$ can conduct, and in region (1) there is no current as the low state $R1$ is still above μ_L.

In Fig. 4b current noise and the Fano factor are presented as functions of V_{sd} for different V_g. It shows that indeed the increase of noise occurs only in region (2) in Fig. 4a. For a quantitative analysis we have to take into account that in our experiment resonant tunnelling via state R exists in parallel with the background hopping. Then the total Fano factor has to be expressed as $F = (F_{RT}I_{RT} + F_B I_B)/(I_{RT} + I_B)$, where F_{RT}, F_B and I_{RT}, I_B are the Fano factors and currents for resonant tunnelling and hopping, respectively. In order to get information about the background hopping we have measured the current and noise at $V_g > -1.681$ V, i.e. away from the resonant tunnelling peak under study in Fig. 4a.

The numerical results have been fitted to the experimental $dI/dV(V_{sd}, V_g)$ and $S_I(V_{sd}, V_g)$, Fig. 4a, b. The fitting parameters are the leak rates of R and M ($\hbar\Gamma_L \simeq 394$ μeV, $\hbar\Gamma_R \simeq 9.8$ μeV, and $\hbar X_e \simeq 0.08$ μeV, $\hbar X_c \simeq 0.16$ μeV), the energy difference between R and M ($\Delta\varepsilon = 1$ meV), and the Fano factor for the background hopping ($F_B = 0.45$). This value of the Fano factor is expected for shot noise in hopping through $N \sim 2-3$ potential barriers (1–2 impurities in series) [7, 13]. The coefficients in the linear relation between the energy levels M, R and V_{sd}, V_g have also been found to match both the experimental data in Fig. 4b and the position of lines $R1$ and $R2$ in Fig. 4a. The Coulomb shift ($U \sim 0.55$ meV) found from Fig. 4a agrees with the estimation for the Coulomb interaction between two impurities not screened by the metallic gate: $U \sim e^2/\kappa d \sim 1$ meV, where $d \sim 1000$ Å is the distance between the gate and the conducting channel.

It is interesting to note that the hopping background effectively hampers the manifestation of the enhanced Fano factor F_{RT}: the largest experimental value of F in Fig. 4b (at $V_g = -1.6975$ V) is approximately 1.5, while a numerical value for RT at this V_g is $F_{RT} \approx 8$.

With further decreasing the temperature down to $T = 70$ mK we have seen a significant increase of the Fano factor up to $F \sim 5$, Fig. 5. This regime corresponds to the quantum case $\hbar\Gamma \gg k_B T$ where the master equation approach is not applicable. We can assume that this increase is also related to the interaction between different resonant tunnelling channels, although more theoretical input is required to understand this effect. There is another interesting feature in Fig. 5: the increase of shot noise slows

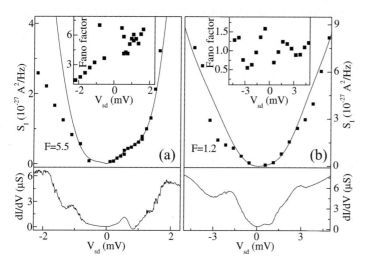

Fig. 5 Shot noise power and differential conductance as a function of the current for the 2D sample with length 0.2 μm at different temperatures: a) $T = 0.07$ K and b) $T = 1.4$ K. Solid lines show a fit for shot noise for resonant tunnelling through a single impurity.

down at the position of the conductance peak, so that the Fano factor shows a decrease at the points, where the new current channel is switched on, Fig. 5 (insets).

4 Conclusion The details of hopping transport in 1D and 2D have been investigated by measuring shot noise. It is shown that suppression of shot noise contains information about the geometry of hopping paths and the distribution of hopping barriers in them. In one-impurity resonant tunnelling we have observed a significant increase of shot noise. It it shown that this increase is caused by the Coulomb interaction between two close resonant tunnelling channels. In general, we have demonstrated that shot noise is a valuable tool for investigations of electron transport in the insulating regime of conduction.

References

[1] Ya. M. Blanter and M. Büttiker, Phys. Rep. **336**, 1 (2000).
[2] M. Reznikov, M. Heiblum, H. Shtrikman, and D. Mahalu, Phys. Rev. Lett. **75**, 3340 (1995).
    A. Kumar, L. Saminadayar, D. C. Glattli, Y. Jin, and B. Etienne, Phys. Rev. Lett. **76**, 2778 (1996).
[3] F. Liefrink, J. I. Dijkhuis, M. J. M. de Jong, L. W. Molenkamp, and H. van Houten, Phys. Rev. B **49**, 14066 (1994).
    A. H. Steinbach, J. M. Martinis, and M. H. Devoret, Phys. Rev. Lett. **76**, 3806 (1996).
    M. Henny, S. Oberholzer, C. Strunk, and C. Schönenberger, Phys. Rev. B **59**, 2871 (1999).
[4] H. Birk, M. J. M. de Jong, and C. Schönenberger, Phys. Rev. Lett. **75**, 1610 (1995).
[5] V. V. Kuznetsov, E. E. Mendez, G. L. Snider, and E. T. Croke, Phys. Rev. Lett. **85**, 397 (2000).
[6] B. I. Shklovskii and A. L. Efros, Electronic Properties of Doped Semiconductors (Springer, Berlin, 1984).
[7] S. H. Roshko, S. S. Safonov, A. K. Savchenko, W. R. Tribe, and E. H. Linfield, Physica E **12**, 861 (2001).
[8] M. E. Raikh and A. N. Ruzin, in: Mesoscopic Phenomena in Solids, edited by B. L. Altshuler, P. A. Lee, and R. A. Webb (Elsevier Science, Amsterdam, 1991).
[9] A. O. Orlov, M. E. Raikh, I. M. Ruzin, and A. K. Savchenko, Sov. Phys. JETP **69**, 1229 (1989).
[10] Y. V. Nazarov and J. J. R. Struben, Phys. Rev. B **53**, 15466 (1996).
[11] S. S. Safonov, A. K. Savchenko, D. A. Bagrets, O. N. Jouravlev, Y. V. Nazarov, E. H. Linfield, and D. A. Ritchie, Phys. Rev. Lett., accepted for publication (2003).
[12] A. van der Ziel, Noise in Solid State Devices and Circuits (Wiley, New York, 1986).
[13] A. N. Korotkov and K. K. Likharev, Phys. Rev. B **61**, 15975 (2000).
[14] Yu. A. Kinkhabwala and A. N. Korotkov, Phys. Rev. B **62**, 7727 (2000).
[15] J. A. Nixon, J. H. Davies, and H. U. Baranger, Phys. Rev. B **43**, 12638 (1991).
    M. J. Laughton, J. R. Barker, J. A. Nixon, and J. H. Davies, Phys. Rev. B **44**, 1150 (1991).
[16] A. B. Fowler, G. L. Timp, J. J. Wainer, and R. A. Webb, Phys. Rev. Lett. **57**, 138 (1986).
    T. E. Kopley, P. L. McEuen, and R. G. Wheeler, Phys. Rev. Lett. **61**, 1654 (1988).
[17] J. R. Clem, Phys. Rep. **75**, 1 (1981).
[18] L. I. Glazman and K. A. Matveev, Zh. Eksp. Teor. Fiz. **94**, 332, (1988) [Sov. Phys. JETP **67**, 1276 (1988)].

phys. stat. sol. (b) **241**, No. 1, 33–39 (2004) / **DOI** 10.1002/pssb.200303651

Magnetic ordering in doped semiconductors near the metal–insulator transition

A. G. Zabrodskii[*]

Ioffe Physico-Technical Institute of the Russian Academy of Sciences, St. Petersburg, 194021, Russia

Received 1 September 2003, revised 3 October 2003, accepted 6 October 2003
Published online 15 December 2003

PACS 71.30.+h, 72.20.My, 76.30.Pk

Experiment clearly shows a certain extent of ordering of localized magnetic moments in doped semiconductors near the metal–insulator transition. An antiferromagnetic spin glass structure is revealed in a number of n-type materials (Ge:As, 6H- and 4H-SiC:N) by Electron Spin Resonance spectroscopy as a sharp decrease in the density of paramagnetic centres, observed when approaching the transition from its insulator side. In a p-type material (Ge:Ga), the macroscopic magnetization and residual magnetization, as well as hole coupling in pairs or clusters with uncompensated moment, follow from the recently discovered hysteresis of variable range hopping magnetoresistance, accompanied by sharp drops in resistance on reversing the magnetization of the system. The location of the magnetically ordered phases is established.

1 Introduction   The present-day experience on the spin structure in n-type semiconductors is mainly based on studies of the Si:P system by Electron Spin Resonance (ESR) spectroscopy ([1] and references therein). For example, a wide background line can be observed at low temperature very far from the critical density $n_c = 3.7 \cdot 10^{18}$ cm^{-3} between two hyperfine lines characteristic of isolated donors [2]. This line can presumably be attributed to spin pairing and clustering, with antiferromagnetic exchange interaction. At higher concentration, $n \approx 0.14\, n_c$,[1] the hyperfine structure disappears and only the background line exists. The line is narrowing towards the metal–insulator (MI) transition under the influence of the motional narrowing resulting from an increase in the electron hopping rate. A conclusion that there occurs an antiferromagnetic interaction between the spins within a cluster has been made [2] on the basis of a weaker, compared with that following from the Curie law for both the hyperfine lines, temperature shift of the background line. According to another point of view, the disappearance of the hyperfine structure is caused by an increase in the diffusion rate of electron spins under the influence of their interaction with different nuclear spins in the ESR. Nevertheless, the formation of the amorphous antiferromagnetic insulating phase near the MI transition in doped semiconductors seems to be widely accepted [1], which is difficult to say about its location.

It is noteworthy that the ESR results cited above refer only to the qualitative form and position of the line, whereas the ESR technique makes it possible, in principle, to estimate the density of paramagnetic centres on the basis of the intensity of the integral ESR absorption signal [5], and thus to verify directly the assumption about spin coupling. Investigations of this kind have been carried out for 6H-SiC:N [6], 4H-SiC:N [7, 8], and Ge:As [9, 10] n-type semiconductors. All of these show a sharp decrease in the density of localized paramagnetic centres in the insulating state, thus indicating directly the occurrence of local antiferromagnetic spin ordering ("spin-glass" structure), but not as far from the MI transition as one could believe previously.

---

[*] e-mail: Andrei.Zabrodskii@mail.ioffe.ru, Phone: +7 812 2472375, Fax: +7 812 2471017
[1] This concentration belongs to a range of strong localization, because the MI transition in the Si:P system is very sharp [3, 4].

In p-type semiconductors, to which the ESR technique is inapplicable, the question of magnetic moment ordering remains open, as far as we know. Several years ago a discovery was reported of the hysteresis of low-temperature magnetoresistance (MR) in the mode of the variable range hopping (VRH) via the Coulomb gap states in a moderately compensated Ge:Ga p-type semiconductor characterized by resistivity drop in some critical field [11–13]. To account for this phenomena, the existence of macroscopic magnetization and residual magnetization was suggested for the Ge:Ga p-type system on the insulator side of the MI transition.

In sections 2 and 3 below, we review and discuss recent results concerning the magnetic ordering near the MI transition in the n- and p-type semiconductor systems [6–13] and establish the location of the magnetically ordered phases.

2 Antiferromagnetic spin glass ordering in n-type 4H-SiC:N and Ge:As systems in the pretransition range

Two of the most appropriate ways to fabricate series of samples suitable for a study of the MI transition were used: (i) variation of the main impurity concentration at nearly constant degree of compensation ("$K \approx 0.2$" series of 4H–SiC:N and "$K \approx 0$" series of uncompensated Ge:As) and (ii) dosed compensation of the initially metallic sample to varied extent ("$N = 3.6 \cdot 10^{17}$ cm^{-3}" series of compensated Ge:As). The 4N–SiC:N sample series, produced by the sublimation sandwich method, covered the electron density range $0.015 \leq n/n_c \leq 1.7$, where the charge density critical for the MI transition was $n_c = N_d - N_a = 1.5 \cdot 10^{19}$ cm^{-3} and was characterized by the natural (uncontrolled) degree of compensation, $K = N_a/N_d$ of about 20 %. The "$K = 0$" series of uncompensated Ge:As was manufactured by the Czochralski growth technique and covered the range $0.06 \leq n/n_c \leq 1.6$, $n_c = 3.7 \cdot 10^{17}$ cm^{-3}. The "$N = 3.6 \cdot 10^{17}$ cm^{-3}" series of compensated Ge:As was produced by introducing a compensated Ga impurity by means of dosed neutron transmutation doping of the initially uncompensated Ge:As with donor concentration $N_d = 3.6 \cdot 10^{17}$ cm^{-3}. The subsequent thermal annealing at $T = 450$ C for 60 h was used to eliminate the radiation defects. The samples produced were characterized by the degree of compensation, $K = 0 \div 0.8$, and electron densities $n = (0.93 \div 0.26) \, n_c$, with $n_c = 3.7 \cdot 10^{17}$ cm^{-3} for $K = 0$.

ESR spectra were measured on a Varian ESR E–112 spectrometer (10 GHz) combined with an ESR–9 cryostat, which allowed recording of spectra in the temperature range $T = 3.2 \div 300$ K. The spin density was calculated by comparing the simultaneously measured signals from a sample under study and a Varian strong pitch reference sample characterized by spin density $n_s = 2.58 \cdot 10^{15}$ cm^{-3} and $g = 2.0028$. In these calculations, we disregarded the skin layer formation because, except in metallic

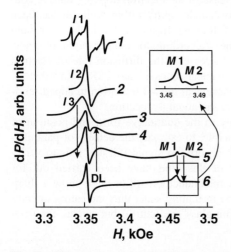

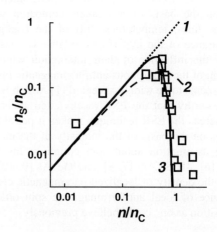

Fig. 1 ESR spectra of 4H-SiC:N samples, n/n_C: 1 – 0.05, 2 – 0.26, 3 – 0.27, 4 – 0.64, 5 – 0.97, 6 – 1.55.

Fig. 2 Relative spin density in 4H-SiC:N samples versus relative electron density: 1 – paramagnetic insulator $n = n_S$, 2 – fit with Eq. (2), 3 – fit with Eq. (4).

4H-SiC:N samples, its depth was no less than the samples thickness. In metallic 4H-SiC:N samples, the corresponding error in the spin density (by a factor of $2 \div 3$) was much smaller than the decrease in the total spin density by $2 \div 3$ orders of magnitude (see below). Even though the ESR measurements were carried out in a wide temperature range $3.2 \div 300$ K, only low-temperature results will be discussed bellow for the sake of brevity. In insulating samples, these data correspond to the hopping region: nearest-neighbour hopping in uncompensated Ge:As and variable range hopping via the Coulomb gap states in compensated Ge:As and 4H-SiC:N, which was confirmed by the known method of resistivity curve derivative analysis (RCDA) [14].

Representative ESR spectra for the 4H-SiC:N series, based on the data from [8], are shown in Fig. 1. It can be seen that a decrease in the doping level results in a number of significant changes, as in Si:P and other systems. A characteristic feature is the three-line hyperfine structure (I1, curve *1*) in lightly doped samples, in accordance with the nuclear spin $S = 1$ of the nitrogen atom in 4H-SiC. With the electron density increasing to $n/n_c \approx 0.2$, the hyperfine structure vanishes and the spectrum is transformed into a single line (I2, curve *2*). With further doping, the Lorentzian line becomes Dysonian (I3, curve *3*) under the influence of the sample conductivity increasing. At still higher doping levels (curves *4–6*), there appears a new narrow line (DL) of constant width, which originates from deep-level paramagnetic centres. At the highest densities, $n/n_c \geq 0.93$ (curve *5*, *6*), two new lines (M1, M2) are observed (seen as a single line only if $\boldsymbol{H} \parallel c$-axis). The intensity of the broad line (M2) is decreasing rapidly when one moves across the critical point $n = n_C$, and that of the M1 line is increasing.

The spin density n_S/n_C, shown as a function of relative Hall concentration $(N_d - N_a)/n_C \equiv n/n_C$ in Fig. 2, was determined by double integration of the narrow weak-field part of the ESR line of the 4H-SiC:N sample, with calibration against a Varian strong pitch reference sample. The dashed line in the figure represents the Curie paramagnetic, for which $n_C = n$. It can be readily seen that the initially linear dependence of the spin density on the doping level becomes weaker at $n/n_C \geq 0.15$ and shows a sharp drop just before the critical point $n = n_C$ of the MI transition. Thus, the ESR absorption becomes negligible at the $n \approx n_C$ and corresponds to the single spin density $n_S \approx 0.004n$, which indicates strong antiferromagnetic spin coupling in the pretransition range with the formation of a magnetically ordered spin-glass structure at low temperatures. This coupling is characterized by the magnitude of the exchange interaction decreasing exponentially on the distance between spins. Evidently, beginning with a certain doping level, the antiferromagnetic spin coupling onsets first for pairs of the nearest neighbours and then for more distant neighbours, and clusters with the number of spins exceeding two.

If the spin pairing at a spin density n is governed only by the probability of finding a neighbouring spin at a distance shorter than some value r_0, the relative concentration of paired spins, n_P, will be determined by the integral

$$n_P/n = \int_0^{r_0} \exp\left(-4 \, r^3 n/3\right) 4\pi r^2 n \, dr = 1 - \exp\left(-4\pi r_0^3 n/3\right), \qquad (1)$$

where $\exp\left(-4\pi r^3 n/3\right) 4\pi r^2 n \, dr$ is the probability of finding a nearest neighbour in a spherical layer of radius r and thickness dr.

Then the relative density of the paramagnetic centres detected by the ESR will be given by an exponent:

$$n_S/n = \exp\left(-4\pi r_0^3 n/3\right) . \qquad (2)$$

Correspondingly, fitting of Eq. (2) to the data in Fig. 2 yields a value $r_0 = 0.7 \, n_C^{-1/3}$ and is shown by the dotted line. The fit fails at the sharp drop near the critical point n_C. It is noteworthy that, in the immediate vicinity of the MI transition, we can suggest another reason for spin pairing or clustering in a volume $4\pi a^3/3$, where a is the localization radius diverging at the critical point with some index ν:

$$a = a_0(1 - n/n_C)^{-\nu} , \qquad (3)$$

with typical values $\nu = 0.5 \div 1.0$ for semiconductors [14].

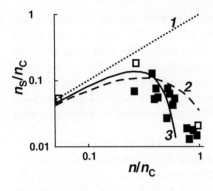 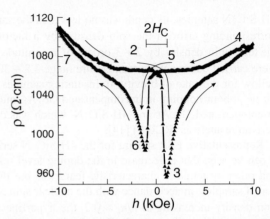

Fig. 3 Relative spin density in compensated (full squares) and uncompensated (open squares) Ge:As samples versus relative electron density; curve numbers correspond to those in Fig. 2.

Fig. 4 MR curves for Ge:Ga sample with $n/n_c = 0.38$ at $T = 0.43$ K; arrows indicate the direction of the field sweep with a rate 3.6 kOe/min.

According to Eq. (3), a starts to exceed r_0 in the vicinity of the MI transition, which will result in a sharper decrease in the paramagnetic spin density n_S with increasing n. To describe this numerically, we substitute a from Eq. (3) for r_0 in Eq. (2):

$$n_S/n = \exp\left(-4\pi a^3 n/3\right). \tag{4}$$

The corresponding fit gives a reasonable value $\nu \approx 0.7$, with $a_0 = 0.35\ n_C^{-1/3}$. The fit is shown in Fig. 2 by a solid line and seems to be much better than that with Eq. (2), thus indicating an important role of clustering in regions of size a exceeding the average spacing between the main impurities just in the vicinity of the critical point.

We now consider briefly the series of uncompensated and compensated Ge:As samples. The spectra of lightly doped samples show 4-line hyperfine structure corresponding to the nuclear spin of As, equal to 3/2. The evolution of the spectra with doping is qualitatively the same, from our standpoint, as that in 4H-SiC:N, Si:P, and other systems. Therefore, let us turn our attention to the resulting behaviour of the paramagnetic spin density with doping level, shown in Fig. 3 for both uncompensated (open circles) and compensated (full circles) samples. The behaviour is not so sharp as that for 4H-SiC:N (Fig. 2) and can be described well enough both by Eq. (2) with $r_0 = 0.8\ n_C^{-1/3}$ and by Eq. (4) with $a_0 = 0.7\ n_C^{-1/3}$ and $\nu = 0.5$.

The common feature of the all the sample series investigated is that a decrease in the paramagnetic centre density occurs in the vicinity of the MI transition within a region much narrower than that stated above: it begins at the doping level $n \approx 0.15\ n_C$, with the density decreasing twice at $n \approx 0.3\ n_C$. It is in this region that the antiferromagnetic spin ordering takes place.

3 Magnetization, residual magnetization and magnetic ordering in the hole Ge:Ga system near the MI transition   Recently, a low temperature hysteresis of MR has been discovered [11–13] in a series of neutron-transmutation-doped p-type Ge:Ga samples near the MI transition in the range of the VRH via the Coulomb gap states. The samples investigated contained Ga acceptors in concentrations $N = 0.26 \div 1.26\ N_c$, where $N_c = 1.85 \times 10^{17}\ cm^{-3}$ is the critical density for the MI transition [15].

The MR hysteresis is accompanied by the characteristic resistivity drop in critical field $H_c = 600 - 850$ Oe and the subsequent relaxation to the equilibrium state, as shown in Fig. 4 for $T = 0.45$ K. Outside the relaxation region, the ordinary quadratic MR behaviour is originating from the well-known shrinkage of the localization radius a in the plane perpendicular to $\boldsymbol{H}$.[2] The drops take place only after the sample is subjected to a field $H \geq 2$ kOe, with subsequent reversal of its sign.

---

[2] It is widely used to determine the value of a in investigations of its critical divergence (see, for example, [14, 15]).

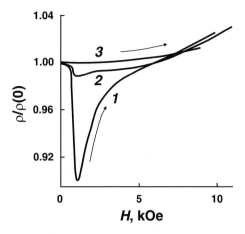

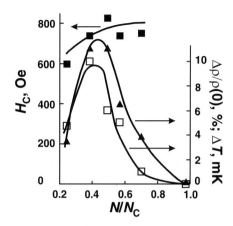

Fig. 5 MR curves for Ge:Ga sample with $n/n_C = 0.38$ at different temperatures, K: 1 – 0.45, 2 – 0.65, 3 – 1.4.

Fig. 6 Critical field (full squares), relative MR drop (open squares) and the corresponding temperature jump (full triangles) versus relative Ga concentration, $T = 0.43$ K.

A quarter of a full MR circle at different temperatures is shown in Fig. 5 for a constant sweep rate. It can be seen that the resistivity drop vanishes at temperatures slightly higher than ~0.7 K, which nearly coincides with the high-temperature boundary of the VRH in the Coulomb gap (so-called "$T^{-1/2}$" law), determined previously for the Ge:Ga system [15]. Thus, the MR hysteresis emerges when the "$T^{-1/2}$" law of the VRH transport becomes dominant and then rapidly becomes more pronounced with decreasing temperature.

The hysteresis described above was not observed in metallic samples. Figure 6 shows the relative drop magnitudes $\Delta\rho/\rho(0)$ and the critical field H_C (determined from the maximum in the derivative $d\rho/dH$) as functions of the relative doping level N/n_c. It can be seen that H_C only slightly increases with doping, whereas $\Delta\rho/\rho(0)$ exhibits a clearly pronounced maximum at $N \approx 0.4N_c$ and vanishes in the vicinity of the critical point of the MI transition.

Our studies have shown that the MR hysteresis described above is virtually independent of the measurement current and **H** orientation (be it transverse or longitudinal) relative to the current direction.

Since the probe and current contacts to the samples studied were typically fabricated by alloying with In, which is superconducting at $T \leq 3.4$ K, a question arises as to whether the destruction of its superconductivity can account for the observed MR hysteresis. In the first place, the superconductivity of In is destroyed by the field $H = 290$ Oe at $T = 0.4$ K and thus bears no relation to MR drops. Direct experimental comparison of the MR behaviour in the samples with superconducting (In) and non-superconducting (Au) contacts shows that the pattern is qualitatively the same in both cases [12].

Could the observed hysteresis drops in MR in the Ge:Ga sample be caused by heat release by an ensemble of localized magnetically ordered moments in the course of its magnetization reversal in the critical field H_C? To confirm the hypothesis qualitatively, two kinds of experiments were done. The first of these was aimed to study the consequences of reducing the heat exchange with the thermostat by coating the sample with stearin. It was suggested that the rate of heat removal from the sample would be lowered and the relaxation to equilibrium would be slowed-down owing to the emergence of additional thermal mass in cooling. This suggestion has been confirmed experimentally [12]. The second experiment was performed to find out whether or not part of the MR hysteresis pattern lying just after the drop has a thermal relaxation nature. To do this, the magnetic-field sweep rate was lowered by a factor of 50 at the point of the minimum in MR, shown in Fig. 7. As a result, the following part of the curve was transformed into a vertical line corresponding to an increase in resistivity in the relaxation process, thereby confirming its cooling nature. Thus, it was confirmed that the MR resistivity drops ob-

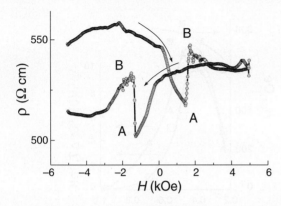

Fig. 7 MR of Ge:Ga sample with $n/n_c = 0.38$; the arrows indicate the direction of field sweep with a rate of 3.6 kOe/min; at the minima (A), the rate is lowered by a factor of 50 times, and after the relaxation ends (B), its initial value is restored.

served are caused by heat release by the magnetization reversal of the system of magnetic moments and are characterized by coercive force $H_C = 600 \div 800$ Oe. It should be emphasized that the existence of the coercive force in the p-type Ge:Ga system clearly shows a residual macroscopic magnetization resulting from a certain extent of magnetic ordering in an external field, characterized by a macroscopic magnetic moment, in contrast to the above-described antiferromagnetic ordering in the n-type 4H-SiC:N and Ge:As systems. However, the ordering takes place in all cases near the MI transition.

We now turn to a qualitative explanation of the MR drops observed. Let us start with the $\Delta\rho/\rho(0)$ behaviour with versus relative Ga concentration, shown in Fig. 6, and take into account that the drop magnitude $\Delta\rho$ is determined by an increase in the sample temperature, ΔT, in the magnetization reversal process:

$$\Delta\rho = (d\rho/dT)\, \Delta T\,, \tag{5}$$

where $\rho = \rho_0 \exp (T_0/T)^{1/2}$ in the VRH mode for the "$T^{-1/2}$" law.

The thus measured ΔT values are shown in Fig 6. It can be seen that the peak in the ΔT curve lies just near the MI transition and is centred at a density $N = 0.5 N_C$.

The temperature jump ΔT, in its turn, results from the energy dissipation ΔQ, which can be estimated on the basis of the thermal balance equation for the sample–thermostat (liquid He^3) system. The rate of thermal exchange between the sample and liquid He^3 depends on the thermal Kapitza resistance. However, as seen in Fig. 4, the rate of sample warming-up during an MR drop is much higher than that of thermal exchange with the He^3 thermostat in the relaxation process. Therefore, we can take into account only the heating of a sample in the simple thermal balance equation:

$$\Delta Q/m \approx c(T)\Delta T\,, \tag{6}$$

where $c = 6 \cdot 10^{-7} \cdot T^{-3}$ (J/g $\cdot$ K) is the specific heat of Ge, and m is the sample mass.

According to Eq. (6), the specific energy dissipation ΔQ is proportional to the temperature jump ΔT shown in Fig. 6, or, in other words, it is peaked at the density $N = 0.5\ N_C$. It was shown in [13] that this energy dissipation can be accounted for by splitting of a shallow acceptor level into 4 ones with quantum numbers $-3/2$, $-1/2$, $1/2$, and $3/2$, and by linear shift of these levels in the external field [16]. At the same time, as demonstrated in the present study, the antiferromagnetic coupling takes place in the system under consideration in pairs and clusters in the pretransition range. A very important difference from the n-type material, in which the antiferromagnetic interaction results in zero moments, is that the pairing in the p-type Ge:Ga system results only in compensation of the magnetic moment component along the pair axis [16]. Thus, its value in the perpendicular plane is finite, and we assume that this may result in residual magnetization phenomena in an external field, reflected by the peaks in Fig. 6.

4 Conclusion   The experiment confirmed the occurrence of a clearly pronounced magnetic ordering in a number of doped n- and p-type semiconductors just near the MI transition. It is in this range that the ESR absorption in n-type 4H-SiC:N and Ge:As shows a sharp decrease in the density of paramagnetic centres and, thus, reveals an antiferromagnetic spin-glass pairing and clustering. At the same time, the residual magnetization and coercive force in the p-type Ge:Ga system assume coupling in pairs or clusters with uncompensated moments, as it follows from a study of the recently discovered VRH MR hysteresis accompanied by sharp resistivity drops and caused by heating of the system in magnetization reversal. Magnetically ordered phases in doped semiconductors exist in the vicinity of the MI transition in a narrow enough density range: $0.3 \leq n/n_C \leq 1$.

Acknowledgements The author is grateful to his co-authors in the related publications, A. I. Veinger, R. V. Parfen'ev, S. E. Egorov, and T. V. Tisnek, and especially to T. V. Tisnek for assistance in manuscript preparation. The study was supported by the Russian Foundation for Basic Research (project 01-02-17873) and by a grant of the President of the Russian Federation (no. 2223.2003.02).

References

[1] N. F. Mott, Metal–Insulator Transition (Tailor & Francis Ltd., London, 1974).
[2] K. Morigaki and S. Moekava, J. Phys. Soc. Jpn. **32**, 462 (1972).
[3] T. F. Rosenbaum, K. Andres, G. A. Thomas, and R. N. Bhatt, Phys. Rev. Lett. **45**, 1723 (1980).
[4] H. F. Hess, K. De Conde, T. F. Rosenbaum, and G. A. Thomas, Phys. Rev. B **25**, 5578 (1982).
[5] C. P. Poole, Electron spin resonance (John Willey & Sons, New York–London–Sydney, 1967).
[6] A. G. Zabrodskii, M. V. Alekseenko, M. P. Timofeev, A. I. Veinger, and V. A. Il'in: Proceedings of the 18th International Conference on the Physik of Semiconductors, Stockholm, 1986 (World Scientific Publishing, 1987), p. 283.
[7] A. I. Veinger, A. G. Zabrodskii, T. V. Tisnek, and E. N. Mokhov, phys. stat. sol. (b) **230**, 113 (2002).
[8] A. I. Veinger, A. G. Zabrodskii, T. V. Tisnek, and E. N. Mokhov, Sov. Phys. – Semicond. **37**, 874 (2003).
[9] A. I. Veinger, A. G. Zabrodskii, and T. V. Tisnek, Sov. Phys. – Semicond. **34**, 46 (2000).
[10] A. I. Veinger, A. G. Zabrodskii, T. V. Tisnek, and S. I. Goloshchapov, phys. stat. sol. (c), **1** (2004) (this conference).
[11] A. G. Andreev, A. V. Chernyaev, S. V. Egorov, R. V. Parfeniev, and A. G. Zabrodskii, phys. stat. sol. (b) **218**, 165 (2000).
[12] A. G. Andreev, S. V. Egorov, A. G. Zabrodskii, R. V. Parfeniev, and A. V. Chernyaev, Sov. Phys. – Semicond. **34**, 768 (2000).
[13] S. V. Egorov, A. G. Zabrodskii, and R. V. Parfeniev, Sov. Phys. – Semicond., in press.
[14] A. G. Zabrodskii, Philos. Mag. **88**, 1131 (2001).
[15] A. G. Zabrodskii, A. G. Andreev, and S. V. Egorov, phys. stat. sol. (b) **205**, 61 (1998).
[16] N. S. Averkiev and S. Yu. Il'inskii, Sov. Phys. – Solid State **36**, 278 (1994).

phys. stat. sol. (b) **241**, No. 1, 40–46 (2004) / **DOI** 10.1002/pssb.200303617

Wigner approach to quantum dynamics simulations of the interacting carriers in disordered systems

V. Filinov[*, 1], **P. Thomas**[2], **M. Bonitz**[3], **V. Fortov**[1], **I. Varga**[4], and **T. Meier**[2]

[1] Institute for High Energy Density, Russian Academy of Sciences, Izhorskay 13/19, Moscow 125412, Russia

[2] Fachbereich Physik, Philipps-Universität Marburg, 35032 Marburg, Germany

[3] Fachbereich Physik, Universität Rostock, Universitätsplatz 3, 18051 Rostock, Germany

[4] Elméleti Fizika Tanszék, Budapesti Műszakiés Gazdaságtudományi Egyetem, 1521 Budapest, Hungary

Received 1 September 2003, revised 17 September 2003, accepted 17 September 2003
Published online 15 December 2003

PACS 05.10.−a, 05.30.−d, 72.15.Rn

The new method for solving Wigner–Liouville's type equations and studying dynamics of quantum particles has been developed within the Wigner formulation of quantum statistical mechanics. This approach combines both molecular dynamics and Monte Carlo methods and computes traces and spectra of the relevant dynamical quantities. Considering, as an application, the quantum dynamics of an ensemble of interacting electrons in an array of random scatterers clearly demonstrates that the many-particle interaction between the electrons can lead to an enhancement of the electrical conductivity.

1 Introduction It is well known that molecular dynamics method due to its highly efficiency is widely used in treatment of dynamic problems of classical statistical physics. The aim of this work is to develop the 'straight generalization' of classical molecular dynamics methods for rigorous consideration of quantum problems. The words 'straight generalization' mean that in classical limit the developed approach should exactly coincide with molecular dynamics method in the phase space. A generalization molecular dynamics method is possible only in the phase space, so in our work it is naturally to use Wigner formulation of quantum mechanics. In 1932 Wigner proposed joint position and momentum (phase space) representation of quantum mechanics and derived quantum analog of the classical distribution function. This representation contains only the values common both for classical and quantum mechanics, which is especially convenient when one of two interacting subsystems is quantum and another − classical. Wigner's paper has given rise to an extensive literature on formal aspects of quantum theory in phase space.

Noninteracting electrons in an array of fixed random scatterers are known to experience Anderson localization at temperature $T = 0$ in one-dimensional systems [1–3]. However, it is expected that the many-particle interaction leads to delocalization tendencies which has been confirmed for simple models [4, 5]. To study the influence of the electron–electron Coulomb interaction on kinetic electron properties in a random environment we have simulated the quantum dynamics in a one-dimensional canonical ensemble at finite temperature for both interacting and noninteracting electrons using the developed Quantum–Dynamics–Monte-Carlo scheme. We discovered that the temporal momentum–momentum correlation functions and their frequency-domain Fourier transforms strongly depend on

---

[*] Corresponding author: e-mail: filinov@ok.ru, Phone: +07 095 9310719, Fax:+07 095 4857990

the electron–electron interaction, clearly demonstrating the delocalizing influence of the many-particle interaction at densities around $R_s = \bar{r}/a_0 = 5$ ($\bar{r}$ is the mean interparticle distance and a_0 the effective Bohr radius) even at finite temperatures. Our approach also treats the positions of the scattering centers as dynamical variables. We are, therefore, able to generate various initial conditions.

2 Wigner representation of quantum mechanics

The basis of our consideration is the Wigner representation of the von Neumann equation – the Wigner–Liouville equation (WLE). To derive the WLE for the full density matrix of the N-particle system $\rho(x_N \mid y_N)$ we introduce center of mass and relative coordinates in standard manner, $q \equiv q_N \equiv (x_N + y_N)/2$ and $r \equiv r_N \equiv x_N - y_N$. The Wigner distribution function (WF) is defined by [6]

$$f(p, q, t) = \frac{1}{(2\pi\hbar)^{6N}} \int \rho\left(q - \frac{r}{2}, q + \frac{r}{2}\right) e^{ipr/\hbar} \, dr . \tag{1}$$

Using this definition it is straightforward to obtain the WLE for the *full* density matrix [6]

$$\frac{\partial f}{\partial t} + \frac{p}{m} \frac{\partial f}{\partial q} - \frac{\partial V(q)}{\partial q} \frac{\partial f}{\partial p} = \int_{-\infty}^{\infty} ds \, f(p - s, q, t) \, \omega(s, q) , \tag{2}$$

where

$$\omega(s, q) = F(q) \frac{d\delta(s)}{ds} + \frac{4}{\hbar(2\pi\hbar)^{6N}} \int d\bar{q} \, V(q - \bar{q}) \sin\left(\frac{2s\bar{q}}{\hbar}\right) , \tag{3}$$

and $F(q) = -\partial V(q)/\partial q$ is the classical force. Obviously, the force term in ω exactly cancels the last term on the *lhs* of Eq. (2). Retaining these terms allows us to write the WLE as the classical Liouville equation [*lhs* of Eq. (2)] plus a quantum correction [all terms on the *rhs* of Eq. (2)] which vanish for $\hbar \to 0$. This form allows us to identically transform Eq. (2) into an integral Eq. (4),

$$f(p, q, t) = f_0(p_0, q_0) + \int_0^t d\tau \int_{-\infty}^{\infty} ds \, f(p_\tau - s, q_\tau, \tau) \, \omega(s, q_\tau) . \tag{4}$$

The first contribution describes quantum dynamics and is given by the initial WF $f_0(p, q) \equiv f(p, q, 0)$, but taken at arguments $p_0 \equiv \bar{p}(0)$ and $q_0 \equiv \bar{q}(0)$, being classical trajectories $p_\tau \equiv \bar{p}(\tau)$ and $q_\tau \equiv \bar{q}(\tau)$ (solutions of the Hamilton equations associated to the WLE and connecting points (p, q) at time t and points (p_0, q_0) at time 0). Notice that even the first term may describe the evolution of a *quantum* many-body state if the initial WF $f_0(p, q)$ is choosen appropriately and contains the all powers of the Plank's constant. The integral term in Eq. (4) describes the perturbation of the classical trajectories due to quantum effects, for details we refer to Ref. [7].

The structure of Eq. (4) suggests to construct its solution iteratively, starting with f_0. Let us, therefore, rewrite Eq. (4) in the following compact form, $f^t = f_0^t + K_\tau^t f^\tau$, where the superscript on the WF denotes the time argument and $K_{\tau_1}^{\tau_2}$ denotes the time integral in Eq. (4). Then, the iteration series has the form:

$$f^t = f_0^t + K_{\tau_1}^t f_0^{\tau_1} + K_{\tau_2}^t K_{\tau_1}^{\tau_2} f_0^{\tau_1} + K_{\tau_3}^t K_{\tau_2}^{\tau_3} K_{\tau_1}^{\tau_2} f_0^{\tau_1} + \dots , \tag{5}$$

where the first term describes the evolution of an initial (classical or quantum) WF f_0 (it may contain any order of Planck's constant). The remaining terms systematically take into account all dynamic quantum corrections [trajectories with momentum jumps arising from the shifted momentum arguments in the WF under the integral in (2)] including e.g. tunneling effects and correctly accounting for the Heisenberg uncertainty principle. Thus, the solution of Eq. (5) can be understood as a properly weighted sum of classical and quantum phase space trajectories [7].

Using the solution f^t we can compute averages of arbitrary operators in standard way and obtain any dynamic macroscopic property of the correlated quantum particles without approximations on the potential interaction. Naturally, the true particle number N is replaced by a greatly reduced number

N_{sim} which is of the order 50–100 in the MC cell with periodic boundary conditions. The solution scheme is a combination of Quantum Monte Carlo and classical Molecular Dynamics methods: Quantum MC is used to generate the correlated initial state, MD generates the p–q trajectories and Monte Carlo methods are applied to perform an importance sampling of the dominant terms of the iteration series (trajectories with momentum jumps).

3 Wigner representation of time correlation functions

According to the Kubo formula the conductivity is the Fourier transform of the current–current correlation function. Our starting point is the general operator expression for the canonical ensemble-averaged time correlation function [8]:

$$C_{FA}(t) = Z^{-1} \operatorname{Tr} \left\{ \hat{F} \, e^{i\hat{H}t_c^*/\hbar} \, \hat{A} \, e^{-i\hat{H}t_c/\hbar} \right\}, \tag{6}$$

where $\hat{H}$ is the Hamiltonian of the system expressed as a sum of the kinetic energy operator, $\hat{K}$, and the potential energy operator, $\hat{U}$. Time is taken to be a complex quantity, $t_c = t - i\hbar\beta/2$, where $\beta = 1/k_{\mathrm{B}}T$ is the inverse temperature with k_{B} denoting the Boltzmann constant. The operators $\hat{F}$ and $\hat{A}$ are quantum operators of the dynamic quantities under consideration and $Z = \operatorname{Tr}\{e^{-\beta\hat{H}}\}$ is the partition function. The Wigner representation of the time correlation function in a v-dimensional space can be written as

$$C_{FA}(t) = (2\pi\hbar)^{-2v} \int\!\int \mathrm{d}\mu_1 \, \mathrm{d}\mu_2 \, F(\mu_1) \, A(\mu_2) \, W(\mu_1; \mu_2; t; i\hbar\beta), \tag{7}$$

where we introduce the short-hand notation for the phase space point, $\mu_i = (p_i, q_i)$, $(i = 1, 2)$, and p and q comprise the momenta and coordinates, respectively, of all particles in the system. $W(\mu_1; \mu_2; t; i\hbar\beta)$ is the spectral density expressed as

$$W(\mu_1; \mu_2; t; i\hbar\beta) = Z^{-1} \int\!\int \mathrm{d}\xi_1 \, \mathrm{d}\xi_2 \, e^{i\frac{p_1\xi_1}{\hbar}} \, 1 \, e^{i\frac{p_2\xi_2}{\hbar}} \times \left\langle q_1 + \frac{\xi_1}{2} \right| e^{i\hat{H}t_c^*/\hbar} \left| q_2 - \frac{\xi_2}{2} \right\rangle \left\langle q_2 + \frac{\xi_2}{2} \right| e^{-i\hat{H}t_c/\hbar} \left| q_1 - \frac{\xi_1}{2} \right\rangle, \tag{8}$$

and $A(\mu)$ denotes Weyl's symbol [6] of operator $\hat{A}: A(\mu) = \int \mathrm{d}\xi \, e^{-i\frac{p\xi}{\hbar}} \left\langle q - \frac{\xi}{2} \right| \hat{A} \left| q + \frac{\xi}{2} \right\rangle$, and similarly for the operator $\hat{F}$. Hence the problem of the numerical calculation of the canonically averaged time correlation function is reduced to the computation of the spectral density.

To obtain the integral equation for W let us introduce a pair of dynamic p, q-trajectories $\{\bar{q}_\tau(\tau; p_1, q_1, t), \bar{p}_\tau(\tau; p_1, q_1, t)\}$ and $\{\tilde{q}_\tau(\tau; p_2, q_2, t), \tilde{p}_\tau(\tau; p_2, q_2, t)\}$ starting at $\tau = t$ from the initial condition $\{q_1, p_1\}$ and $\{q_2, p_2\}$ propagating in 'negative' and 'positive' time direction:

$$\frac{\mathrm{d}\bar{p}_\tau}{\mathrm{d}\tau} = \frac{1}{2} F[\bar{q}_\tau(\tau)] \, ; \quad \frac{\mathrm{d}\bar{q}_\tau}{\mathrm{d}\tau} = \frac{\bar{p}_\tau(\tau)}{2m} \, ,$$

with

$$\bar{p}_t(\tau = t; p_1, q_1, t) = p_1 \, ; \quad \bar{q}_t(\tau = t; p_1, q_1, t) = q_1 \, ,$$

$$\frac{\mathrm{d}\tilde{p}_\tau}{\mathrm{d}\tau} = -\frac{1}{2} F[\tilde{q}_\tau(\tau)] \, ; \quad \frac{\mathrm{d}\tilde{q}_\tau}{\mathrm{d}\tau} = -\frac{\tilde{p}_\tau(\tau)}{2m} \, ,$$

with

$$\tilde{p}_t(\tau = t; p_2, q_2, t) = p_2 \, ; \quad \tilde{q}_t(\tau = t; p_2, q_2, t) = q_2 \, ,$$

where $F(q) \equiv -\nabla\tilde{U}$ with $\tilde{U}$ being the total potential, i.e. the sum of all pair interactions U_{ab}. Then, as has been proven in [9], W obeys the following integral equation

$$W(\mu_1; \mu_2; t; i\hbar\beta) = \bar{W}(\bar{p}_0, \bar{q}_0; \tilde{p}_0, \tilde{q}_0; i\hbar\beta) + \frac{1}{2} \int_0^t \mathrm{d}\tau \int \mathrm{d}s \, W(\bar{p}_\tau - s, \bar{q}_\tau; \tilde{p}_\tau, \tilde{q}_\tau; \tau; i\hbar\beta) \, \varpi(s, \bar{q}_\tau)$$

$$- \frac{1}{2} \int_0^t \mathrm{d}\tau \int \mathrm{d}s \, W(\bar{p}_\tau, \bar{q}_\tau; \tilde{p}_\tau - s, \tilde{q}_\tau; \tau; i\hbar\beta) \, \varpi(s, \tilde{q}_\tau), \tag{9}$$

where $\varpi(s, q) = \dfrac{4}{(2\pi\hbar)^\nu \hbar} \int dq'\, \tilde{U}(q - q') \sin\left(\dfrac{2sq'}{\hbar}\right) + F(q)\, \nabla\delta(s)$, and $\delta(s)$ is the Dirac delta function. Equation (9) has to be supplemented by an initial condition for the spectral density at $t = 0$: $W(\mu_1; \mu_2; 0; i\hbar\beta) = \bar{W}(\mu_1; \mu_2; i\hbar\beta) \equiv \bar{W}$. The τ-integrals connect the points $\bar{p}_\tau, \bar{q}_\tau; \tilde{p}_\tau, \tilde{q}_\tau$ at time τ of the mentioned above dymamic p, q-trajectories with the points $p_1, q_1; p_2, q_2$ at time t whereas in $\bar{W}$ the trajectories are to be taken at $\tau = 0$. The function $\bar{W}$ can be expressed in the form of a finite difference approximation of the path integral [7, 9, 10]:

$$\bar{W}(\mu_1; \mu_2; i\hbar\beta) \approx \int\int d\tilde{q}_1 \ldots d\tilde{q}_n \int\int dq'_1 \ldots dq'_n\, \Psi(\mu_1; \mu_2; \tilde{q}_1, \ldots, \tilde{q}_n; q'_1, \ldots, q'_n; i\hbar\beta), \qquad (10)$$

with

$$\Psi(\mu_1; \mu_2; \tilde{q}_1, \ldots, \tilde{q}_n; q'_1, \ldots, q'_n i\hbar\beta)$$

$$\equiv \frac{1}{Z} \left\langle q_1 \left| e^{-\epsilon\hat{K}} \right| \tilde{q}_1 \right\rangle e^{-\epsilon U(\tilde{q}_1)} \left\langle \tilde{q}_1 \left| e^{-\epsilon\hat{K}} \right| \tilde{q}_2 \right\rangle e^{-\epsilon U(\tilde{q}_2)} \ldots e^{-\epsilon U(\tilde{q}_n)} \left\langle \tilde{q}_n \left| e^{-\epsilon\hat{K}} \right| q_2 \right\rangle \varphi(p_2; \tilde{q}_n, q'_1)$$

$$\times \left\langle q_2 \left| e^{-\epsilon\hat{K}} \right| q'_1 \right\rangle e^{-\epsilon U(q'_1)} \left\langle q'_1 \left| e^{-\epsilon\hat{K}} \right| q'_2 \right\rangle e^{-\epsilon U(q'_2)} \ldots e^{-\epsilon U(q'_n)} \left\langle q'_n \left| e^{-\epsilon\hat{K}} \right| q_1 \right\rangle \varphi(p_1; q'_n, \tilde{q}_1),$$

$$(11)$$

where $\varphi(p; q', q'') \equiv (2\lambda^2)^{\nu/2} \exp\left[-\dfrac{1}{2\pi}\left\langle \dfrac{p\lambda}{\hbar} + i\pi\dfrac{q'-q''}{\lambda} \left| \dfrac{p\lambda}{\hbar} + i\pi\dfrac{q'-q''}{\lambda} \right. \right\rangle\right]$, and $\langle x \mid y \rangle$ denotes the scalar product of two vectors x, y. In this expression the original (unknown) density matrix of the correlated system $e^{-\beta(\hat{K}+\hat{U})}$ has been decomposed into $2n$ factors, each at a $2n$ times higher temperature, with the inverse $\epsilon = \beta/2n$ and the corresponding high temperature DeBroglie wave length squared $\lambda^2 \equiv 2\pi\hbar^2\epsilon/m$. This leads to a product of known high-temperature (weakly correlated) density matrices, however, at the price of $2n$ additional integrations over the intermediate coordinate vectors (over the "path"). This representation is exact in the limit $n \to \infty$, and, for finite n, an error of order $1/n$ occurs. The function Ψ has to be generalized to properly account for spin-statics effects. This gives rise to an additional spin part of the density matrix and antisymmetrization of one off-diagonal matrix element. To improve the accuracy of the obtained expression, we will replace $U_{ab} \to U_{ab}^{\mathrm{eff}}$ where U_{ab}^{eff} is the proper effective quantum pair potential. For more details on the path integral concept, we refer to Refs. [11]–[13].

Let us now come back to the integral Eq. (9). For the discussion we note that the integral Eq. (9) can be exactly converted into an iteration series (which is obtained by successively replacing $W \to \bar{W}$ under the integrals). This series is, however, not a perturbative expansion in the interaction, neither in the electron–scatterer nor in the electron–electron interaction. It rather is an expansion in terms of corrections to classical trajectories of fully interacting electrons and electrons with scatterers. So multiple scattering effects are fully included. Physically the second order and other terms of the iteration series include corrections to the classical electron trajectories (momentum jumps related to the uncertainty principle between momentum–coordinate and energy-time). A detailed investigation of the conditions for which the contribution of the these terms of the iteration series should be taken into account is presented in [9, 14–16].

As mentioned above, the first term $\bar{W}$ describes propagation of a correlated quantum initial state along the characteristics of the classical Wigner–Liouville equation. This term, containing all powers of Planck's constant, is the coherent sum of complex-valued contributions of a trajectory ensemble related to $\bar{W}$. This term allows to describe quantum coherent effects such as Anderson localization, while other terms of iteration series describe deviations from the classical trajectories: the trajectories are perturbed by a finite momentum jump s occuring at arbitrary times τ, $0 \leq \tau \leq t$ [7]. These terms are essential for the recovery of tunneling effects, we expect that they do not give dominant contributions to coherence and localization phenomena. With increasing quantum degeneracy (i.e. decreasing temperature or/and increasing density) the magnitude of these terms will grow. The time correlation functions are linear functionals of the spectral density, for them the same series representation holds,

$$C_{FA}(t) = (2\pi\hbar)^{-2v} \int\int d\mu_1 \, d\mu_2 \, \phi(\mu_1; \mu_2) \, W(\mu_1; \mu_2; t; i\hbar\beta) \equiv (\phi \mid W^t)$$

$$= (\phi \mid \bar{W}^t) + \left(\phi \mid K^t_{\tau_1} \bar{W}^{\tau_1}\right) + \left(\phi \mid K^t_{\tau_2} K^{\tau_2}_{\tau_1} \bar{W}^{\tau_1}\right) + \left(\phi \mid K^t_{\tau_3} K^{\tau_3}_{\tau_2} K^{\tau_2}_{\tau_1} \bar{W}^{\tau_1}\right) + \ldots \qquad (12)$$

where $\phi(\mu_1; \mu_2) \equiv F(\mu_1) A(\mu_2)$ and the parentheses $(\ldots \mid \ldots)$ denote integration over the phase spaces $\{\mu_1; \mu_2\}$, as indicated in the first line of the equation.

Our numerical results below refer to finite temperature and moderate degeneracy $n_e\lambda_e = 0.2 \ldots, 7$. We, therefore, will include in the following numerical analysis only the first term in this series.

4 Quantum dynamics

As an application, in this work we will consider a system composed of heavy particles (called scatterers) with mass m_s and negatively charged electrons with mass m_e. To avoid bound state effects due to attraction we consider in this case study only negatively charged scatterers, assuming a positve backgroud for charge neutrality. The influence of electron–scatterer attraction will be studied in a further publication.

The possibility to convert a iteration series into a form convenient for probabilistic interpretation allows us to apply Monte Carlo methods to its evaluation. According to the general theory of Monte Carlo methods for solving linear integral equations, e.g. [17], one can simultaneously calculate all terms of the iteration series. Using the basic ideas of [17] we have developed a Monte Carlo scheme, which provides domain sampling of the terms giving the main contribution to the iteration series cf. [9]. For simplicity, in this work, we take into account only the first term of iteration series, which is related to the propagation of the initial quantum distribution according to the Hamiltonian equation of motion. This term, however, does not describe pure classical dynamics but accounts for quantum effects [14] and, in fact, contains arbitrarily high powers of Planck's constant. The remaing terms of the iteration series describe momentum jumps [9, 15] which account for higher–order corrections to the classical dynamics of the quantum distribution, which are expected to be relevant in the limit of high density.

This approach allows us to generate, in a controlled way, various kinds of quantum dynamics and initial conditions of the many-body system, in particular (i) those which are characteristic of the fully interacting system [i.e. including scatterer–scatterer (s–s), electron–scatterer (e–s), and electron–electron (e–e)] and (ii) those which result if some aspects of these interactions are ignored.

5 Numerical results

We now apply the numerical approach explained above to the problem of an interacting ensemble of electrons and disordered scatterers in one dimension. In all calculations times, frequencies and distances are measured in atomic units. The average distance between electrons, $R_s = 1/(n_e a_0)$, was varied between 12.0 and 0.55, with the densities of electrons and heavy scatterers taken to be equal. The results obtained were practically insensitive to the variation of the whole number of the particles in MC cell from 30 up to 50 and also of the number of high temperature density matrices (determined by the number of factors n), ranging from 10 to 20. Estimates of the average statistical error gave the value of the order 5–7%. We studied two different temperatures: $k_B T/|V_0^{es}| = 0.45$ and 0.28, corresponding to $\lambda_{ee}/a_0 \sim 2.2$ and $\lambda_{ee}/a_0 \sim 3.5$, respectively.

According to the Kubo formula [8] our calculations include two different stages: (i) generation of the initial conditions (configuration of scatterers and electrons) in the canonical ensemble with probability proportional to the quantum density matrix and (ii) generation of the dynamic trajectories on the time scale t' in phase space, starting from these initial configurations. The results presented below are related to two different cases: 1. *with* e–e interaction included in the dynamics ("interacting dynamics") and 2., *without* e–e interaction ("noninteracting dynamics"). In both cases, the initial state fully includes all interactions.

Figure 1 presents for our model the real part of the diagonal elements of the electrical conductivity tensor versus frequency (real part of the Fourier transform of the temporal momentum–momentum correlation functions) which characterizes the Ohmic absorption of electromagnetic energy and has the

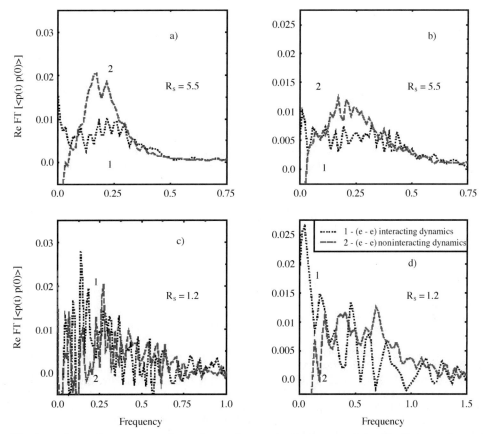

Fig. 1 (online colour at: www.interscience.wiley.com) Real part of the Fourier transform of the temporal momentum–momentum correlation functions for dynamics with (1) and without (2) e–e interaction. Figure parts are for two densities (a, b: $R_s = 5.5$; c, d: $R_s = 1.2$) and temperatures (a, c: $k_B T / |V_0^{es}| = 0.28$; b, d: $k_B T / |V_0^{es}| = 0.45$).

physical meaning of electron conductivity. To compare influence of electron interactions conductivities are given on Fig. 1 in the same arbitrary units. The first observation is that, in all cases, the conductivity for the non-interacting dynamics (2) has a maximum at some finite frequency related to the coherent oscillations in the time domain and vanishes at low frequency [18]. The latter clearly indicates Anderson localization. The effect of the e–e interaction is, as shown by curves 1, a reduction of the maximum (damping of the coherent oscillations) and, in most cases, an increase of the zero-frequency conductivity. Thus, our calculations confirm the delocalizing effect of the interactions (Figs. 1a, b, d) at the considered densities. Interestingly, Fig. 1c is an exception: even with interactions included, the localization behavior persists. The large oscillations in Fig. 1c are not result of numerical noise, they exist inspite of very long simulation duration.

The reason for the observed behavior is an interplay of varying strength of the e–e-interaction (which is weakened with reducing R_s, i.e. from top to bottom figures) and of the magnitude of quantum effects (which grow with temperature reduction, i.e. from right to left figures). Thus, the delocalization tendency observed from Figs. 1c to d is due to thermal activation which, similarly as the interaction, destroys coherence phenomena.

Our simulations qualitatively confirm analytical predictions for the low-frequency and zero temperature limit of the 1D conductivity [19]. Yet our computer power allows us to generate dynamic trajectories up to times t' equal $100 \ldots 200$ in atomic units. Thus for small frequencies of the order 10^{-2}, large fluctuations of the conductivity appear [18], and the accuracy is not yet sufficient to extract an asymptotic frequency behavior. On the other hand, the advantage of our computational method is that it allows to study systematically the influence of finite temperature and of electron correlation effects on localization phenomena in a wide range of densities. We note that we have also performed simulations at lower densities and found that the delocalizing effect of the e−e-interaction has also been observed at lower density up to $R_s = 12$. At even lower densities, we expect that future simulations will yield a pinned electron Wigner crystal at weak disorder [20] and Coulomb glass behavior at strong disorder.

In summary, we have presented numerical results on the influence of Coulomb interaction on Anderson localization in a one-dimensional system. At low density ($R_s = 5.5$) the interaction is comparatively strong and localization is destroyed. With increasing density $R_s = 1.2$, localization is found to persist even in the presence of Coulomb interaction. For a full understanding of the physical processes additional investigations are needed which are presently under way.

Acknowledgments   The author thank B. L. Altshuler for stimulating discussions and valuable notes and the NIC Jülich for computer time. V. Filinov acknowledges the hospitality of the Graduate College "Optoelectronics of Mesoscopic Semiconductors" and the Department of Physics of the Philipps-Universität Marburg. This work is partly supported by the Max-Planck Research Prize of the Humboldt and Max-Planck Societies, by the Deutsche Forschungsgemeinschaft (DFG) through the Quantenkohärenz Schwerpunkt, and by the Leibniz Prize. I.V. acknowledges financial support from the Hungarian Research Fund (OTKA) under T029813, T032116 and T034832.

References

[1] P. A. Lee and T. V. Ramakrishnan, Rev. Mod. Phys. **57**, 287 (1985).

[2] D. Belitz and T. R. Kirkpatrik, Rev. Mod. Phys. **64**, 261 (1994).

[3] E. Abrahams, P. W. Anderson, D. C. Licciardello, and T. V. Ramakrishnan, Phys. Rev. Lett. **42**, 637 (1979).

[4] A. Schmid, Phys. Rev. Lett. **60**, 80 (1991).
   F. von Oppen and E. K. Riedel, Phys. Rev. Lett. **60**, 84 (1991).
   B. L. Altshuler, Y. Gefen, and Y. Imry, Phys. Rev. Lett. **60**, 88 (1991).

[5] M. Pollak, Philos. Mag. **42**, 799 (1980).

[6] V. Tatarskii, Sov. Phys. Uspekhi **26**, 311 (1983).

[7] V. Filinov, P. Thomas, I. Varga, T. Meier, M. Bonitz, V. Fortov, and S. Koch, Phys. Rev. B **65**, 165124, (2002).

[8] D. N. Zubarev, Nonequilibrium Statistical Thermodynamics (Plenum Press, New York/London, 1974).

[9] V. S. Filinov, J. Mol. Phys. **88**, 1517, 1529 (1996).

[10] R. P. Feynman and A. R. Hibbs, Quantum Mechanics and Path Integrals (McGraw-Hill, New York, Moscow, 1965).

[11] V. M. Zamalin, G. E. Norman, and V. S. Filinov, The Monte Carlo Method in Statistical Thermodynamics (Nauka, Moscow, 1977) (in Russian).

[12] B. V. Zelener, G. E. Norman, and V. S. Filinov, Perturbation theory and Pseudopotential in Statistical Thermodynamics (Nauka, Moscow, 1981) (in Russian).

[13] Details on our direct fermionic path integral Monte Carlo simulations are given in V. S. Filinov, M. Bonitz, W. Ebeling, and V. E. Fortov, Plasma Phys. Contr. Fusion **43**, 743 (2001).

[14] G. Ciccotti, C. Pierleoni, F. Capuani, and V. Filinov, Comp. Phys. Commun. **121–122**, 452−459 (1999).

[15] V. Filinov, Yu. Medvedev, and V. Kamskyi, J. Mol. Phys. **85**, 717 (1995).

[16] Yu. Lozovik and A. Filinov, Sov. Phys. JETP, **88**, 1026 (1999).

[17] I. M. Sobol and R. Messer (Translator), Monte Carlo Methods (Univ. Chicago Publisher, Chicago, 1975).

[18] In fact, we observe negative values, although, the real part of the conductivity has to be positive. The reason are weakly damped oscillations with a period exceeding the scale t' used in the calculation of the dynamics. To overcome this deficiency of our model one has to increase the time t' and/or to take into account the slow motion of the heavy particles, which will destroy the coherent oscillations of the light electrons trapped by the heavy particles. Additional calculations with increased t' lead to decreasing negative contributions for low frequencies, as expected.

[19] V. L. Berezinskii, Zh. Eksp. Teor. Fiz. **65**, 1251 (1973) [Sov. Phys. JETP **38**, 620 (1974)].

[20] In the absence of disorder, Wigner crystallization is clearly found in path-integral MC simulations, see A. Filinov, M. Bonitz, and Yu. Lozovik, Phys. Rev. Lett. **86**, 3851 (2001) and references therein.

phys. stat. sol. (b) **241**, No. 1, 47–53 (2004) / **DOI** 10.1002/pssb.200303634

Are the interaction effects responsible for the temperature and magnetic field dependent conductivity in Si-MOSFETs?

V. M. Pudalov[*,1,2], **M. E. Gershenson**[2], **H. Kojima**[2], **G. Brunthaler**[3], and **G. Bauer**[3]

[1] P. N. Lebedev Physics Institute, Moscow 119991, Russia
[2] Department of Physics and Astronomy, Rutgers University, New Jersey 08854, USA
[3] Institut für Halbleiterphysik, Johannes Kepler Universität, 4040 Linz, Austria

Received 1 September 2003, revised 15 September 2003, accepted 18 September 2003
Published online 15 December 2003

PACS 71.27.+a, 71.30.+h. 73.40.Qv

We compare the temperature and in-plane magnetic field dependences of resistivity $\rho(T, B_\parallel)$ of Si MOSFETs with the recent theories. In the comparison we use the effective mass m^* and g^*-factor determined independently. An anomalous increase of ρ with temperature, which has been considered a signature of the "metallic" state, for high conductivities ($\sigma > e^2/h$) can be described quantitatively by the interaction effects in the ballistic regime. $\rho(B_\parallel)$ is consistent with the theory only qualitatively; it is found to be more susceptible than $\rho(T)$ to details of disorder, possibly associated with magnetic moments of localized states. Description of the transport in the "critical" regime of $\rho \sim h/e^2$ is still lacking.

1 Introduction One of the outstanding problems of modern solid state physics is the behavior of strongly-interacting and disordered electron systems. The theoretical description of 2D systems has far not reached a predictive "first principle" stage and a number of effects observed experimentally at high interaction strength remains puzzling. A well-known example is anomalous low-temperature increase of the conductivity in 2D systems with cooling, which has been observed in the beginning of 90's in high-mobility Si-MOSFETs. This "metallic" behavior transforms eventually to the "insulating" behavior as electron density decreases below a critical value n_c [1, 2]. Because of the apparent disagreement with the single-particle theory, the phenomenon of the "metallic" conduction and apparent metal-insulator transition in 2D attracts a great deal of interest. The similar behavior was found later for many other high-mobility and low-density 2D systems (p- and n-type GaAs/AlGaAs, n-type AlAs, p-type Si/SiGe, etc.) [1, 2]. The interactions must play a significant role in this phenomenon, because the ratio of the Coulomb to Fermi energy r_s increases $\propto n^{-1/2}$ and reaches a factor of ~ 10 as electron density n decreases.

2 Prior key results Recently, a considerable progress has been achieved in understanding the "metallic" conduction in the dilute regime. Firstly, a theory has been developed [3] which allows calculation of the interaction corrections to the conductivity beyond the diffusive regime, in terms of the Fermi-liquid (FL) coupling constants. Secondly, FL constants have been determined from the study of the renormalization of the effective spin susceptibility χ^*, mass m^*, and g^*-factor of mobile electrons over a wide range of densities. The corresponding data in Fig. 1 show a strong increase in χ^*, m^*, and g^* with r_s, which significantly affects transport at low densities. In the current paper we make a comparison with the available theories, and conclude that the "metallic" drop of conductivity

---

[*] Corresponding author: e-mail: pudalov@lebedev.ru, Phone: +7 095 135 4278, Fax: +7 095 132 6780

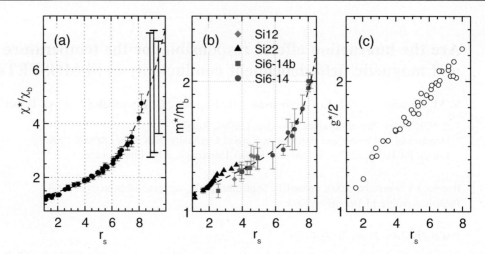

Fig. 1 (online colour at: www.interscience.wiley.com)   a) Renormalized spin susceptibility, b) effective mass, and c) g^*-factor [4].

with cooling in the regime $\sigma \gg e^2/h$ can be accounted for by the interaction effects in electron "liquid" at temperatures which correspond to the ballistic regime. Several experimental teams came independently to a similar conclusion for p-GaAs/AlGaAs [5] and Si MOSFETs [6, 7]. These observations suggest that the anomalous "metallic" conduction in 2D, at least for densities not too close to the critical density, is indeed the finite-temperature phenomenon rather than the signature of a new quantum ground state.

The theory introduced recently by Zala et al [3] considers interference between the electron waves scattered by the short-range scattering centers and by the associated Friedel oscillations of the electron density. This interference gives rise to the quantum corrections, which have been calculated to the higher orders in the interaction strength and to the leading order in the temperature. The quantum corrections to the Drude conductivity $\sigma_D = e^2 n\tau/m^*$ (τ is the momentum relaxation time, m^* is the effective mass of carriers), can be expressed (in units of $e^2/\pi\hbar$) as follows [3]:

$$\sigma(T, B_\|) - \sigma_D = \delta\sigma_C + 15\,\delta\sigma_T + 2[(\sigma(E_Z, T) - \sigma(0, T)] + 2[\sigma(\Delta_v, T) - \sigma(0, T)]$$

$$+ [\sigma(E_Z + \Delta_v, T) - \sigma(0, T)] + [\sigma(E_Z - \Delta_v, T) - \sigma(0, T)] + \delta\sigma_{\mathrm{loc}}(T). \quad (1)$$

Here $\delta\sigma_C \approx x[1 - (3/8)f(x)]$ and $\delta\sigma_T \approx A(F_0^a)\,x[1 - (3/8)\,t(x, F_0^a)]$ are the interaction contributions in the singlet and triplet channels, respectively; $x \equiv T\tau/\hbar$, $\delta\sigma_{\mathrm{loc}}(T) = (1/2\pi)\ln(\tau/\tau_\varphi(T))$ is the weak localization contribution. The terms $\sigma(Z, T) - \sigma(0, T)$ reduce the triplet contribution when the Zeeman energy ($Z = E_Z \equiv 2\mu_B B_\|$), the valley splitting ($Z = \Delta_v$), or combination of these factors ($Z = E_Z \pm \Delta_v$) exceed the temperature. The prefactor 15 to $\delta\sigma_T$ reflects enhanced number of triplet components due to valley-degenerate electron spectrum in (100) Si MOSFETs [8]. Because of this enhancement, the "negative" correction to the conductivity due to the triplet channel, $d\,\delta\sigma_T/dT < 0$, overwhelms the "positive" correction due to the singlet channel and weak localization, $d(\delta\sigma_C + \delta\sigma_{\mathrm{loc}})/dT > 0$. Equation (1) predicts the linear dependence $\sigma(T)$ in the ballistic regime $T\tau \gg 1$ and the logarithmic dependence in the diffusive regime $T\tau \ll 1$; the crossover between the two regimes occurs at $T \approx (1 + F_0^a)\hbar/(2\pi\tau)$ [3].

The terms in Eq. (1) are functions of $x = T\tau/\hbar$, Z, and F_0^a; their explicit expressions are given in Ref. [3]. The FL interaction constant $F_0^a \equiv F_0^\sigma$ which controls the renormalization of the g^*-factor [$g^* = g_b/(1 + F_0^a)$, where $g_b = 2$ for Si], has been independently determined in Ref. [4]. The momentum relaxation time τ is found from the Drude resistivity $\rho_D \equiv \sigma_D^{-1}$ using the renormalized effective mass m^* (controlled by another coupling parameter F_1^s) determined in Ref. [4]. Thus, in principle, one can compare the experiment and the theory with no adjustable parameters; such comparison is the main goal of our paper. Earlier, there were attempts to apply the theory [3] for fitting the experimental data on $\rho(T)$ and $\rho(B)$ in p-type GaAs [5] and Si-MOSFETs [6, 7]; in these attempts, a number of fitting parameters were used. In contrast, our approach provides a rigorous test of applicability of the theory [3] to high-mobility Si-inversion layers, because we determined the two FL coupling constants in independent measurements.

3 Experimental results The ac (13 Hz) measurements of the $\rho(T)$ and $\rho(B_\parallel)$ dependences have been performed on six (100) Si-MOS samples from different wafers: Si15 (peak mobility $\mu^{peak} = 4.0 \, m^2/$ Vs), Si2Ni ($3.4 \, m^2/Vs$), Si22 ($3.3 \, m^2/Vs$), Si6-14 ($2.4 \, m^2/Vs$), Si43 ($1.96 \, m^2/Vs$), and Si46 ($0.15 \, m^2/$ Vs); more detailed description of the samples can be found in Ref. [9].

3.1 Transport at high densities and high conductivities $n \gg n_c$, $\sigma \gg e^2/h$ Figure 2 shows resistivity versus normalized temperature in a wide range of densities. Two important borderline in Fig. 2, T/τ (dashed curve), depicts the ballistic regimes in electron–electron interactions. One can see that the strong metallic-like drop in $\rho(T)$ (i) is characteristic for a wide range of densities (rather than for a critical regime only) and (ii) corresponds mostly to the ballistic regime of interactions. We focus firstly on the high-temperature/high densities ballistic regime, where $\sigma \gg e^2/h$ and the weak localization contribution $\delta\sigma_{loc}(T) = (1/2\pi) \ln(\tau/\tau_\varphi(T))$ can be safely neglected.

Figure 3 illustrates the central result of this paper: for high conductance $\sigma > e^2/h$, the experimental data $\rho(T)$ can be *quantitatively* described by the theory of electron-electron ($e-e$) interaction corrections in the ballistic regime $T\tau \gg 1$ [3]. The solid lines show the $\rho(T)$ dependences calculated according to Eq. (1) with $F_0^a(n)$ and $m^*(n)$ values determined in SdH measurements [4]. Throughout the paper, we assume $\Delta_v = 0$; small values $\Delta_v \leq 1$ K do not affect the theoretical curves at intermediate temperatures.

In the comparison, the Drude resistivity is needed both for calculating the magnitude of ρ and for determining τ in Eq. (1). The theory [3] suggests a recipe for finding the classical ρ_D value by extrapolating the high-temperature quasi-linear $\rho(T)$ dependence in the ballistic regime to $T = 0$. One needs to do more accurate non-linear extrapolation, according to Eq. (1), to account for the contribution of the $\ln(T/E_F)$ terms and non-linear crossover functions $t(T\tau)$ and $f(T\tau)$ [12]. The obtained ρ_D values differ from the result of the simplified linear extrapolation by ~ 1 to 10% as n changes from 40 to 2×10^{11} cm^{-2}. We note that this difference is important only for finding the magnitude of ρ, whereas the slope of the $\rho(T)$ dependence in the ballistic regime is not sensitive to such small variations in ρ_D.

The measured and calculated resistivities are in a good agreement over a broad intervals of temperatures and densities. For some samples the agreement with the theory holds up to such high tempera-

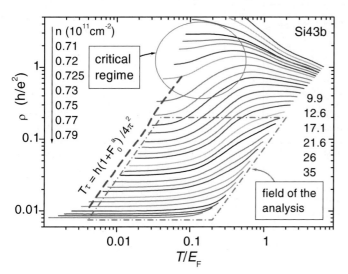

Fig. 2 (online colour at: www. interscience.wiley.com) Resistivity vs. normalized temperature in a wide range of densities. The dashed line depicts diffusive/ballistic border for electron–electron interactions.

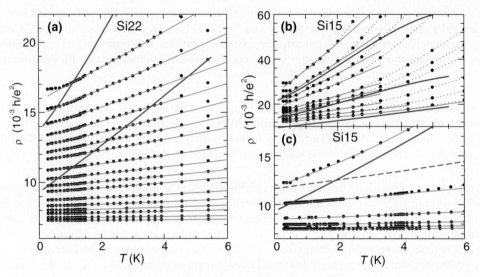

Fig. 3 (online colour at: www. interscience.wiley.com) ρ vs. T for two samples. Dots show the data, thin solid lines correspond to Eq. (1). The densities, from top to bottom, are (in units of 10^{11} cm^{-2}): on panel a) 5.7, 6.3, 6.9, 7.5, 8.1, 8.7, 9.3, 10.5, 11.7, 12.9, 14.1, 16.5, 18.9, 21.3, 23.7, 28.5, 35.7; on panel b) 2.23, 2.46, 2.68, 2.90, 3.34, 3.56, 3.78, 4.33, 4.88, 5.44; on panel c) 5.44, 10.45, 15.95, 21.4, 26.9, 32.4. Thick lines reproduce calculated $\rho(T)$ dependences for sample Si-15 from Fig. 1b of Ref. [11], dashed line – full RPA calculations for $n = 5.6$ from Ref. [13].

tures ($T \sim 0.3E_F$) that $\delta\rho/\rho \sim 1$ (see Fig. 4). In this case, which is beyond the applicability of the theory [3], we still calculated the corrections to the resistivity according to $\delta\rho = -\delta\sigma\,\rho_D^2$, keeping in mind that the corrections pertain primarily to the scattering rate. For much more disordered sample Si46, the agreement with the theory is worse: the theoretical $\rho(T)$ curves are consistent with the data only at $T < 10$ K.

The temperature range, in which ρ varies quasi-linearly with T extends for approximately a decade up to $T \approx 0.1E_F$; it shrinks, however, towards low densities, $n \sim 1 \times 10^{11}$ cm^{-2}, and high densities, $n \sim 4 \times 10^{12}$ cm^{-2}. The linear $\rho(T)$ dependence is only a part of the overall non-linear $\rho(T)$ dependence [2, 10]. At higher temperatures $T \sim T_F$, the $\rho(T)$ data depart from the theory (see Figs. 3, 4). It

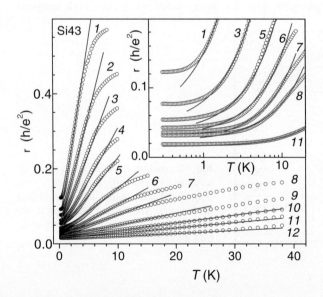

Fig. 4 (online colour at: www. interscience. wiley.com) $\rho(T)$-dependences at $B = 0$ for sample Si43. Inset shows the same data in log T-scale. Dots show the data, lines correspond to Eq. (1) with F_0^a from Ref. [4]. The densities are (in units of 10^{11} cm^{-2}): $1 - 1.49$, $2 - 1.67$, $3 - 1.85$, $4 - 2.07$, $5 - 2.30$, $6 - 2.75$, $7 - 3.19$, $8 - 3.64$, $9 - 4.54$, $10 - 5.43$, $11 - 6.33$, $12 - 8.13$.

is worth mentioning, that the numerical calculations to all orders in T [11, 13] do not provide better fit than the linear T-corrections [3]. At $T \to E_F$, the deviation from the theories might be caused by thermal activation of the sample-dependent interface localized states, which are ignored in the theory.

On the low-temperature side, the dependences $\rho(T)$ tend to saturate for all samples, in contrast to the theoretical prediction Eq. (1). Weakening of the $\rho(T)$ dependence is not caused by valley splitting, because the saturation temperature is too high (e.g., $1-8\,\mathrm{K}$ for sample Si43 – see Figs. 2, 4). One of the reasons for diminishing of the interaction contribution might be strong (and sample-specific) inter-valley scattering. The theory which takes the inter-valley scattering into account is currently unavailable.

3.2 Magnetotransport (MR) in the in-plane field Figure 5 shows a typical behavior of $\rho(B)$. At low densities $n \sim n_c$, as field increases, transport becomes temperature activated [9]; the dashed line $n_c(B)$ in Fig. 5 depicts the border between activated and "metallic" transport regimes, above and below the border, respectively. When n approaches n_c, by definition, $n_c(B)$ tends to $n_c(B=0) \equiv n_c$, therefore, the range of magnetic fields where MR may be studied in the "metallic state", shrinks to zero.

According to the theory of interaction corrections [3],

$$\delta\sigma(B_{\|}) \equiv \sigma(0,T) - \sigma(E_Z, T) \approx \frac{e^2}{\pi\hbar} f(F_0^a) \frac{T\tau}{\hbar} \left[K_b \left(\frac{E_z}{2T}, F_0^a \right) \right] \tag{2}$$

where $K_b \approx (E_Z/2T)^2 f(F_0^a)$ in the low field limit $E_Z/2T \ll 1$, and $K_b \approx (E_Z/T) f_2(F_0^a)$ in the limit $E_Z/2T \gg 1$ limit. Correspondingly, as field increases, $\delta\sigma(B)$ should increase initially $\propto B^2/T$, and than $\propto B$.

Experimental data, in general, show similar behavior [7, 12]; however, in contrast to the temperature dependences of ρ, the agreement with theory [3] is only qualitative. In high fields, $g^*\mu B \gg T$, the

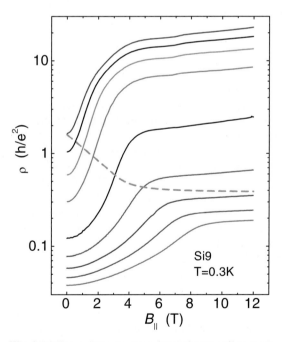

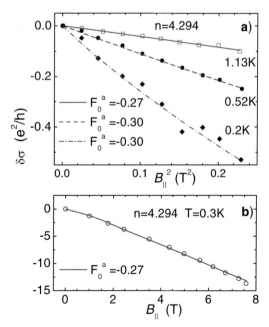

Fig. 5 (online colour at: www. interscience. wiley.com) Resistivity versus in-plane field for 9 densities, from $n = 0.75$ to $2.1 \times 10^{11}\,\mathrm{cm}^{-2}$. Dashed line depicts a border between the activated and metallic transport regimes [9]. Sample Si9.

Fig. 6 (online colour at: www. interscience. wiley.com) a) Magnetoconductivity for sample Si6-14 vs. B^2 in low fields, and b) vs. B in high fields, for different temperatures and at the same density. Lines show dependences, calculated according to Eq. (1).

© 2004 WILEY-VCH Verlag GmbH & Co. KGaA, Weinheim

magnetoresistance deviates substantially from the theory, and the deviations are sample dependent [12]. The discrepancy is most dramatic for low densities [9, 12]. To quantify the deviations from the theory, F_0^a was treated in Ref. [12] as an adjustable parameter in fitting the MR data. In the low field regime, $g^* \mu B < T$, for some samples, $\sigma(B, T)$ seems to be in a reasonable agreement with the theory (see Fig. 2): the fitted F_0^a values are close to those determined from SdH measurements (see Fig. 1). However, more detailed consideration reveals disagreement with the theory even for relatively high densities and high conductivities $\sigma \gg e^2/h$: $\sigma(B)$ data can not be fitted with a given F_0^a value overall temperature range. An example of such disagreement is shown in Fig. 6. From the MR fitting, one could draw a conclusion that F_0^a varies by 10% with temperature. However, direct measurements of $\chi^*(T)$ from SdH effect [4] reveal much weaker monotonic T-dependence of χ^*, (2–3% at $T \leq 1$ K), which has the opposite sign. Furthermore, as density decreases and field increases, the deviations of the measured MR from the theory grow and reach a factor of two at $n = 1.5n_c$ [12]. The T-dependence of the fitting parameter F_0^a therefore signals that the measured $\sigma(B, T)$ scales with temperature somewhat different from that given by Eq. (2), which predicts $\delta\sigma \propto -(B^2/T)$ in low fields. Indeed, the experimental data in low fields scales as $\delta\sigma \propto -(B^2/T^p)$, where p varies, e.g. from 1.1 to 1.6 as density decreases from $5n_c$ to $1.2n_c$.

4 Transport in the critical regime at $\sigma \sim e^2/h$ In view of the encouraging comparison with the theory of interaction corrections in the high density regime, it seems attractive to extend the comparison to the critical regime, where $\sigma \sim e^2/h$ (see Fig. 2). The renorm-group (RG) theory by Finkel'stein [15] is an extension of the same diagrams of the interaction theory to the case of the strong disorder and interaction; it suggests a basis for such comparison. The RG equations describe renormalization of both, disorder (i.e. ρ) and interaction parameters (in particular, F_0^a) as the length scale varies with T [15, 16]. Such comparison has been recently performed in Ref. [8]; it demonstrated a certain similarity between the data and the theory. Application of the RG-theory to the data is based on the following assumptions: (i) the critical domain of strong $\rho(T)$ variation (see Fig. 2) belongs to the diffusive regime of interaction, and (ii) the $\rho(T)$ maximum signifies a turning point from localization to delocalization behavior due to the renormalization of the interaction parameter (F_0^a) with T. Both assumptions need to be verified. The first assumption does not agree well with the empirical diagram in Fig. 2; therefore, its application would require to find a different interpretation for the $\tau(n)$ or $F_0^a(n)$ data in the critical regime. Furthermore, the $\rho(T)$ data in the critical regime, as a rule, show a strong sample-dependent and non-universal behavior [17, 18]. In Ref. [8], the assumption (ii) was presumed to be fulfilled in the most clean samples; however, some high mobility Si-MOS samples do not exhibit a pronounced $\rho(T)$ maximum in the critical regime [17, 19].

As mentioned above, the magnetoresistance studies in the critical regime are restricted to the fields vanishing to zero as $n \to n_c$. Nevertheless, down to $n \approx 1.2n_c$ the magnetotransport can be safely studied in the regime $g^* \mu B_\| < T$. Comparing the empirical scaling law $\delta\sigma \propto -B^2/T^{1.6}$ with that predicted by the RG theory [16], $\delta\sigma \propto \gamma(B/T)^2$, we conclude that the interaction parameter $|\gamma|$ (roughly, $\propto |F_0^a|$) *decreases* as temperature decreases. This result contradicts the main idea of the two-parameter scaling, where $|\gamma|$ is expected to increase as length scale increases (i.e. T decreases). We note also that the direct SdH measurements in low $B_\|$ fields do not confirm strong T-dependence of F_0^a [14]. We suggest therefore, that the transport in the critical regime is determined not only by universal effects of interactions between the itinerant electrons, but is also affected by interactions of localized states with itinerant ones.

5 Conclusions   Answering the question raised in the title, we conclude that there is a strong evidence in support of the anomalous metallic-like transport in the 2D electron liquid, far away from the critical regime ($\sigma \gg e^2/h$), to be caused by $e-e$ interaction effects at intermediate temperatures. The MR is controlled, in addition to the interaction effects, by a complicated behavior of the localized electrons. An adequate theory of the transport in the critical regime is still lacking; such theory must include both, interaction of itinerant electrons with each other and with localized states. Despite a serious disagreement between the experiment and the theory in the critical regime, there is an interesting encouraging observation that the sample-dependent effects of disorder (which strongly affect transport at finite temperatures) vanish as $T \to 0$ [18]. This suggests that the transition from "metallic" to insulating regime at $T \to 0$ may have a universal character, consistent with the picture of a disordered electron crystal [21, 22]. Despite the transition occurs at 4 times lower r_s value than that predicted for the quantum solidification in the pure system, the electron transport in the insulating phase, at densities very close to but less than n_c, shows pronounced collective effects; the collective transport behavior transforms to a single particle one far away from n_c, or with disorder enhancement [20].

Acknowledgements The work was supported by the NSF, ARO MURI, FWF Austria, NWO, NATO, INTAS, and the Russian programs RFBR, "The State Support of Leading Scientific Schools", and "Integration".

References

[1] for a review, see: E. Abrahams, S. V. Kravchenko, and M. P. Sarachik, Rev. Mod. Phys. **73**, 251 (2001).

[2] for a review, see: B. L. Altshuler, D. L. Maslov, and V. M. Pudalov, Physica E **9**(2), 209 (2001).

[3] G. Zala, B. N. Narozhny, and I. L. Aleiner., Phys. Rev. B **64**, 214204 (2001); Phys. Rev. B **65**, 020201 (2001).

[4] V. M. Pudalov, M. Gershenson, H. Kojima, N. Butch, E. M. Dizhur, G. Brunthaler, A. Prinz, and G. Bauer, Phys. Rev. Lett. **88**, 196404 (2002).

[5] Y. Y. Proskuryakov, A. K. Savchenko, S. S. Safonov, M. Pepper, M. Y. Simmons, and D. A. Ritchie, Phys. Rev. Lett. **89**, 076406 (2002).

[6] A. A. Shashkin, S. V. Kravchenko, V. T. Dolgopolov, and T. M. Klapwijk, Phys. Rev. B **66**, 073303 (2002).

[7] S. A. Vitkalov, K. James, B. N. Narozhny, M. P. Sarachik, and T. M. Klapwijk, Phys. Rev. B **67**, 113310 (2003).

[8] A. Punnoose and A. M. Finkelstein, Phys. Rev. Lett. **88**, 016802 (2002).

[9] V. M. Pudalov, G. Brunthaler, A. Prinz, and G. Bauer, cond-mat/0103087; Phys. Rev. Lett. **88**, 076401 (2002).

[10] G. Brunthaler, A. Prinz, G. Bauer, and V. M. Pudalov, Phys. Rev. Lett. **87**, 096802 (2001).

[11] S. Das Sarma and E. H. Hwang, Phys. Rev. Lett. **83**, 164 (1999).

[12] V. M. Pudalov, M. Gershenson, H. Kojima, N. Butch, E. M. Dizhur, G. Brunthaler, A. Prinz, and G. Bauer, Phys. Rev. Lett. **91**, 126403 (2003).

[13] S. Das Sarma and E. H. Hwang, cond-mat/0302047.

[14] V. M. Pudalov, M. Gershenson and H. Kojima, to be published elsewhere.

[15] For a review, see A. M. Finkelstein, Sov. Sci. Rev. A **14**, 3 (1990).

[16] C. Castellani, C. Di Castro, P. A. Lee, M. Ma, Phys. Rev. B **30**, 527 (1984).
C. Castellani, C. Di Castro, H. Fukuyama, P. A. Lee, and M. Ma, Phys. Rev. B **33**, 7277 (1986).
C. Castellani, C. Di Castro, and P. A. Lee, Phys. Rev. B **57**, R9381 (1998).
C. Castellani, G. Kotliar, and P. A. Lee, Phys. Rev. Lett. **59**, 323 (1987).

[17] V. M. Pudalov, G. Brunthaler, A. Prinz, and G. Bauer, JETP Lett. **68**, 442 (1998).

[18] V. M. Pudalov, M. E. Gershenson, and H. Kojima, cond-mat/0201001.

[19] S. V. Kravchenko and T. M. Klapwijk, Phys. Rev. Lett. **84**, 2909 (2000).

[20] V. M. Pudalov, J. Phys. IV (France) **12**, Pr9-331 (2002).

[21] S.-T. Chui and B. Tanatar, Phys. Rev. Lett. **74**, 458 (1995).

[22] B. Spivak, Phys. Rev. B **64**, 085317 (2001); Phys. Rev. B **67**, 125205 (2003).

phys. stat. sol. (b) **241**, No. 1, 54–60 (2004) / **DOI** 10.1002/pssb.200303626

Electric-field-induced hopping transport in superlattices

P. Kleinert[*,1] and **V. V. Bryksin**[2]

[1] Paul-Drude-Institut für Festkörperelektronik, Hausvogteiplatz 5–7, 10117 Berlin, Germany
[2] Physical Technical Institute, Politekhnicheskaya 26, 194021 St. Petersburg, Russia

Received 1 September 2003, accepted 9 September 2003
Published online 15 December 2003

PACS 72.10.Bg, 72.20.Ee, 72.20.Ht, 73.61.–r

A strong electric field applied parallel to the superlattice axis leads to Wannier–Stark localization. There are two formally equivalent pictures for the drift velocity and the diffusion coefficient, the hopping model, which is most suitable for localized states, and the band picture, which is applicable to extended states. Complete localization of carriers is achieved by strong electric and magnetic fields. In these molecule-like systems, there is a strong interdependence between the carrier spectrum and its statistical properties, the description of which requires a new approach. The related double-time character of the transport gives rise to interesting quantum effects in the current–voltage characteristics.

1 Introduction The physics of carrier transport in solids has attracted a great deal of interest since the development of microstructured devices, which rely on unique transport properties. The character of the transport mechanism depends on the character of the eigenstates. Spatially localized charge carriers, which occur in disordered solids due to deep potential fluctuations, give rise to hopping. On the contrary, extended states in crystalline solids constitute the band-transport regime. Both mechanisms are quite differently affected by inelastic scattering, e.g., on phonons. In the band-transport regime, scattering acts as the mechanism limiting the mobility of the carriers. However, carge carrier motion via hopping requires inelastic scattering. This striking discrepancy between the two regimes appears to require the development of two quite different approaches. The question can be raised: Do we really need two different approaches or can we derive a general theory that is applicable both to disordered and ordered systems? In this contribution, we show that there is one general description of transport, the formulation of which starts either from the hopping or from the band picture.

Let us take a closer look at crystalline solids. Even in these ordered systems, there exist various mechanisms that lead to spatially localized carriers, such as the formation of small polarons [1] in crystals with a narrow conduction band and strong interaction with phonons, the impurity hopping conduction and weak localization [2] in doped and compensated semiconductors, and the quantum-Hall effect [3] in a two-dimensional electron gas subject to a quantizing magnetic field. In the present paper, we focus on the localiztion of originally extended electronic states of a crystal by a strong electric field. Within the one-band approximation, the extended eigenstates of the ordered semiconductor become localized in the direction, in which the strong electric field is applied (Wannier–Stark localization). High-electric-field effects are most pronounced in layered semiconductor structures, which consist of a periodic array of alternating materials so that a one-dimensional superlattice (SL) is formed with a SL period d much larger than the lattice constant a of the constituent materials. Due to this large SL constant, high-field phenomena and quantum effects caused by field-induced localization of carriers are observed at much lower field strengths than in bulk material. Many interesting effects

---

[*] Corresponding author: e-mail: kl@pdi-berlin.de, Phone: +49 30 203 77 350, Fax: +49 30 203 77 515

have been reported [4] such as negative differential conductivity (NDC), stationary and traveling field domains, Bloch oscillations, and sequential resonant tunneling.

When an additional magnetic field is applied perpendicular to the SL layers, the energy spectrum becomes completely discrete due to the simultaneous influence of Wannier–Stark and Landau quantization. For high enough fields, for which the period of the cyclotron motion is smaller than the collision time, a quantum-box SL is formed, the building blocks of which are tunable by varying the electric and magnetic field strengths. The field-induced spatial localization of eigenstates leads to hopping transport, which exhibits an interesting quantum nature. In such a field-induced molecule-like solid, lifetime effects are very important, because without any collisional broadening the current-voltage characteristics becomes completely singular with δ-like peaks at certain resonance positions. Scattering effects drastically change both the energy spectrum and the carrier statistics. The interdependence between the energy spectrum and the carrier statistics is constituted at each instant due to the time dependence of scattering events so that even the description of the stationary quantum transport would require the determination of a time-dependent nonequilibrium distribution function.

The application of an ac field in the THz domain opens up an additional transport channel due to delocalization of carriers by the irradiation field. The ac field can suppress the formation of field domains in the NDC regime and stabilize, therefore, a homogeneous electric field. When photon absorption becomes more efficient than photon emission, the carriers use the energy gain to travel against the field direction. This may lead to absolute negative current. The double-time character of the quantum transport gives rise to a photon-induced phononless current contribution, which dominates when inelastic scattering is suppressed.

As mentioned above, quantum transport and quantum diffusion can be treated by the hopping version of the theory or by the quantum Boltzmann equation. In this paper, we will focus on the hopping picture, which is most efficient in the regime of quantizing electric and magnetic fields.

2 Wannier–Stark localization   Let us treat a SL with one occupied miniband at low electron concentration so that the electron gas is nondegenerate and Boltzmann statistics applies. The SL energy dispersion relation $\varepsilon(\mathbf{k})$ describes carrier propagation along the SL axis by a tight-binding model and considers free electron motion with an effective mass m^* for all lateral directions

$$\varepsilon(\mathbf{k}) = \frac{\hbar^2 k_\perp^2}{2m^*} + \frac{\Delta}{2}(1 - \cos k_z d). \tag{1}$$

$\hbar k_\perp$ denotes the transverse quasi-momentum and Δ the miniband width. The k_z dependence of the energy dispersion relation is schematically illustrated in Fig. 1. When the Bloch frequency $\Omega = eEd/\hbar$ (with E being the electric field) exceeds a characteristic scattering rate $1/\tau$ ($\Omega\tau \gg 1$), the drift velo-

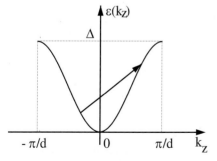

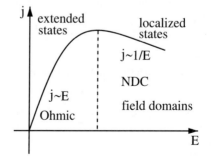

Fig. 1 Schematic diagram of the cosine-shape tight-binding dispersion relation of the SL. Δ denotes the miniband width. The arrow illustrates the immediate change of the carrier state by inelastic scattering.

Fig. 2 Schematic current–voltage characteristics of the SL. Two distinct transport regimes are identified. The transport behaviour at low fields is governed by extended states. In the hopping regime at high fields NDC occurs.

city $v_z = \partial \varepsilon / \partial \hbar k_z$ oscillates due to the continuous time dependence of the wave-vector component according to $k_z(t) = k_0 + eEt/\hbar$. The quantization of these oscillations leads to the Wannier–Stark ladder. The oscillatory motion with frequency Ω is averaged out so that the stationary current vanishes, unless the mean electron positions are suddenly changed by scattering events as shown in Fig. 1. In this hopping transport regime, the carriers jump along the field direction from layer to layer. When, in contrast, the period of the Bloch oscillation is much larger than the collision time ($\Omega\tau \ll 1$), the electrons are scattered so quickly that their energy remains at the bottom of the subband $\varepsilon(k_z) \sim k_z^2$, where the states are spatially extended. This leads to the band-transport regime. The field-induced change of the transport behaviour is illustrated in Fig. 2. At low electric fields, i.e., in the Ohmic regime, the current density increases with increasing electric field $j \sim E$ and scattering hinders the carrier transport. Due to the localization of carriers at high electric fields, the character of the transport changes drastically. The current density is induced and not hindered by inelastic scattering. It decreases with increasing electric field $j \sim 1/E$ (NDC regime). Despite the striking difference between both transport regimes, a general theory can be derived, which is applicable for all electric field strengths. Such an approach can start either from the hopping or band-transport picture. We will focus on the hopping description of quantum transport and quantum diffusion in biased SLs.

3 Hopping transport picture   Let us now specify the basis of our approach. As was stated a long time ago [5], a microscopic one-electron theory for the drift velocity and the diffusion coefficient can be derived from the conditional probability $P_{m'm}^{m'm}(\boldsymbol{k}'_\perp, \boldsymbol{k}_\perp \mid t)$ to find a particle at time t in the m'-th layer with the in-plane quasi-momentum $\boldsymbol{k}'_\perp$, provided it occupied the layer m with $\boldsymbol{k}_\perp$ at an earlier time $t = 0$. Within the Laplace representation, this probability function satisfies the rate equation [6]

$$s P_{m',m}^{m',m}(\boldsymbol{k}'_\perp, \boldsymbol{k}_\perp \mid s) = \delta_{m,m'}\delta_{k_\perp,k'_\perp} + \sum_{m''}\sum_{k''_\perp} P_{m',m''}^{m',m''}(\boldsymbol{k}'_\perp, \boldsymbol{k}''_\perp \mid s)\, \widetilde{W}_{m'',m}^{m'',m}(\boldsymbol{k}''_\perp, \boldsymbol{k}_\perp \mid s)\,, \qquad (2)$$

in which an effective transition probability $\widetilde{W}$ appears. It is calculated from

$$\widetilde{W}_{m_2,m_4}^{m_1,m_3}(\boldsymbol{k}'_\perp, \boldsymbol{k}_\perp \mid s)$$

$$= W_{m_2,m_4}^{m_1,m_3}(\boldsymbol{k}'_\perp, \boldsymbol{k}_\perp \mid s) + \sum_{k''_\perp}\sum_{m_5 \neq m_6} \widetilde{W}_{m_2,m_6}^{m_1,m_5}(\boldsymbol{k}'_\perp, \boldsymbol{k}''_\perp \mid s)\, \frac{\hbar}{ieEd(m_6 - m_5)}\, W_{m_6,m_4}^{m_5,m_3}(\boldsymbol{k}''_\perp, \boldsymbol{k}_\perp \mid s)\,. \qquad (3)$$

It is understood that the bare scattering probability W (which also depends on the electric field) enters the integral equation for the hopping rate $\widetilde{W}$. Eq. (3) generates a power series in $1/E$, which terminates rapidly in the ultra-quantum limit, where both quantities $\widetilde{W}$ and W almost coincide. Within the hopping picture, the drift velocity is expressed by the effective scattering rate $\widetilde{W}$ multiplied by the hopping length md and the normalized lateral distribution function $n(\boldsymbol{k}_\perp)$ [6]

$$v_z = \sum_{k_\perp, k'_\perp}\sum_m (md)\, n(\boldsymbol{k}_\perp)\, \widetilde{W}_{0,m}^{0,m}(\boldsymbol{k}'_\perp, \boldsymbol{k}_\perp)\,. \qquad (4)$$

The scattering-induced heating of the lateral electron motion is characterized by $n(\boldsymbol{k}_\perp)$, which solves the integral equation

$$\sum_{m, k'_\perp} n(\boldsymbol{k}'_\perp)\, \widetilde{W}_{0,m}^{0,m}(\boldsymbol{k}'_\perp, \boldsymbol{k}_\perp) = 0\,, \qquad \sum_{k_\perp} n(\boldsymbol{k}_\perp) = 1\,. \qquad (5)$$

Quantum diffusion along the z-axis is also induced by inelastic scattering. It is expressed by half of the squared hopping length $(md)^2/2$ multiplied by the effective scattering rate $\widetilde{W}$ and the lateral distribution function $n(\boldsymbol{k}_\perp)$ [7, 8]

$$D_{zz} = \frac{1}{2}\sum_{k_\perp, k'_\perp}\sum_{m=-\infty}^{\infty} (md)^2\, n(\boldsymbol{k}'_\perp)\, \widetilde{W}_{0,m}^{0,m}(\boldsymbol{k}'_\perp, \boldsymbol{k}_\perp)\,. \qquad (6)$$

Taking into account Eqs. (4) and (6) together with the principle of detailed balance, we obtain in the ultra-quantum limit ($\Omega\tau \gg 1$) for the mobility $\mu = v_z/E$ the asymptotic form [7]

$$\mu = \frac{eD_{zz}}{k_BT}\frac{\tanh\left(eEd/2k_BT\right)}{eEd/2k_BT} , \tag{7}$$

which reproduces the Einstein relation for sufficiently high temperatures ($\hbar\Omega < 2k_BT$). In the hopping regime ($2k_BT < \hbar\Omega$), we obtain $\mu = 2D_{zz}/(Ed)$ and $D_{zz} = v_zd/2$.

With decreasing electric field strength, the eigenstates become more and more delocalized, and the band-transport regime is approached. To derive an appropriate theory, which covers also this field region, we can switch in an exact manner from the Houston representation, used above to express the scattering probabilities, to the quasi-momentum representation [9]. In a first step, let us summarize the k-representations of the transport coefficients for high electric fields ($\Omega\tau \gg 1$)

$$v_z = \frac{1}{eE}\sum_{k,k'}\left[\varepsilon(k'_z) - \varepsilon(k_z)\right]n(k'_\perp)\,\widetilde{W}(k', k) , \tag{8}$$

$$D_{zz} = \frac{1}{2}\frac{1}{(eE)^2}\sum_{k,k'}\left[\varepsilon(k_z) - \varepsilon(k'_z)\right]^2 n(k'_\perp)\,W(k', k) . \tag{9}$$

These equations allow a clear physical interpretation based on a quasi-classical picture [7, 9]. For classically high fields, the scattering probability W does not depend on the electric field so that we obtain $v_z \sim 1/E$ and $D_{zz} \sim 1/E^2$. When the field dependence of the scattering probability is accounted for, non-analytic electro-phonon resonances [10, 11] are predicted to occur whenever the frequency ω_0 of polar-optical phonons matches a multiple of the Bloch frequency $l\Omega = \omega_0$. Figure 3 shows an example of this intra-collisional field effect. The positions of electro-phonon resonances are marked by vertical lines. At intermediate field strengths, v_z decreases more slowly than the diffusion coefficient D_{zz}, whereas both quantities have the same high-field asymptotics in the ultra-quantum limit [7].

We can switch from the hopping picture as expressed by Eqs. (4) and (6) to the band-transport picture in an exact manner by expressing the scattering probability in k-space [9]. The result for the drift velocity has the well-known form

$$v_z = \sum_k v_{\text{eff}}(k)\,f(k) , \quad v_{\text{eff}}(k) = \frac{1}{\hbar}\frac{\partial\varepsilon(k)}{\partial k_z} - i\sum_{k'}\frac{\partial}{\partial\kappa_z}W(k, k', \kappa)|_{\kappa=0} , \tag{10}$$

where the distribution function $f(k)$ is the solution of the quantum Boltzmann equation [9]. The scattering-induced contribution to the effective velocity v_{eff} vanishes when the dipole operator commutes with the Hamiltonian. This is the case, e.g., for Fröhlich coupling. However, within the theory of

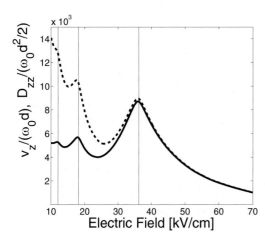

Fig. 3 The relative diffusion coefficient $D_{zz}/(\omega_0d^2/2)$ [dashed line] and the relative drift velocity $v_z/(\omega_0d)$ [solid line] as a function of the electric field for $T = 77$ K, $\Delta = 20$ m eV, and $d = 10$ nm, for a phenomenological damping parameter of 1 m eV.

small polarons or for systems with discrete energy levels, it is just this scattering-induced contribution, which dominates the average drift velocity. The related band picture of the diffusion coefficient has been recently derived [7].

4 Localization by parallel electric and magnetic fields The main ingredients in the hopping picture of quantum transport are the effective field-dependent scattering probability in the Houston representation and the lateral nonequilibrium distribution function. The latter quantity describes the bandlike motion of carriers in the plane. Applying in addition to the electric field a strong magnetic field perpendicular to the layers, the lateral energy spectrum becomes also quantized, giving rise to a set of discrete eigenstates. A hopping transport picture emerges, in which scattering plays a fundamental role. Both the energy spectrum and the statistical properties of the carrier ensemble are drastically changed by scattering. Moreover, it turns out that there is a strong interdependence between the spectrum and the statistics, which is accounted for by a specific explicit time dependence of the distribution function. This reveals the double-time nature of the problem, the adequate description of which is only possible beyond the Kadanoff–Baym (KB) ansatz [12]. An appropriate theoretical description of this retroaction provides the double-time Kadanoff–Baym–Keldysh nonequilibrium Green function technique. In this approach, the stationary carrier transport is described by two time-dependent distribution functions $f^{\lessgtr}(\boldsymbol{k}, t)$ [13], which refer to the carrier spectrum and its statistical properties. Most essential is the time dependence, which constitutes at each instant a scattering-mediated interdependence between the spectrum and the carrier statistics. Note that this effect is not accounted for by the quantum Boltzmann equation, which describes the steady state by a time-independent nonequilibrium distribution function $f(\boldsymbol{k})$. Indeed, it has been shown that calculations within this framework carried out on the basis of the KB ansatz yield results that deviate even qualitatively from the correct data (cf. Refs. [13] and [14]). We conclude that there are additional quantum effects due to the finite duration of scattering events that cannot be taken into account by the quantum Boltzmann equation. Figure 4 shows an example for the current component induced by inelastic scattering on polar-optical phonons. As in Fig. 3, pronounced electro-phonon resonances appear at resonance positions $l\Omega = \omega_0$ (indicated by vertical lines). Between the main resonances, current gaps occur, which are not reproduced by data derived from the quantum Boltzmann equation (dashed line) [14]. Most interesting is the change of

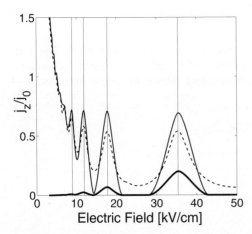

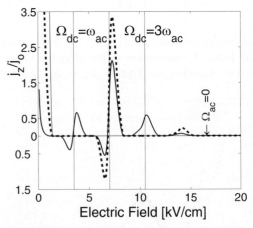

Fig. 4 The electric-field dependence of the relative current density j_z/j_0 for $\Delta/\hbar\omega_0 = 1$ (thin solid line) and $\Delta/\hbar\omega_0 = 0.5$ (heavy solid line). The calculation has been carried out with the parameters $B = 20$ T, $\hbar\omega_0/k_BT = 5$, and $d = 10$ nm. The dashed line has been calculated on the basis of the quantum Boltzmann equation.

Fig. 5 Relative stationary current density j_z/j_0 as a function of the dc electric field for $E_{ac} = 10$ kV/cm, $T = 4$ K, and $d = 10$ nm. The solid and dashed lines have been calculated for $\omega_{ac}/\omega_0 = 0.1$ and 0.2, respectively. Positions of electro-photon resonances are marked by vertical lines.

the current–voltage characteristics at intermediate field strengths as a function of the miniband width. Depending on the strength of the coupling between the SL quantum wells, there is a crossover from the quasi–classical ($j \sim 1/E$) to the activated, hopping transport regime. For weakly coupled SLs ($\Delta/\hbar\omega_0 < 1$), phonon-induced hopping prevails also at intermediate field strengths so that the current decreases with decreasing electric field (heavy solid line). This is in accordance with the activated nature of hopping transport. This peculiarity cannot be derived using an approach, which rests on the KB ansatz.

5 Phononless current contribution The experimental identification of quantum effects, which are related to the double-time nature of the transport, is complicated by the occurrence of electric–field domains in the NDC region. An ac field additionally applied parallel to the dc field can suppress field domains and leads, therefore, to a stabilization of a homogeneous electric field within the SL sample. The irradiation field opens up a new transport channel by delocalizing the eigenstates. The energy provided by the ac field can be used by the carriers to travel against the field direction giving rise to an absolute negative current. To determine quantum effects, which emerge beyond the KB ansatz, the Dyson equations for the Wigner-transformed double-time Green functions $\widetilde{G}^{\gtrless}(k \mid t, t')$ are treated within the self-consistent Born approximation [15]. Within this non–perturbative approach, an infinite set of diagrams has to be summed up, which collects terms depending non–analytically on the coupling constant of scattering. A second peculiarity refers to the double-time character of the nonequilibrium distribution functions $f^{\gtrless}(k \mid t, t')$, which is due to the interdependence between the carrier spectrum and its statistics. Most theoretical studies (for an example see Ref. [16, 17]) did not consider this retroaction by treating only functions $f^{\gtrless}(k, T)$ [with $T = (t + t')/2$], which are periodic in T with a period $2\pi/\omega_{ac}$ (where ω_{ac} denotes the frequency of the ac field). However, in general, one has to cope with an additional time dependence of the distribution functions $f^{\gtrless}(k \mid T, t_-)$, which is determined from a kinetic equation that emerges only beyond the KB ansatz. An interesting result of this sophisticated quantum approach is the appearance of a photon-induced phononless current [15]. Figure 5 shows an example for the relative current contribution induced by an ac electric field in the THz domain as a function of the dc electric field. The photon-induced current indeed vanishes for all dc electric fields, when the irradiation field is switched off. Besides pronounced electron–photon resonances at $\Omega_{dc} = m\omega_{ac}$ ($m = 1, 2, 3$), current gaps appear, whenever the joint density of states becomes zero. These features are resolved, when the collisional broadening drops below $\hbar\omega_{ac}$. In addition, the stationary current, which is induced by the THz irradiation, may flow against the field direction ($j_z < 0$). A measurement of this specific phononless current contribution would stimulate further studies of the double-time quantum nature of SL transport.

6 Summary Biased SLs with a very large periodicity along the SL axis are particularly suited to study quantum transport and quantum diffusion. Treating a one-band model, the electronic states become localized by a strong electric field. This localization gives rise to a Wannier–Stark ladder of discrete electronic eigenstates. Based on the Houston representation, both the drift velocity and the diffusion coefficient are expressed by an effective field-dependent scattering probability and the lateral distribution function. From this hopping picture of transport, we can recover in an exact manner the band model, which is most appropriate for extended states at low field strengths. Both the hopping and band model are formally equivalent, but from the numerical point of view essentially restricted to high and low electric fields, respectively. The appearance of electron–phonon resonances is the most interesting quantum effect of the electric field-induced carrier localization.

The energy spectrum of the carriers becomes completely discrete by applying strong electric and magnetic fields parallel to the SL axis. In this field-driven "metal–insulator transition", inelastic scattering plays a fundamental role. There is a permanent interdependence between the carrier spectrum and its statistical properties so that even for the stationary transport a time-dependent nonequilibrium function has to be determined. To adequately describe these quantum effects, one needs a new approach, which takes into account the double-time character of the transport. From a more fundamental point of view, such an approach provides an interesting example of quantum statistics, in which the main physics of transport appears just beyond the KB ansatz. An interesting quantum effect of this kind is given by a photon-induced phononless current component, in which electron–photon resonances appear.

References

[1] H. Böttger and V. V. Bryksin, Hopping Conduction in Solids (Akademie Verlag, Berlin, 1985).
[2] D. Vollhardt and P. Wölfle, Electronic Phase Transitions. edited by W. Hanke and Y. V. Kopaev (Elsevier Science Publishers, Amsterdam, 1992).
[3] Eds.: R. E. Prange and S. M. Girvin, The Quantum Hall Effect (Springer, New York, 1990).
[4] Ed.: H. T. Grahn, Semiconductor Superlattices (World Scientific, Singapore, 1995).
[5] E. K. Kudinov and Y. A. Firsov, Zh. Eksper. Teor. Fiz. **49**, 867 (1965), [Soviet Phys. JETP **22**, 603 (1966)].
[6] V. V. Bryksin and Y. A. Firsov, Fiz. Tverd. Tela **13**, 3246 (1971), [Soviet Phys. Solid State **13**, 2729 (1972)]; Erratum: Fiz. Tverd. Tela **14**, 1857 (1972), [Soviet Phys. Solid State **14**, 1615 (1972)].
[7] V. V. Bryksin and P. Kleinert, J. Phys.: Condens. Matter **15**, 1415 (2003).
[8] P. Kleinert and V. V. Bryksin, Phys. Lett. A **317**, 315 (2003).
[9] V. V. Bryksin and Y. A. Firsov, Zh. Eksper. Teor. Fiz. bf 61, 2373 (1971) [Soviet Phys. JETP **34**, 1272 (1971)].
[10] V. V. Bryksin and Y. A. Firsov, Solid State Commun. **10**, 471 (1972).
[11] V. V. Bryksin and P. Kleinert, J. Phys.: Condens. Matter **9**, 7403 (1997).
[12] V. V. Bryksin and P. Kleinert, J. Phys. A: Math. Gen. **33**, 233 (2000).
[13] P. Kleinert and V. V. Bryksin, Int. J. Mod. Phys. B **15**, 4123 (2001).
[14] V. V. Bryksin and P. Kleinert, Physica B **269**, 163 (1999).
[15] V. V. Bryksin and P. Kleinert, Zh. Eksper. Teor. Fiz. **119**, 1235 (2001) [Soviet Phys. JETP **92**, 1072 (2001)].
[16] V. V. Bryksin and P. Kleinert, Phys. Rev. B **59**, 8152 (1999).
[17] P. Kleinert and V. V. Bryksin, J. Phys.: Condens. Matter **11**, 2539 (1999).

phys. stat. sol. (b) **241**, No. 1, 61–68 (2004) / **DOI** 10.1002/pssb.200303624

On the mechanism of superionic conduction in the zero-dimensional hydrogen-bonded crystals $M_3H(XO_4)_2$ with M = K, Rb, Cs and X = S, Se

Hiroshi Kamimura[*], **Yasumitsu Matsuo**, **Seiichiro Ikehata**, **Takuo Ito**[**], **Masaru Komukae**, and **Toshio Osaka**

Department of Applied Physics, Faculty of Science, Tokyo University of Science, 1-3 Kagurazaka, Shinjuku-ku, Tokyo 162-8601, Japan

Received 1 September 2003, revised 18 September 2003, accepted 18 September 2003
Published online 15 December 2003

PACS 66.30.Dn, 76.60.Es

In this paper we first present recent experimental results related to superionic conduction in the zero-dimensional hydrogen-bonded crystals of alkali-hydrosulfates or hydroselenates, $M_3H(XO_4)_2$ [M = K, Rb, Cs and X = S, Se]. Then a novel approach to the mechanism of proton conduction in the paraelastic phase is described. The key features of the conduction mechanism in the high temperature paraelastic phase which we call the superionic phase are: (1) Two kinds of ionic states, $H_2XO_4^{(+e)}$ and $XO_4^{(-e)}$ are formed by breaking a hydrogen-bond thermally; (2) $H_2XO_4^{(+e)}$ and $XO_4^{(-e)}$ ionic states move coherently from an XO_4 tetrahedron to a distant XO_4 as the result of successive proton tunneling among the hydrogen bonds. We calculate the density of states for the coherent motions of these ionic states by the recursion formula. From this result we show that a characteristic feature of the band-like states obtained from itinerancy of $H_2XO_4^{(+e)}$ and $XO_4^{(-e)}$ ionic states is that of the Bethe lattice; that is the appearance of the twin peak structure due to self-similarity. The calculated conductivity is very high such as the order of 10^{-2} S/cm at and above T_c, consistent with experiment.

1 Introduction Recently the zero-dimensional hydrogen-bonded crystals of $M_3H(XO_4)_2$ type [M = K, Rb, Cs and X = S, Se] have been extensively studied, in particular with regard to the phenomenon of superionic conduction [1–10]. $M_3H(XO_4)_2$ type dielectric crystals exhibit a ferroelastic phase transition at high temperature such as 400 K. For example, the values of T_c for $K_3H(SeO_4)_2$ and $Rb_3H(SeO_4)_2$ are 390 K and 449 K, respectively. Figure 1 shows the crystal structure of rubidium hydroselenate in the ferroelastic phase ($T < T_c$), where Fig. 1a shows its projection on the $a - c$ plane while Fig. 1b shows its projection on the $a - b$ plane [11]. From Fig. 1a we note that the top and bottom oxygen of the neighboring tetrahedrons lie nearly at the same height along the c-axis and that a hydrogen bond is formed in between these top and bottom oxygen whose distance is about 2.5 Å. These hydrogen bonds are isolated and thus do not form a network. Such isolation can be seen clearly in Fig. 1b. Because of this localized nature of the hydrogen bonds, $M_3H(XO_4)_2$ type crystals are called "zero-dimensional hydrogen-bonded crystals". In the paraelastic phase ($T > T_c$), each tetrahedron is tilted so as for Rb, Se and O to stand in line perpendicular to the $a - b$ plane (i.e. c^*-axis). For top or bottom oxygen of each tetrahedron there are three equivalent positions for tilting. Since each tetrahedron thermally rotates very fast at high temperatures above T_c, three equivalent positions

---

[*] Corresponding author: e-mail: kamimura@rs.kagu.tus.ac.jp, Phone: +81 3 3260 4272, Fax: +81 3 3260 4021
[**] Present address: Fujitsu Corporation, Akiruno-Technology Center, Akiruno, Tokyo 197-0833

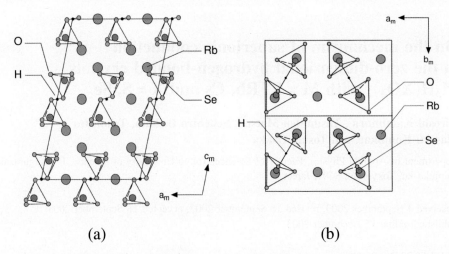

Fig. 1 Projection of the crystal structure of the $Rb_3H(SeO_4)_2$ crystal in the ferroelastic phase on the $a - c$ plane (a) and the $a - b$ plane (b).

appear equally. As a result a crystal structure in the paraelastic phase has the three-fold axes along the c^*-axis and the length of the hydrogen bonds becomes equal. The space group of the *superionic phase* is $R\bar{3}m$.

In these crystals the anomalous increase of electrical conductivity near the phase transition temperature T_c has been observed. Further the conductivity in the paraelastic phase ($T > T_c$) is very high, and it is of the order of 10^{-4} to 10^{-2} S/cm. Because of this fact we call the paraelastic phase as "superionic phase". In Fig. 2 we present the experimental results on the temperature dependence of the electrical conductivity in $M_3H(SeO_4)_2$ crystals (M: K, Rb, Cs). In the superionic phase it is evident that all crystals show the high electrical conductivity of 10^{-4} to 10^{-2} S/cm, although the phase transition temperature depends on the kind of the alkaline metal and on the direction of the electrical conduction. The electrical conductivities along the a-axis are about 10^2 times larger than those along the c^*-axis. It is also noted that in the superionic phase, the conductivity obeys the Arrhenius law. The activation energy of the $M_3H(SeO_4)_2$ crystals along the a-axis becomes about 0.2 eV. This value is 2 times smaller than that along the c^*-axis, indicating that the $M_3H(SeO_4)_2$ crystals display the quasi-two dimensional conductivity. Therefore, it is deduced that the electrical conductivity is closely related to the hopping motion of protons in the $a - b$ plane accompanied by the breaking of the

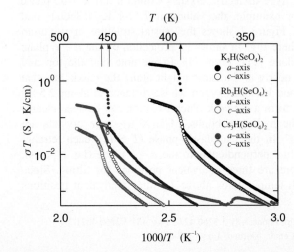

Fig. 2 Temperature dependence of the electrical conductivity in $M_3H(SeO_4)_2$ crystals.

hydrogen bond between SeO_4^{2-} ions. In addition to the behaviour of the electrical conductivity in the superionic phase, the considerable increase of the electrical conductivity with increasing temperature even in the ferroelastic phase is also one of the characteristic features in $M_3H(SeO_4)_2$ crystals. The conductivity just below T_c seems to increase following the $(T_c - T)^{-1/2}$ power law when the temperature approaches T_c from the lower temperature.

As regards a mechanism of superionic conduction, Plakida and Salejda developed a theory of the superionic conduction in $Rb_3H(SeO_4)_2$, based on the phenomenological theory of the ferroelastic transition [12]. Later Gordon discussed the effect of phase transition on the ionic conductivity in these materials [13]. In 1998 Ito and Kamimura [1] developed the first principles theory of superionic conduction using the theory of Matsubara and Toyozawa [14] for impurity conduction in semiconductors and by the Kubo formula [15]. In 2001 Kamimura and Watanabe proposed a new approach to the mechanism of ionic conductivity below and at the ferroelastic phase transition and predicted that the temperature dependence of ionic conductivity follows the $(T_c - T)^{-1/2}$ power law just below T_c [2]. In the present paper we will describe the mechanism of superionic conduction in the superionic phase $(T > T_c)$ based on the theory by Ito and Kamimura [1]. In addition, we will also present new experimental evidence by the proton-NMR for superionic conduction due to proton in $Rb_3H(SeO_4)_2$.

2 Mechanism of ionic conductivity In this section we describe a mechanism of ionic conductivity of the $M_3H(XO_4)_2$ proposed by Ito and Kamimura [1]. In the superionic phase the tetrahedrons rotate as stated above, and the geometrical arrangement of the hydrogen-bonds becomes random. In order to describe the correlated behavior of the protons and of the rotation of the tetrahedrons, let us start with a certain arrangement of hydrogen-bonds shown in Fig. 3a, where $XO_4 - H - XO_4$ dimmers are represented by the thick solid lines. Suppose a hydrogen-bond between two tetrahedrons marked by i and j is broken thermally, by the rotational displacement of top and bottom oxygen of neighboring tetrahedrons as shown by the short arrows in Fig. 3a. Then, when j is tilted and displaced toward a neighboring tetrahedron, say k, as shown by the dotted lines in this figure, a proton in the hydrogen-bond denoted by ij hops to an interstitial position between the tetrahedrons j and k as shown by a long arrow. Then tetrahedron k is tilted to j so as for the energies before and after a hop to be equal with each other. Simultaneously an electron separated from the proton is accommodated in the tetrahedron i. As a result two kinds of ionic states, $H_2XO_4^{(+e)}$ and $XO_4^{(-e)}$, are formed in the superionic phase, as indicated in the dotted circles as shown in Fig. 3b. The formation process of ionic states is of the

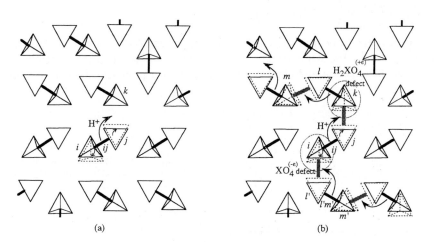

(a)   (b)

Fig. 3 a) Schematic view for the breaking of a hydrogen-bond by the rotational displacement of two neighboring tetrahedrons. b) Schematic view for the formation of two kinds of ionic states, $H_2XO_4^{(+e)}$ and $XO_4^{(-e)}$, and the coherent motion of these ionic states due to successive proton tunneling.

thermal activation type, because the breaking of the hydrogen-bonds is caused by the thermal rotational motion of the tetrahedrons. In the $H_2XO_4^{(+e)}$ ionic state, each of the two hydrogen bonds denoted by the thick solid and gray lines in Fig. 3b is attached to the top and bottom oxygen, while in the $XO_4^{(-e)}$ ionic state there are no hydrogen-bonds. The $(+e)$ and $(-e)$ represent the extra charges yielded in the respective one of the two tetrahedrons when a hydrogen-bond is broken.

First we shall pay attention to the $H_2XO_4^{(+e)}$ ionic state. When a new hydrogen-bond is formed between j and k, this brings about the breaking of the hydrogen-bond between the tetrahedrons k and l. Then, when l is tilted to a neighboring tetrahedron, say m, an extra proton in the hydrogen-bond kl tunnels to an interstitial position between the tetrahedrons l and m, because the energies before and after the tunneling of a proton is equal. As a result the $H_2XO_4^{(+e)}$ ionic state which was originally located at k moves to the tetrahedron m. In the same way, the $XO_4^{(-e)}$ ionic state moves from i to a neighboring tetrahedron, say m', when a proton in the hydrogen-bond $l'm'$ tunnels to an interstitial position between the tetrahedrons i and l' as shown by the arrow in Fig. 3b. In this way, the successive breaking and formation of hydrogen-bonds at different positions appear. Since the thermal rotation of the tetrahedrons is very fast at higher temperatures above T_c of about 400 K, compared with the time of the proton hopping, the coherent protonic transport phenomenon takes place. The remarkable feature of the itinerant motion of two ionic states shown in Figs. 3a and b reflects the characteristics of the Bethe lattice (Cayley tree).

We consider that the transfer interactions of the $H_2XO_4^{(+e)}$ and $XO_4^{(-e)}$ ionic states, Γ and Γ', are of the same order of magnitudes as the rotational energy of an AB_4 type molecule, that is, the absolute magnitudes of Γ and Γ' are the order of 10^{-4} eV. Since the energy for the thermal orientation of the tetrahedrons is of the order of k_BT with k_B being the Boltzmann constant, the time of the hopping motion of a proton between the neighboring sites, $\hbar/\Gamma$ (or $\hbar/\Gamma'$), is much longer than the time of the reorientation of tetrahedrons, $\hbar/k_BT$, for $T = 400$ K, where $\hbar = h/2\pi$, and h is the Planck constant. As a result the coherent motions of both the $H_2XO_4^{(+e)}$ and $XO_4^{(-e)}$ ionic states take place. Recently Waplak et al. [16] reported experimental evidence for coherent motion of protons by EPR measurements for $K_3H(SO_4)_2$ doped with Cr^{5+} ion. In this context, choosing the site energy of the ionic state as the origin of energy, the model Hamiltonian to describe the coherent motion of the $H_2XO_4^{(+e)}$ and $XO_4^{(-e)}$ ionic states are expressed, respectively, by

$$H = \sum_k \sum_j \sum_i \Gamma a_{jk}^{\dagger} a_{ij} \left(\sum_{j'=1}^{4} a_{ij'}^{\dagger} a_{ij'} - 1 \right) \tag{1a}$$

and

$$H = \sum_k \sum_j \sum_i \Gamma' a_{jk}^{\dagger} a_{ij} \left(1 - \sum_{j'=1}^{3} a_{ij'}^{\dagger} a_{ij'} \right), \tag{1b}$$

where Γ and Γ' represent the resonant integral of the $H_2XO_4^{(+e)}$ and $XO_4^{(-e)}$ ionic states, respectively: In other words, those are the transfer interaction of a proton between adjacent available sites. The symbols $a_{ij}^{\dagger}$ and a_{ij} are the creation and annihilation operators of a proton in the site ij, respectively.

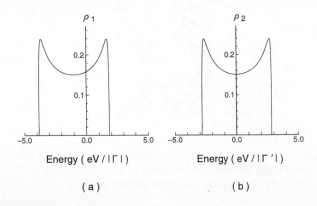

Fig. 4   a) Density of states for the itinerant motion of the $H_2XO_4^{(+e)}$ ionic state where Γ is the transfer interaction of the $H_2XO_4^{(+e)}$ state. b) Density of states for the $XO_4^{(-e)}$ ionic state, where Γ' is the transfer interaction of the $XO_4^{(-e)}$ ionic state.

Here the summation over j' in the parentheses is taken over the nearest neighbor sites around the site i. It should be noted that the numbers of the nearest neighboring sites are 4 and 3, respectively, for $H_2XO_4^{(+e)}$ and $XO_4^{(-e)}$. The factor $\left(\sum_{j'=1}^{4} a_{ij'}^{\dagger} a_{ij'} - 1 \right)$ in Eq. (1a) is 1 when $H_2XO_4^{(+e)}$ ionic state lies at the i-position while otherwise zero, because of two hydrogen bonds attached to $H_2XO_4^{(+e)}$. On the other hand, $\left(1 - \sum_{j'=1}^{3} a_{ij'}^{\dagger} a_{ij'} \right)$ in Eq. (1b) is 1 when $XO_4^{(-e)}$ ionic state lies at the i-position while otherwise zero, because an $XO_4^{(-e)}$ state does not have hydrogen bonds.

Based on this model we will calculate the density of states for the two kinds of ionic states and the mobility for the coherent motions of these ionic states. On the basis of the model Hamiltonians (1a) and (1b), Ito and Kamimura calculated the density of states (DOS) using the recursion formula by Haydock et al [1, 17, 18]. We use their calculated results for DOS which are shown in Figs. 4a and b.

3 Mobility

Ito and Kamimura [1] calculated the mobility of the $H_2XO_4^{(+e)}$ and $XO_4^{(-e)}$ ionic states by using the Kubo formula [15]. According to them, we characterize an ionic state as α, β, etc. by its position and orientation. Then the mobility μ is given by

$$\mu = \left(\frac{\pi \hbar}{2qkT} \right) \int e^{-E/kT} \, \Xi(E) \, dE \Big/ \int e^{-E/kT} \, \rho(E) \, dE \,, \tag{2}$$

where

$$\Xi(E) = -\frac{1}{N} \sum_{\alpha} \sum_{\beta} \sum_{\gamma} \sum_{\delta} G_{\alpha\delta}(E) H_{\delta\gamma} R_{\delta\gamma} \times G_{\gamma\beta}(E) H_{\beta\alpha} R_{\beta\alpha} \,. \tag{3}$$

Here k is the Boltzmann constant, T the absolute temperature, N the total number of the ionic states, and $R_{\beta\alpha}$ represents the displacement vector of a proton associated with the change of an ionic state from α to β, which is equal to the distance between sites i and j, R_{ij}, in Fig. 3b.

As seen above, the calculations of μ are ascribed to calculating the Green's function $G_{\beta\alpha}(E)$, which is defined by

$$G_{\beta\alpha}^{(+)}(E) = \left\langle \beta \left| \frac{1}{E + i\epsilon - H} \right| \alpha \right\rangle$$

$$G_{\beta\alpha}(E) = -\frac{1}{\pi} \operatorname{Im} G_{\beta\alpha}^{(+)}(E) \,, \tag{4}$$

where E is the energy of a system, and ϵ a positive infinitesimal. Ito and Kamimura calculated $\Xi(E)$ in Eq. (3) by using the method of Matsubara and Toyozawa [14], and obtained the mobility for the coherent transfer motion of $H_2XO_4^{(+e)}$ and $XO_4^{(-e)}$ states as follows;

$$\mu = \frac{q\Gamma R^2}{\hbar kT} \times 1.07 \qquad \text{for} \quad H_2XO_4^{(+e)} \text{ state} \,, \tag{5a}$$

$$\mu' = \frac{|q'| \, |\Gamma'| \, R^2}{\hbar kT} \times 1.066 \quad \text{for} \quad XO_4^{(-e)} \text{ state} \,, \tag{5b}$$

where R is the average hopping distance of the $H_2XO_4^{(+e)}$ or $XO_4^{(-e)}$ states, and q and q' the effective charges of an $H_2XO_4^{(+e)}$ and $XO_4^{(-e)}$ states, respectively. Although the proton motion itself is coherent in the quantum-mechanical sense, the distribution function appeared in Eq. (2) is given by the Boltzmann statistics at high temperatures. Thus we have obtained the Einstein relation for the mobility as seen in Eqs. (5a) and (5b). Here we should mention that, if one adopts a classical hopping motion for a proton, the magnitude of mobility is smaller than those of Eqs. (5a) and (5b) at least one order of magnitude.

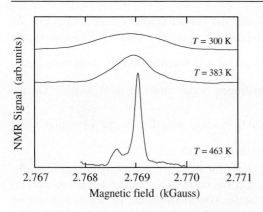

Fig. 5 ^{1}H-NMR absorption lines at various temperatures in the $Rb_3H(SeO_4)_2$ crystal, where $T_c = 449$ K.

Now, let us calculate the ionic conductivity. In doing so we recall that two kinds of ionic states are formed thermally by breaking a hydrogen-bond within a layer. Thus the concentration of ionic states, n, is expressed in the form of the Boltzmann distribution,

$$n = N\,e^{-E_a/kT}\,, \tag{6}$$

where E_a is the formation energy for creating simultaneously an $H_2XO_4^{(+e)}$ ionic state and an $XO_4^{(-e)}$ ionic state thermally. Thus, from Eqs. (5a), (5b) and (6) the static ionic conductivity is obtained as follows;

$$\sigma = qn(\mu + \mu')$$
$$= \frac{q^2 R^2 N(1.07|\Gamma| + 1.066|\Gamma'|)}{\hbar kT}\,e^{-E_a/kT}\,, \tag{7}$$

where we have taken $q = |q'|$. From Eq. (7) one can see clearly that the obtained conductivity obeys the Arrhenius law, which is consistent with experimental results.

We estimate the magnitude of σ for the case of $Rb_3H(SeO_4)_2$. From experimental results in Figs. 1 and 2 we obtain $E_a = 0.26$ eV, $R = 5.8 \times 10^{-8}$ cm, and $T_c = 449$ K. Then, assuming that $q = |q'| = e$, $N = 10^{22}$ cm^{-3}, and $|\Gamma| = |\Gamma'| = 10^{-4}$ eV, we estimate the mobility as 2.6×10^{-3} cm^2/V sec. Thus the conductivity has the order of a magnitude of 10^{-2} S/cm, consistent with experimental results shown in Fig. 2.

4 Experimental evidence for the coherent motion of a proton: ^{1}H-NMR results

Figure 5 shows the observed ^{1}H-NMR absorption lines at various temperatures in the $Rb_3H(SeO_4)_2$ crystal. At room temperature, we have observed the ^{1}H-NMR absorption line with the full width at half maximum of 1.1 G. With increasing temperature, the ^{1}H-NMR absorption line shape becomes sharper even below

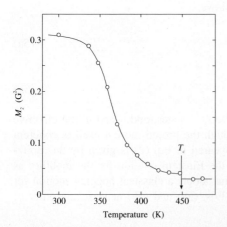

Fig. 6 Temperature dependence of the second moment M_2.

the superionic phase transition. In the superionic phase, we observe the much sharp NMR absorption line with the line width of 0.3 G. This line width is considerably shaper than those in the ferroelastic phase. This result indicates that the motional narrowing of ^{1}H-NMR absorption line is observed above the room temperature in the $Rb_3H(SeO_4)_2$ crystal. Moreover, we note that the ^{1}H-NMR absorption line shows the fine structure with the two peaks (especially in the superionic phase). This result seems to be due to the appearance of some magnetic inequivalent sites accompanied by the proton motion.

The temperature dependence of the second moment M_2 is shown in Fig. 6. The second moment M_2 is closely related with the NMR absorption line width and is calculated from the measured NMR absorption line $f(H)$ using the following equation,

$$M_2 = \frac{\int_{-\infty}^{\infty} (H - H_0)^2 f(H - H_0)\, dH}{\int_{-\infty}^{\infty} f(H - H_0)\, dH}, \tag{8}$$

where H and H_0 mean the external magnetic field and the resonance magnetic field, respectively. As shown in Fig. 6, M_2 begins to decrease at around 340 K and decreases drastically with increasing temperature even below the superionic phase transition, $T_c = 449$ K, and becomes a constant in the superionic phase. The decrease of M_2 is due to the motional narrowing of ^{1}H-NMR absorption line. Therefore, these results imply that the hopping motion of protons appears not only in the superionic phase but also even in the ferroelastic phase.

It is deduced that the appearance of the mobile proton observed even in the ferroelastic phase is caused by the precursor effect of the superionic phase transition. This result is consistent with the theoretical model reported by Kamimura and Watanabe [2]. Moreover, it is also noted that M_2 changes discretely at the superionic phase transition. This result might be due to the drastic increase of the mobility of proton at the first order phase transition. From these results it is deduced that the drastic change of the mobility at T_c leads to the drastic increase of the electrical conductivity at the superionic phase transition. In addition, we also note that M_2 becomes constant in the superionic phase. This indicates that in the superionic phase the hopping motion of protons is fast enough to narrow the NMR absorption line in comparison with those in the ferroelastic phase (that is, the correlation time of the hopping motion of proton is sufficiently shorter than that of relative precession of ^{1}H nuclei). This experimental result is consistent with the theoretical result predicting the drastic increase of the mobility due to the coherent motion of protons in the superionic phase as described in Sections 2 and 3.

5 Summary and concluding remarks In this paper we have described a novel mechanism of ionic conductivity in the superionic phase of $M_3H(XO_4)_2$ type dielectric crystals, based on the theory by Ito and Kamimura. The key features of the conduction mechanism in the superionic phase are the following: (1) Two kinds of ionic states, $H_2XO_4^{(+e)}$ and $XO_4^{(-e)}$ are formed by breaking a hydrogen-bond thermally; (2) $H_2XO_4^{(+e)}$ and $XO_4^{(-e)}$ ionic states move coherently from an XO_4 tetrahedron to a very distant XO_4 as the result of successive proton tunneling among the hydrogen bonds. We have showed that, although the hopping motion of the ionic states is coherent, the mobility has the form of the Einstein relation in its temperature dependence. Finally a remark is made on the mechanism of ionic conductivity just below the phase transition, based on the theory by Kamimura and Watanabe [2]. Just below $T < T_c$, Kamimura and Watanabe assumed that, due to fluctuation effects near the phase transition, the ferroelastic phase consists of a mixture of two regions; a "pseudo-paraelastic region" and an "intervening region". The "pseudo-paraelastic region" is defined as a region in which the the distances between XO_4 tetrahedrons are the same, while the "intervening region" is defined by a region in which XO_4's form $XO_4 - H - XO_4$ dimers connected by hydrogen-bonds. Thus the conduction mechanism developed in the paraelastic phase is applied to the "pseudo-paraelastic regions". As a result they showed that the temperature dependence of ionic conductivity follows the $(T_c - T)^{-1/2}$ power law just below T_c. Their prediction is consistent with experimental results shown in Fig. 2.

References

[1] T. Ito and H. Kamimura, J. Phys. Soc. Jpn. **67**, 1999 (1998).

[2] H. Kamimura and S. Watanabe, Philos. Mag. B **81**, 1011 (2001).

[3] Y. Matsuo, K. Takahashi, and S. Ikehata, Ferroelectrics **272**, 199 (2002).

[4] A. Bohn, R. Melzer, T. Sonntag, R. E. Lechner, G. Schuck, and K. Langer, Solid State Ion. **77**, 111 (1995).

[5] B. V. Merinov, Solid State Ion. **84**, 89 (1996).

[6] M. Komukae, K. Sakata, T. Osaka, and Y. Makita, J. Phys. Soc. Jpn. **63**, 1009 (1994).

[7] Y. Matsuo, K. Takahashi, and S. Ikehata, J. Phys. Soc. Jpn. **70**, 2934 (2001).

[8] S. M. Haile, D. A. Boysen, C. R. I. Chisholm, and R. B. Merle, Nature **410**, 910 (2001).

[9] T. Norby, Nature **410**, 877 (2001).

[10] Y. Matsuo, K. Takahashi, and S. Ikehata, Solid State Commun. **120**, 85 (2001).

[11] I. Makarova, I. A. Verin, and N. M. Shagina, Kristalografiya **31** 178 (1986).

[12] N. M. Plakida and W. Salejda, phys. stat. sol. (b) **148**, 473 (1988).

[13] A. Gordon, Phys. Rev. (b) **52**, R6999 (1995).

[14] T. Matsubara and Y. Toyozawa, Prog. Theor. Phys. **26**, 739 (1961).

[15] R. Kubo, J. Phys. Soc. Jpn. **12**, 570 (1957).

[16] S. Waplak, W. Bednarski, and A. Ostrowski, J. Phys. Chem. Solids **64**, 229 (2003).

[17] R. Haydock, V. Heine, and M. J. Kelly, J. Phys. C **5**, 2846 (1972).

[18] R. Haydock, V. Heine, and M. J. Kelly, J. Phys. C **8**, 2591 (1975).

phys. stat. sol. (b) **241**, No. 1, 69–75 (2004) / **DOI** 10.1002/pssb.200303603

Charge transfer and charge transport on the double helix

N. P. Armitage[*], **M. Briman,** and **G. Grüner**

Department of Physics and Astronomy, University of California, Los Angeles, CA 90095, USA

Received 1 September 2003, accepted 3 September 2003
Published online 15 December 2003

PACS 72.80.Le, 87.14.Gg

We present a short review of various experiments that measure charge transfer and charge transport in DNA. Some general comments are made on the possible connection between various 'chemistry-style' charge transfer experiments that probe fluorescence quenching and remote oxidative damage and 'phys-style' measurements that measure transport properties as defined typically in the solid-state. We then describe measurements performed by our group on the millimeter wave response of DNA. By measuring over a wide range of humidity conditions and comparing the response of single strand DNA and double strand DNA, we show that the appreciable AC conductivity of DNA is not due to photon assisted hopping between localized states, but instead due to dissipation from dipole motion in the surrounding water helix.

1 Overview The electrical conductivity of DNA has been a topic of much recent interest and controversy [1]. Measurements from different groups have reached a variety of conclusions about the nature of charge transfer and transport along the double helix. Although there has been a flurry of recent activity, the subject has long history. Eley and Spivey in 1962 [2] were the first to note that the unique structure of DNA with $\pi-\pi$ orbital stacking separated by 3.4 Å resembled high mobility aromatic crystals and suggested it as efficient structure for electron transfer.

Charge transfer is one of the most fundamental chemical processes, driving such disparate reactions as corrosion and photosynthesis. The semi-classical Marcus [3] theory predicts a charge transfer efficiency in large molecules that falls off as $e^{-\beta r}$ with $\beta \approx 1.5$ Å^{-1}. These considerations seemed borne out by two decades of experiments on proteins and other σ-bonded network bridges between photoexcited metal complexes and electron acceptors. Hence, initial experiments [4] probing the π-bond stack of DNA that showed the possibility of longer range charge transfer were surprising. In these first experiments, fluorescent molecules bound to calf thymus DNA were quenched by the addition of electron acceptors to the strands. They suggested a transfer efficiency $e^{-\beta r}$ with $\beta \approx 0.2$ Å. The expectations of Eley and Spivey not-withstanding this was counter to the prevailing paradigm of transfer efficiency $\beta \approx 1.5$ Å from the Marcus theory. Such long range mobile electrons raised the possibility of interesting electronic effects on the double helix. Transfer along this supposed π-way was referred to as wire-like. This work prompted many other experiments to be done, both within the chemistry community and within the solid-state physics community the latter attempting to measure the transport properties of DNA directly. The activity has lead to new theories, such as polaron transport [5] and conformational gating [6], regarding charge transfer and transport in molecular stacks and biological systems.

---

[*] Corresponding author: e-mail: npa@physics.ucla.edu

Additional experiments showed that the value of β obtained seemed to depend on the details of the strand sequences and donor-acceptor complex used. In the initial experiments Murphy et al. [4] tethered a ruthenium intercalator to end of a single DNA strand and a rhodium intercalator to a complementary stand. When annealed, ruthenium luminescence was completely quenched by the rhodium intercalator positioned almost 40 Å down the π-stack. With the organic intercalator ethidium [7] as the photoexcited donor and rhodium as the acceptor similar quenching behavior was shown over distances of 20 to 30 Å. However other organic donor-acceptor complexes showed $\beta \approx 1$ Å^{-1} [8]. Lewis et al. [9], using stilbene as fluorescence at the end of an A–T chain, systematically moved a G–C pair (functioning as an acceptor) away from the stilbene. They found that quenching rate decreased quickly until about 4 separating A–T's and then more slowly after that.

So-called 'chemistry-at-a-distance' by electron transfer was shown by radical induced strand cleavage. Meggers et al. [10] formed a highly oxidizing radical guanine cation at one end of a DNA strand that had a GGG unit on the other end. The GGG unit is purported to have a lower ionization potential than a single G and hence can accept the hole which neutralizes the radical G. The strand was then treated to cleave at the resulting oxidation site. The length dependence of the electron transfer could be found by varying the number of intervening bridge states and performing electrophoresis to find the number and lengths of cleaved strands. The measurements showed exquisite sensitivity to intervening T–A bases. The efficiency was found to be determined by the longest bottleneck i.e. the longest hopping T–A step.

Strong evidence that the charge transfer was truly happening through base–base hopping via the π–π overlap was given by measurements that probed changes in oxidized guanine damage yield with response to base perturbations [11, 12]. Overall the efficacy of charge transfer through the mismatch was found to correlate with how well bases in the mismatch were stacked. This gave strong evidence that charges are transferred through the π–π stack directly.

These measurements, taken as a whole, gave an emerging picture where a hole has its lowest energy on the G–C sites and for short distances moves from one G–C pair to the next by coherent tunnelling through the A–T sites. The overall motion from the donor to the acceptor site is an incoherent hopping mechanism i.e. the charged carrier is localized on G–C sites along the path. For longer distances between G–C base pairs the picture was that thermal hopping onto A–T bridges becomes the dominant charge-transfer mechanism which gave the weaker distance dependence above four separating A–T pairs of Lewis et al. [9]. Under such circumstances β becomes a poor parametrization of the transfer efficiency as the distance dependence is no longer exponential. Such a picture has been supported from the quantum-mechanical computation models of Burin, Berlin, and Ratner [13].

These 'chemistry-style' experiments give convincing evidence that electron or holes can delocalize over a number of base pairs and that the extent of the delocalization is governed by strand sequence among other aspects. Although such experiments have motivated the direct measure of transport properties via DC and AC techniques, the information gained from luminescence quenching measurements and the like is not directly related to their conductivity i.e. the ability to behave as a molecular wire. Although the descriptor 'wire-like' has been applied to sequences where a small β has been found, such terminology is misleading.

Luminesce quenching is an excited state property. Under appreciated by the solid state community working in this field is the relatively large energy scale (≈ 2 eV) of the typical redox potentials for a luminence quenching reaction (stilbene*/stilbene: 1.75 eV and Rh-complex + 3/Rh-complex + 2: 2 eV). In solid-state physics jargon these are very high energy electron–hole excitations. Perhaps a good solid-state analog of this phenomenon is the luminescence quenching of fluorescent atoms doped into semiconductors, as for instance in Si:Er or ZnSe:Cu [14–16]. In erbium doped silicon a photoexcited electron–hole pair is captured by an impurity level on Er. Decay of this level imparts energy to the 4f Er^{+3} system, which then decays and emits a fluorescent photon. Such a fluorescence can be quenched by detrapping of the captured electron pair on the Er level into the conduction band. In this case, the detrapping is into an orbital which is completely invisible to DC transport. Such an experiment tells us only that there is finite overlap between the localized level and some further extended states. We learn nothing directly applicable to the material's ability to conduct electricity. Likewise luminescence quenching in DNA can be viewed as the detrapping of a hole into the HOMO orbital, whereupon even weak orbital overlap allows it to make its way to the acceptor under the influence of the driving redox potential. Charge transfer experiments confirm delocalization of hole over a few bases, but we learn little about materials ability to behave as wire. In this regard these charge transfer experiments merely reflect the strong effects of disorder in 1-D i.e. the localization of all states. A small β is not synonymous with 'wire-like' behavior.

Although there is relative agreement among chemists regarding the charge transfer properties of DNA, the physics community has not reached a similar détente with respect to measurements of its direct charge transport properties. DNA has been reported to be metallic [17], semiconducting [18], insulating [19, 20], and even a proximity effect induced superconductor [21]. However, questions have been raised in many papers with regards to length effects, the role played by electrical contacts, and the manner in which electrostatic damage, mechanical deformation by substrate–molecule interaction, and residual salt concentrations and other contaminants may have affected these results. Some recent measurements, where care was taken to both establish a direct chemical bond between λ-DNA and Au electrodes and also control the excess ion concentration, have given compelling evidence that the DC resistivity of the DNA double helix over long length scales ($<10\,\mu$m) is very high indeed ($\rho > 10^6\,\Omega$ cm) [22]. These results were consistent with earlier work that found flat $I-V$ characteristics and vanishingly small conductances [20], but contrast with other studies that found a substantial DC conductance that was interpreted in terms of small polaron hopping [5]. DC measurements that show DNA to be a good insulator are also in apparent contradiction with recent contactless AC measurements that have shown appreciable conductivity at microwave and far-infrared frequencies [23, 24] the magnitude of which approaches that of a well–doped semiconductor [25].

In previous finite frequency studies, the AC conductivity in DNA was found to be well parameterized as a power-law in ω [23, 24]. Such a dependence can be a general hallmark of AC conductivity in disordered systems with photon assisted hopping between random localized states [26] and led to the reasonable interpretation that intrinsic disorder, counterion fluctuations, and possibly other sources created a small number of electronic states on the base pair sequences in which charge conduction could occur. However, such a scenario would lead to thermally activated hopping conduction between these localized states and is thus inconsistent with a very low DC conductivity [22]. To the end of resolving some of these matters, we have extended our previous AC conductivity experiments in the millimeter wave range to a wide range of humidity conditions. We have shown that the appreciable AC conductivity of DNA in the microwave and far infrared regime should not be viewed as some sort of hopping between localized states and is instead likely due to dissipation in the dipole response of the water molecules in the surrounding hydration layer.

2 Experimental details Double stranded DNA films were obtained by vacuum drying of 7 mM PBS solution containing 20 mg/ml sodium salt DNA extracted from calf thymus and salmon testes (Sigma D1501 and D1626). In order to improve the DNA/salt mass ratio we used a high concentration of DNA, but it was found that the limit was 20 mg/ml. Higher concentrations makes it difficult for DNA fibers to dissolve and the solution becomes too viscous, which prevents producing the flat uniform films which are of paramount importance for the quasi-optical resonant technique. It was found that as long as the excess salt mass fraction is kept between 2–5% the final results were not significantly affected. Single stranded DNA films were prepared from the same original solution as the double stranded ones. The solution was heated to 95 °C for 30 minutes and the quickly cooled to 4 °C. We checked the conformational state of both double-strand DNA (dsDNA) and single-strand DNA (ssDNA) by fluorescent microscope measurements. Films, when dry, were 20 to 30 microns thick and were made on top of 1 mm thick sapphire windows. Immediately after solution deposition onto the sapphire substrates the air inside the viscous solution was expelled by vacuum centrifuging at 500 g, otherwise the evaporation process causes the formation of air bubbles that destroy the film uniformity.

The AC conductivity was measured in the millimeter spectral range. Backward wave oscillators (BWO) in a quasi-optical setup (100 GHz–1 THz) were employed as coherent sources in a transmission configuration. This frequency range, although difficult to access experimentally, is particularly

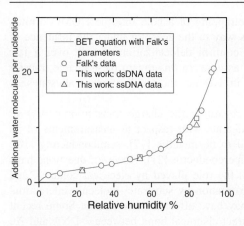

Fig. 1 Adsorbtion of water molecules per nucleotide as a function of humidity. The data represented by the open circles is taken from Falk et al. [28].

relevant as it corresponds to the approximate expected time scale for relaxation processes in room temperature liquids (0.1–10 ps). Importantly, it is also below the energy range where one expects to have appreciable structural excitations. The technique and analysis are well established [27].

3 Results We measured samples at room temperature at several fixed humidity levels. They were maintained in a hermetically sealed environment with a saturated salt solution [28] that kept moisture levels constant. The mass of the DNA films and changes in thickness were tracked by separate measurements within a controlled environment for each sample in a glove box. The total number of water molecules per nucleotide A can be correlated to the relative humidity x ($x = 0 - 1$) through the so-called Branauer–Emmett–Teller (BET) equation [29]

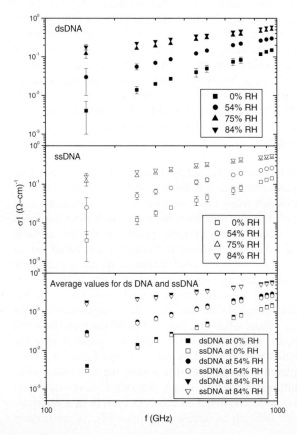

$$A = \frac{BCx}{(1-x)(1-x+Cx)} . \tag{1}$$

The constant B denotes the maximum number of water molecules in the first layer sites. Mobile water molecules within the double helix can be characterized as 2 types according to the statistical formulation of the BET equation by Hill [30]. The first are those within the initial hydration layer, which are directly attached to DNA and have a characteristic binding energy ϵ_1. Water molecules of the second and all other layers can be approximated as having a binding energy ϵ_L. To a good approximation this ϵ_L can be taken to be that of bulk water. These parameters enter into the BET equation through the expression for C which equals $De\left(\frac{\epsilon_1 - \epsilon_L}{kT}\right)$

Fig. 2 AC conductivity of calf thymus DNA at different relative humidity levels. a) Double stranded DNA, b) single stranded DNA. c) A comparison of conductivity between single and double stranded DNA.

© 2004 WILEY-VCH Verlag GmbH & Co. KGaA, Weinheim

where D is related to the partition function of water. Also we should note that there is, in actuality, a structural 0-th layer of water molecules, containing 2.5–3 water molecules per nucleotide that cannot be removed from the helix under typical conditions [31].

Falk et al.'s [28] first established that the adsorption of mobile water layers of DNA can be modelled by distinguishing 2 different types of water parameters by use of the BET equation to describe the hydration of sodium and lithium DNA salts. They found good agreement between experimental data and theory with constants $B = 2.2$ and $C = 20$. We performed a similar hydration study of our dsDNA and ssDNA films; as shown in Fig. 1 the hydration of our films is perfectly consistent with the results of Falk. Moreover, we found no appreciable difference in the hydration between dsDNA and ssDNA.

In Fig. 2 data is presented for the extracted $\sigma_1(\omega)$ of both dsDNA and ssDNA thin films. In both cases, the conductivity is an increasing function of frequency. Since the conductivity also increases with humidity, one may wish to try to separate the relative contributions of charge motion along the DNA backbone from that of the surrounding water molecules.

First, one can consider that there should be two main effects of hydration in our dsDNA films. There is the hydration itself, where water molecules are added in layers. Additionally, the conformational state of dsDNA changes as a function of adsorbed water. Although water molecules can certainly contribute to the increase in conductivity, at high humidities there is the possibility that some of the conduction might be due to an increase in electron transfer along the dsDNA helix in the ordered B form. However since such an effect would be much reduced in disordered and denaturalized single strand DNA and since Fig. 2 shows that to within the experimental uncertainty the conductivity of dsDNA and ssDNA in the millimeter wave range is indistinguishable, it is most natural to suggest that water is the major contribution to the AC conductivity. From this comparison of dsDNA and ssDNA, we find no evidence for charge conduction along the DNA between bases.

4 Discussion In Fig. 3 we plot the conductivity σ_1 of the DNA films normalized by the expected volume fraction of water molecules including both the hydration layers plus the structural water. Although this normalization reduces the spread in the thin film conductivity at the lowest frequencies it does not reduce it to zero, showing that if the observed conductivity comes from water, the character of its contribution changes as a function of humidity.

The complex dielectric constant of bulk water has been shown to be well described by a biexponential Debye relaxation model [32–34], where the first relaxation process [32], characterized by a time scale $\tau_D = 8.5$ ps, corresponds to the collective motion of tetrahedral water clusters, and the second from faster single molecular rotations [35] with a time scale $\tau_F = 170$ fs. For bulk water, the contribution of each relaxation process is determined by the static dielectric constant $\epsilon_S(T) \approx 80$, $\epsilon_1 = 5.2$,

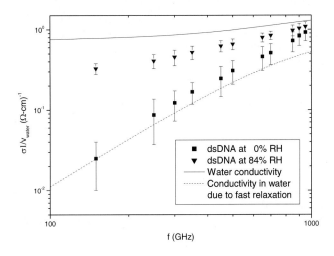

Fig. 3 Conductivity of dsDNA and ssDNA films normalized by the volume fraction of all water molecules (structural plus hydration layer). For clarity, only 0% and 84% humidities are shown. The solid line represents the conductivity of pure water as modelled by the biexponential Debye model using the parameters of Ronne et al. The dashed line shows just the contribution from single water molecule relaxation.

and the dielectric constant at high frequencies $\epsilon_\infty = 3.3$.

$$\widehat{\epsilon}(w) = \epsilon_\infty + \frac{\epsilon_S - \epsilon_1}{1 + i\omega\tau_D} + \frac{\epsilon_1 - \epsilon_\infty}{1 + i\omega\tau_F} \qquad (2)$$

Eq. 2 gives us insight into the conduction and loss processes occurring in the water layers. For high hydration levels, where multiple water layers exist around the dipole helix, the relaxation losses may approach those of bulk water. The above equation can be compared, using the independently known values [32] for τ_D, ϵ_S, τ_F and ϵ_1, to the experimental data normalized to the expected volume fraction of the water. The conductivity of well hydrated DNA is seen to approach that of bulk water.

One expects that the contribution to the loss of cluster relaxation processes to decrease as the number of water layers decreases. As the structural water is not tetrahedrally coordinated, it is reasonable that first term of Eq. 2, which is due to the collective motion of water clusters, cannot contribute at low humidity. Remarkably, the 0% humidity conductivity appears to be described by a model that only includes the fast single molecule rotation of bulk water. This is notable because such behavior is at odds with many systems that find longer average relaxation times in thin adsorbed gas layers than in the corresponding bulk systems [36].

In Fig. 3, along with the experimental data at two representative humidity levels, two theoretical curves for 0% and 100% humidity are plotted. With the only two assumptions being that at 0% humidity, the sole relaxational losses come from singly coordinated water molecules in the structural water layer and that it is only at higher humidity levels where the collective losses can gradually play a greater role, the theoretical curves provide a very good bound to the data over almost all of the measured frequency range.

The only large difference between the experiment and theory is the high frequency data at low humidity, where the model underestimates the conductivity. There are a number of possibilities for these discrepancies. It may be that at higher frequencies for low hydration samples, the weak restoring force from charge-dipole interaction in the structural water layer becomes more significant and our biexponential Debye model is less applicable. Alternatively, it is possible that at very low relative humidities for the ionic phosphate groups on the DNA backbone to form stable dihydrates which may give their own contribution to relaxation losses through their additional degree of freedom [28]. We should also note that one advantage of working in the millimeter spectral range is the known weak contribution of ionic conduction in this regime [37]. The motion of the surrounding relatively large mass counterions only becomes appreciable at lower frequencies [38].

5 Conclusion   In conclusion, we have found that the considerable AC conductivity of DNA can be attributed largely to relaxational losses of the surrounding water dipoles. The AC conductivity of ssDNA and dsDNA was found to be identical to within the experimental error. As ssDNA has significantly different base–base orbital overlap than dsDNA, this indicates the absence of charge conduction along the DNA backbone itself. The conclusion that the observed conductivity derives from the water layer is supported by the fact that, over much of the range, it can be well described by a biexponential Debye model, where the only free parameter is the relative contributions of single water molecule and tetrahedral water cluster relaxation modes.

Acknowledgements   We would like to thank K. Greskoviak for help with sample preparation and E. Helgren for assistance at various stages of the measurements. The research at UCLA was supported by the National Science Foundation grant DMR-0077251.

References

[1] C. Dekker and M. Ratner, Physics World **14**, 29 (2001).
[2] D. D. Eley and D. I. Spivey, Trans. Faraday Soc. **58**, 411 (1962).
[3] R. A. Marcus and N. Sutin, BioChim. BioPhys. Acta **811**, 265 (1985).
[4] C. J. Murphy, M. R. Arkin, Y. Jenkins, N. D. Ghatlia, S. H. Bossmann, N. J. Turro, and J. K. Barton, Science **262**, 1025 (1993).

[5] Yoo et al., Phys. Rev. Lett. **87**, 198102 (2002).

[6] R. Bruinsma, G. Grüner, M. R. D Orsogna, and J. Rudnick, Phys. Rev. Lett. **85**, 4393 (2000).

[7] S. O. Kelley and J. K. Barton, Science **273**, 475 (1996).

[8] A. M. Brun and A. Harriman, J. Am. Chem. Soc. **116**, 10383 (1994).

[9] F. D. Lewis, T. Wu, Y. Zhang, R. L. Letsinger, S. R. Greeneld, and M. R.Wasielewski, Science **277**, 673 (1997).

[10] E. Meggers, M. E. Michel-Beyerle, and B. Giese, J. Am. Chem. Soc. **120**, 12950 (1998).

[11] P. K. Bhattacharyaa and J. K. Barton, J. Am. Chem. Soc. **123**, 8649 (2001).

[12] M. A. Rosen, L. Shapiro, and D. J. Patel, Biochemistry **31**, 4015 (1992).

[13] Y. A. Berlin, A. L. Burin, and M. A. Ratner, J. Phys. Chem. A **104**, 443 (2000).

[14] Jung H. Shin et al., App. Phys. Lett. **68**, 997 (1996).

[15] F. Priolo et al., J. App. Phys. **49**, 16313 (1994).

[16] J. F. Suyver et al., App. Phys. Lett. **79**, 4222 (2001).

[17] H. W. Fink and C. Schonenberger, Nature **398**, 407 (1999).

[18] D. Porath et al., Nature **403**, 635 (2000).

[19] E. Braun et al., Nature **391**, 775 (1998).

[20] P. J. de Pablo et al., Phys. Rev. Lett. **85**, 4992 (2000).

[21] A. Y. Kasumov et al., Science **291**, 280 (2001).

[22] Zhang et al., Phys. Rev. Lett. **89**, 198102 (2002).

[23] P. Tran, B. Alavi, and G. Grüner, Phys. Rev. Lett. **85**, 1564 (2000).

[24] E. Helgren et al., cond-mat/0111299.

[25] E. Helgren, N. P. Armitage, and G. Grüner, Phys. Rev. Lett. **89**, 246601 (2002).

[26] A. L. Efros and B. I. Shklovskii, J. Physics C **8**, L49 (1975).

[27] A. Schwartz et al., Rev. Sci. Instrum. **66**, 2943 (1995).

[28] M. Falk, K. Hartman, and R. Lord, J. Am. Chem. Soc. **84**, 3843 (1962).

[29] S. Braunauer, P. Emmett, and E. Teller, J. Am. Chem. Soc. **80**, 309 (1938).

[30] T. L. Hill, J. Chem. Phys. **14**, 263 (1946).

[31] N. Tao and S. Lindsay, Biopolymers **28**, 1019 (1989).

[32] C. Ronne et al., J. Chem. Phys. **107**, 5319 (1997)

[33] J. Kindt and C. Schmuttenmaer, J. Phys. Chem. **100**, 10373 (1996).

[34] J. Barthel and R. Buchner, Pure Appl. Chem. **63**, 1473 (1991).

[35] N. Agmon, J. Phys. Chem. **100**, 1072 (1996).

[36] G. P. Singh et al., Phys. Rev. Lett. **47**, 685 (1981).

[37] J. D. Jackson, Classical Electrodynamics, 2nd ed. (John Wiley and Sons, New York, 1975), p. 291.

[38] Z. Kutnjak et al., Phys. Rev. Lett. **90**, 98101 (2003). Kutnjak et al. have observed an extension of the region of apparent power-law conductivity to frequencies below that of the lower relaxation time. Within our interpretation this is reasonably attributed to counterion motion.

phys. stat. sol. (b) **241**, No. 1, 76–82 (2004) / **DOI** 10.1002/pssb.200303627

Hopping transport in 1D chains (DNA vs. DLC)

K. Kohary[*,1,2], **H. Cordes**[3], **S. D. Baranovskii**[2], **P. Thomas**[2], and **J.-H. Wendorff**[4]

[1] Department of Materials, University of Oxford, Parks Road, Oxford, OX1 3PH, UK
[2] Department of Physics and Material Sciences Center, Philipps University Marburg, 35032, Marburg, Germany
[3] Department of Physics, Princeton University, New Jersey, NJ 08544, USA
[4] Department of Chemistry and Material Sciences Center, Philipps University Marburg, 35032, Marburg, Germany

Received 1 September 2003, revised 16 September 2003, accepted 16 September 2003
Published online 15 December 2003

PACS 72.20.Ee, 72.80.Ng, 73.63.–b, 87.14.Gg

We discuss charge transport in one-dimensional organic solids (DNA and discotic liquid-crystalline glass (DLC)), focusing on the effects of static and dynamic disorder. In the presence of static disorder it can be shown that the temperature dependence of the low-field mobility is $\mu \propto \exp\left[-(T_0/T)^2\right]$, with characteristic temperature T_0 depending on the scale of the energy distribution of localized states responsible for transport. In the case of both static and dynamic disorder the situation is different. We obtain a temperature independent mobility in our molecular dynamics calculations in the case of large static and dynamic disorder compared to the energy overlap integral between the neighbouring sites. The theoretical results are in good agreement with experimental data.

1 Introduction It was already suggested in the 1960's that DNA is a possible candidate for a one-dimensional transport material because of the π-overlap between the neighbouring bases. Recent experiments have demonstrated a long-range charge migration along the DNA double helix, indicating that DNA can serve as a one-dimensional molecular wire. However, the actual magnitude of DNA conductivity as well as its physical mechanism is still under debate. A number of conflicting experimental results have been reported and a variety of theoretical models have been suggested [1–17], but a consensus on charge transport in DNA has not yet been reached. For example, it was published by different groups that DNA is an insulator [3, 4], or a semiconductor [5, 6], or has very good conduction properties [7], and there are even suggestions for a superconducting behaviour [8]. It is a fundamental task to understand charge transport mechanism along the DNA chain. It will enable us to better understand the damage and repair mechanism present in biological systems. In addition, we can get a better insight into the possible electronic and medical applications. Charge (electron or hole) transport in DNA can be modelled as a quasi-one-dimensional transport along the helical chain. This transport takes place along the main building elements of DNA (guanine, cytosine, thymine, and adenine), which can be considered as sublattice sites for the charge transport. In such a system, static and dynamic disorder can play a crucial role.

Numerous chemical studies have shown that after inserting a charge carrier into a donor-DNA-acceptor system, the carrier moves along the DNA chain via hopping between DNA bases [18]. For example, holes tend to hop via guanine bases, which provide essentially lower energies for such carriers compared to other DNA bases. The rate of hopping processes strongly depends on the number of adenine-thymine bases bridging the neighbouring guanines for short bridges evidencing the tunnelling hop-

---

[*] Corresponding author: e-mail: krisztian.kohary@materials.ox.ac.uk, Phone: +44 1865 283325, Fax: +44 1865 273789

ping transitions. Unfortunately, being based on the reaction yields and not on the direct measurements of the transient times, the study of transport phenomena in such "chemistry at a distance" experiments does not reveal much physical information. Therefore, studies of hopping transport in organic one-dimensional systems have been concentrated so far mostly on systems which are widely believed to be similar to the DNA chain, namely, on discotic liquid-crystalline glasses (DLC) [10, 19, 20]. Both DNA and DLC consist of quasi-one-dimensional chains of flexible organic units separated by distances of about 0.35 nm. In contrary to DNA chains, the DLC systems allow a straightforward experimental study of transport phenomena including the dependencies of the carrier mobility on essential parameters like temperature and electric field in wide ranges of the corresponding parameter values.

Columnar discotic liquid crystalline glasses have recently caused a particular interest because of their potential application for electrography and for transport materials in light-emitting diodes [21]. The proximity (0.35 nm) of molecular units leads to a strong overlap of electron wave functions on neighbouring units, which can provide high charge carrier mobility values favourable for technical applications. These systems enable researchers to test different transport approaches. Regarding the temperature dependence of the mobility, DLC can be separated into two groups. In the first group, materials have temperature dependent mobility proportional to $\mu \propto \exp[-(T_0/T)^2]$ with some characteristic temperature T_0 [22, 23]. This type of behaviour of the mobility can be described in the framework of a hopping theory in a quasi-one-dimensional chain [24–26]. To the second group belong materials which exhibit almost temperature independent mobilities [27]. This latter behaviour cannot be described by simply assuming the presence of static disorder and a dynamic disorder should be also taken into account.

In this paper, we show principles of numerical algorithms and analytic calculations suitable for describing the charge transport in quasi-one-dimensional systems, where the effects of static and dynamic disorder play an important role. This paper is organised as follows. In the next section the effects of static disorder are studied. Section 3 describes the effects of both static and dynamic disorder. In section 4 their relevance to charge transport in DNA and DLC systems is discussed.

2 Static disorder   It was shown that in chain systems, such as DLC, the charge transport is extremely anisotropic. The mobility along the chains is more than three orders of magnitudes larger than in the perpendicular direction. This allows one to model the transport in a quasi-one-dimensional environment. Experimentally observed dependences of the conductivity on the frequency, on the strength of the applied electric field, and on the temperature evidence that an incoherent hopping process is the dominant transport mechanism in such materials [22, 23, 28]. It is widely believed that energy levels which are responsible for charge transport in such systems have a Gaussian density of states

$$d(\varepsilon) = \frac{1}{\sqrt{2\pi\sigma^2}} \exp\left(-\frac{\varepsilon^2}{2\sigma^2}\right), \tag{1}$$

where σ is the width of the distribution [21]. In these quasi-one-dimensional chains the charge carriers hop via molecular sites. The transition rate from the occupied localised state i to an unoccupied site j is usually described by the Miller-Abrahams rates $\Gamma_{ij} = \Gamma_0 \exp\left(-2R_{ij}/\alpha\right) \exp\left(-\frac{\varepsilon_j - \varepsilon_i + |\varepsilon_j - \varepsilon_i|}{2kT}\right)$, where Γ_0 is the attempt-to-escape frequency, R_{ij} is the distance between sites i and j, k is the Boltzmann constant and ε_i and ε_j are the corresponding site energies.

The temperature and electric field dependences of the mobility in such a system were extensively studied by Cordes et al. [24–26]. It was shown by analytical calculations and Monte Carlo simulations that the temperature dependence of mobility has a form of

$$\mu(T) = \mu_0 \exp\left(-A\left(\frac{\sigma}{kT}\right)^2\right), \tag{2}$$

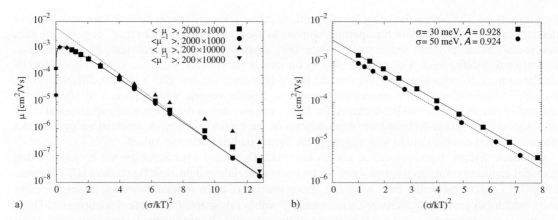

Fig. 1   In the left figure, the temperature dependence of the low-field mobility for $\sigma = 50$ meV is shown. Different symbols show different types of averaging and different system sizes. Solid line represents the solution for infinite chain. Dashed line is a guide for the eyes with a slope equal to −1. In the right figure, the results of Monte Carlo simulations are shown.

with some coefficient A in a good agreement with experiments [22, 23]. Figure 1 shows such a strong exponential dependence on temperature both for analytical calculations and for Monte Carlo simulations. It was found that at large temperatures, the nearest neighbour hopping is the dominant process in these systems. At lower temperatures tunnelling to second-, third-nearest neighbours, etc. gets more and more important. This affects the value of the coefficient A in Eq. (2), which changes from 1 (at large temperatures) to $(n + 1)/2n$ (low temperatures), where n is the n-nearest neighbour essential for hopping processes. However, in the experimentally studied systems this change is only relevant for A being between about 0.7 and 1 [26].

Much attention has been paid in the recent years to the dependence of the drift mobility in organic systems on the electric field. In particular, the reported increase of the drift mobility with decreasing field [21, 29–31] at low electric fields caused a lot of discussion in the scientific literature. In their theoretical calculations, Cordes et al. [24] have shown that this increase at $F \to 0$ was erroneously interpreted and can be explained as follows. In the experiments the mobility is calculated by $\mu = b/(Ft_{\mathrm{tr}})$, where b is the length of the sample, F is strength of the electric field and t_{tr} is the transit time. If this expression is used in the theoretical calculations the above mentioned anomaly can be observed [24]. Figure 2 illustrates such an increase of the mobility at low fields. However, the charge carrier would pass through the system even at zero electric field solely due to diffusion. Therefore, at very low fields it is more appropriate to calculate the mobility via the diffusion formula using the Einstein relation $\mu = (eD)/(kT)$, with calculating the diffusion constant by $D = b^2/2t_{\mathrm{tr}}$. In Fig. 2 it is shown that this latter calculation coincides at

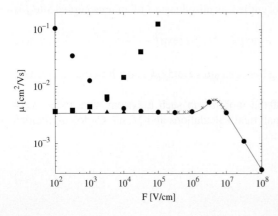

Fig. 2   Field dependence of the carrier mobility at $kT = 25$ meV. The solid line represents the exact solution for infinite chain. Data shown by circles and squares were calculated in finite systems via drift and diffusion formulas, respectively. Crosses show the results of the Monte Carlo simulation.

low electric fields with the results for the mobility calculated by drift formula at higher fields. This agreement between two approaches confirms the application of the diffusion formula at $F \to 0$. It was therefore concluded that the anomalous behaviour of the mobility at low electric fields found in the experiments is only an artefact and has no physical meaning. The same conclusion has been found earlier on the basis of experimental studies of the mobility in organic semiconductors at low electric fields [32], though it could not be claimed for sure before the theoretical calculations confirmed this result [24].

3 Static and dynamic disorder   The results with static disorder are really important, because they are able to explain such puzzling features as the field dependence of the mobility and the temperature dependence at low temperatures. However, in some DLC the temperature dependence of the mobility is not exponential and temperature independent mobility values were reported above room temperature [27]. In such systems dynamic disorder also should be taken into account. Palenberg et al. used a model in which only dynamic disordered effects were considered [33]. However, their model only provides temperature independent parts of the mobility in a small temperature range.

   We used a tight-binding model Hamiltonian for describing static and dynamic energy disorder in organic solids. We consider a one-dimensional chain where every site belongs to a molecular sequence of the related material. Following Vekhter et al. [34], a charge carrier is generated at a particular site. The motion of this charge carrier is determined by the Hamiltonian

$$H = H_c + H_{ph} + H_{c-ph} \,. \tag{3}$$

The first part H_c describes the on-site energies and the first-neighbour hopping terms

$$H_c = \sum_n \varepsilon_n a_n^+ a_n - \sum_n \beta(a_n^+ a_{n+1} + a_{n+1}^+ a_n) \,, \tag{4}$$

where n is the site label in the chain, $a_n^+ (a_n)$ denotes the creation (annihilation) operator of the charge carrier, the variables ε_n correspond to the on-site energies (static disorder), and β is the energy overlap integral between neighbouring sites. The motion of the site molecules (phonon bath) is described by a classical harmonic oscillator Hamiltonian

$$H_{ph} = \sum_n \left\{ \frac{p_n^2(t)}{2M} + \frac{1}{2} M \omega^2 \left[x_{n+1}(t) - x_n(t) - x_{eq} \right]^2 \right\}, \tag{5}$$

where $p_n(t)$ and $x_n(t)$ are the momenta and the coordinates of the molecular sites, M is their mass , x_{eq} is the equilibrium distance between the sequential sites on the chain, and ω is the vibration frequency. The interaction of a charge carrier with molecular sites (electron–phonon interaction) in the chain is described by the

$$H_{c-ph} = \sum_n g\left(x_{n+1}(t) - x_{n-1}(t) - 2x_{eq} \right) a_n^+ a_n \,, \tag{6}$$

where g is the coupling constant. The last equation provides the time dependent site energy values ($\tilde{\varepsilon}_n(t)$), which fluctuate around the static disorder values: $\tilde{\varepsilon}_n(t) = \varepsilon_n + g\left(x_{n+1}(t) - x_{n-1}(t) - 2x_{eq} \right)$. The wave function of the charge carrier in the above second quantization formalism is $|\Psi(t)\rangle = \sum_n c_n(t)|n\rangle$.

   We insert the particle into the system at time $t = 0$, by taking the coefficients in this expansion as $c_0(t = 0) = 1$ and $c_{n \neq 0}(t = 0) = 0$. Then the time evolution of the system can be described by calculating the new charge carrier wave function at time $t + \Delta t$ using the electronic part of the total Hamiltonian (Eq. (3)) $i\hbar \frac{\partial}{\partial t} |\Psi(t)\rangle = H|\Psi(t)\rangle$, followed by the calculation of the positions and velocities of the oscillators at time $t + \Delta t$ with the help of the effective potential derived from the electronic contribution.

Once it is done, the procedure can be repeated. To study the transport, we calculated the mean square displacement of the charge carrier by $\langle \Psi(t) | n^2 | \Psi(t) \rangle = \sum_n |c_n(t)|^2 n^2$. From this quantity one can evaluate the diffusion constant and the mobility can be calculated from the diffusion constant via the Einstein relation.

In our computer simulations before the injection of the charge carrier into the system, the molecular complex was equilibrated by the following method. At the starting stage, every molecular site was placed at its equilibrium position. The velocities were distributed according to a given temperature by the Maxwell–Boltzmann distribution. Then the system was let to evolve for 2.5 ps. After this time delay we had an equilibrated system with kinetic energy fluctuating at half of the starting temperature. We took this kinetic energy as the temperature of the system. For the next step, a charge carrier was generated in the middle of the chain at $t = 0$. The time evolution of the wave function was approximated by the symmetric Cayley formula:

$$|\Psi(t + \Delta t)\rangle = \exp\left[-iH(t)\Delta t\right]|\Psi(t)\rangle \approx \frac{1 - \frac{1}{2}iH(t)\Delta t}{1 + \frac{1}{2}iH(t)\Delta t}|\Psi(t)\rangle. \qquad (7)$$

The time step Δt was chosen as 0.1 fs. We checked that smaller time steps do not bring any effect to the physical picture. The separation distance between the molecular sites was 0.35 nm. Following previous studies [19, 34], we have chosen the M mass of the oscillators as 60 proton mass, the oscillation frequency ω as 1.4×10^{13} Hz, and the coupling constant g as 5.15 eV/nm. The charge carriers were injected in the middle of the chains. We checked that no size effects are present by choosing long enough chains during the 150 ps long simulations. Our systems typically consisted of 1001 sites and the averages were calculated over 100 realisations of each system. In our computer simulations the site energies ε_n were distributed according to a Gaussian formula (Eq. (1)), with $\sigma = 0, 25, 50,$ and 75 meV. The overlap parameter β was taken as 10, 15, and 25 meV. This choice leads to the range of the ratio σ/β between 1.0 and 7.5.

Figure 3 summarises our results showing the temperature dependence of the mobility in systems having large static disorder ($\sigma = 75$ meV) with the overlap parameter β changing from 10 to 25 meV. The results clearly show that dynamic disorder can really suppress the strong temperature dependence of the mobility calculated in the systems with only static disorder. In a wide temperature range we found almost temperature independent mobility values if the $\sigma/\beta > 4$. However, if this ratio gets smaller, the typical temperature dependence can be described as $1/T$.

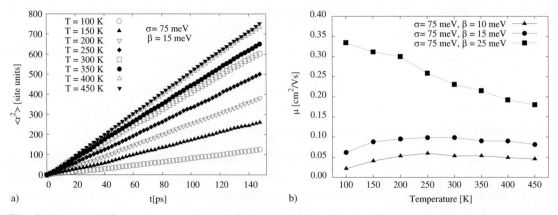

Fig. 3 In the left figure, the mean square displacement is shown for different temperatures for parameters $\sigma = 75$ meV and $\beta = 15$ meV. In the right figure, the temperature dependence of the mobility with static and dynamic disorder is shown.

4 Discussion   Organic disordered systems (such DLC and DNA) can be very much different from inorganic disordered solids. It is believed that in the latter materials only static disorder is dominant, whereas it is "expected" that in organic systems dynamic disorder also plays a crucial role for transport and other physical properties. Both DLC and DNA have flexible organic units along the chain axis. In the previous two sections we have presented theoretical descriptions of different approaches to such quasi-one-dimensional systems. It was shown that the temperature dependence of the mobility can change drastically from an exponential to an almost temperature independent behaviour, depending on the presence on dynamic disorder.

Chemical studies have shown that there is hopping in DNA. In such experiments, small donor-DNA-acceptor bridges are constructed and the transport properties were extracted from reaction yield measurements. Unfortunately, in those experiments the dependence of the transport properties on temperature or electric field has not been studied. Physical experiments tried to address such issues, but the reported results are contradictory and almost each group reported different transport properties. The difficulty lies in the fact that the transport properties depend on several circumstances, which are really hard to control in the experiments. Aiming and hoping in the future to apply algorithms to DNA molecules, we concentrated so far on DLC, a system similar to DNA, in which interesting experiments are abundant.

Surprising and interesting results were reported for DLC. In one set of experiments strong exponential temperature dependence of mobility was reported, whereas there are other systems where temperature independent mobility values were measured. We claim that these variations can be explained by the strength of dynamic disorder in different systems. Our molecular dynamics simulations show that dynamic disorder can suppress the exponential temperature dependence of the mobility, which was found in the case of "pure" static disorder. Therefore, we suggest that in these latter systems (with strong exponential dependence on temperature) the effect of dynamic disorder on transport is small. On the other hand, in systems where temperature independent mobility was reported the effect of dynamic disorder seems to be crucial.

5 Conclusion   The temperature dependence of mobility in quasi-one-dimensional organic systems, such as discotic liquid-crystalline glasses (DLC), has been studied with taking into account static and dynamic disorder effects. We have shown that in the case of static disorder the temperature dependence of the mobility is exponential. In the case of static and dynamic disorder temperature independent mobility values can be obtained with reasonable parameters. Speculating about the strong structural similarity between DLC and DNA, we believe that our theoretical approach would also be valid for the description of charge carrier transport in DNA molecules.

Acknowledgements   Financial support of the Fonds Dder Chemischen Industrie, of the Deutsche Forschungsgemeinschaft, the Optodynamic Center Marburg and that of the European Graduate College "Electron–electron Interactions in Solids" Marburg–Budapest is gratefully acknowledged. K. K. also acknowledges the financial support of the Lockey Committee (University of Oxford).

References

[1] F. D. Lewis, T. Wu, Y. Zhang, R. L. Letsinger, S. R. Greenfield, and M. R. Wasielewski, Science **277**, 673 (1997).

[2] P. J. Dandliker, R. E. Holmlin, and J. K. Barton, Science **275**, 1465 (1997).

[3] E. Braun, Y. Eichen, U. Sivan, and G. Ben-Yoseph, Nature **391**, 775 (1998).

[4] P. J. de Pablo, F. Moreno-Herrero, J. Colchero, J. Gomez Herrero, P. Herrero, A. M. Baro, P. Orderjon, J. M. Soler, and E. Artacho, Phys. Rev. Lett. **85**, 4992 (2000).

[5] H.-W. Fink and C. Schonenberger, Nature **398**, 407 (1999).

[6] D. Porath, A. Bezryadin, S. de Vries, and C. Dekker, Nature **403**, 635 (2000).

[7] A. Rakitin, P. Aich, C. Papadopoulos, Y. Kobzar, A. S. Vedeneev, J. S. Lee, and J. M. Xu, Phys. Rev. Lett. **86**, 3670 (2001).

[8] A. Y. Kasumov, M. Kociak, S. Gueron, B. Reulet, V. T. Volkov, D. V. Klinov, and H. Bouchiat, Science **291**, 280 (2001).

[9] J. Jortner, M. Bixon, T. Langenbacher, and M. E. Michel-Beyerle, Proc. Natl. Acad. Sci. USA **95**, 12759 (1998).
[10] F. C. Grozema, Y. A. Berlin, and L. D. A. Siebbeles, Int. J. Quantum Chem. **75**, 1009 (1999).
[11] Y. A. Berlin, A. L. Burin, and M. A. Ratner, J. Am. Chem. Soc. **123**, 260 (2001).
[12] P. Tran, B. Alavi, and G. Gruner, Phys. Rev. Lett. **85**, 1564 (2000).
[13] Bruinsma, G. Gruner, M. R. D'Orsogna, and J. Rudnick, Phys. Rev. Lett. **85**, 4393 (2000).
[14] Z. G. Yu and X. Song, Phys. Rev. Lett. **86**, 6018 (2001).
[15] K.-H. Yoo, D. H. Ha, J.-O. Lee, J. W. Park, J. Kim, J. J. Kim, H.-Y. Lee, T. Kawai, and H. Y. Choi, Phys. Rev. Lett. **87**, 198102 (2001).
[16] Y.-J. Ye, R. S. Chen, F. Chen, J. Sun, and J. Ladik, Solid State Commun. **119**, 175 (2001).
[17] Z. Kutnjak, C. Filipic, R. Podgornik, L. Norddenskiold, and N. Korolev, Phys. Rev. Lett. **90**, 098101 (2003).
[18] B. Giese, Annu. Rev. Biochem. **71**, 51 (2002), and references therein.
[19] L. D. A. Siebbeles and Y. A. Berlin, Chem. Phys. **238**, 97 (1998).
[20] F. C. Grozema, Y. A. Berlin, and L. D. A. Siebbeles, J. Am. Chem. Soc. **122**, 10903 (2000).
[21] H. Bassler, phys. stat. sol. (b) **175**, 15 (1993), and references therein.
[22] A. Ochse, A. Kettner, J. Kopitzke, J. H. Wendorff, and H. Bassler, Phys. Chem. Chem. Phys. **1**, 1757 (1999).
[23] I. Bleyl, C. Erdelen, H.-W. Schmidt, and D. Haarer, Philos. Mag. B **79**, 463 (1999).
[24] H. Cordes, S. D. Baranovksi, K. Kohary, P. Thomas, S. Yamasaki, F. Hensel, and J-H. Wendorff, Phys. Rev. B **63**, 094202 (2001).
[25] K. Kohary, H. Cordes, S. D. Baranovski, P. Thomas, S. Yamasaki, F. Hensel, and J.-H. Wendorff, Phys. Rev. B **63**, 094202 (2001).
[26] I. P. Zvyagin, S. D. Baranovskii, K. Kohary, H. Cordes, and P. Thomas, phys. stat. sol. (b) **230**, 227 (2002).
[27] T. Kreouzis, K. J. Donovan, N. Boden, R. J. Bushby, O. R. Lozman, and Q. Liu, J. Chem. Phys. **114**, 1797 (2001).
[28] N. Boden, R. J. Bushby, J. Clements, B. Movaghar, K. J. Donovan, and T. Kreouzis, Phys. Rev. B **52**, 13274 (1995).
[29] P. M. Borsenberger, L. Pautmeier, and H. Bässler, J. Phys. Chem. **94**, 5447 (1991).
[30] P. M. Borsenberger, E. H. Magin, M. van der Auweraer, and F. C. de Schryver, phys. stat. sol. (a) **140**, 9 (1993).
[31] H. Bassler, in: Semiconducting Polymers, edited by G. Hadziioannou and P. E. van Hutten (Wiley-VCH, Weinheim, 2000) p. 365.
[32] A. Hirao, H. Nashizawa, and M. Sugiuchi, Phys. Rev. Lett. **75**, 1787 (1995).
[33] M. A. Palenberg, R. J. Silbey, M. Malagoli, and J.-L. Bredas, J. Chem. Phys. **112**, 1541 (2000).
[34] B. G. Vekhter and M. A. Ratner, J. Chem. Phys. **101**, 9710 (1994).

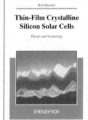

59313064_gu All prices are approx prices and subject to change / All € prices are only valid for Germany

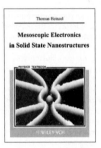
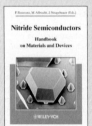

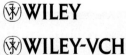

physica **p s s** solidi c

www.physica-status-solidi.com

conferences and critical reviews

WILEY-VCH

phys. stat. sol. (c) **1**, No. 1, 9–12 (2004) / **DOI** 10.1002/pssc.200303622

Slow conductance relaxation in granular aluminium films

T. Grenet[*]

LEPES CNRS, BP 166, 38042 Grenoble cedex 9, France

Received 12 September 2003, revised 17 September 2003, accepted 17 September 2003
Published online 28 November 2003

PACS 72.80.Ng, 73.23.Hk, 73.61.At

Conductance measurements of thin granular aluminium films are reported. They exhibit a symmetrical field effect anomaly and slow relaxation of the conductance after the samples are put out of equilibrium by abruptly changing the temperature or the gate voltage. The results are identical to the ones reported in InO_x thin films, suggesting that these phenomena are quite general in disordered insulators. The study of granular systems may help to elucidate the origin of the glassy behaviour.

1   Introduction

This work was motivated by the studies of indium oxide (InO_x) thin films performed by Z. Ovadyahu and his co-workers [1] which showed in the insulating phase (i) a symmetrical minimum in the field effect characteristics $G(V_{gate})$ (where G is the InO_x channel conductance and V_{gate} the gate voltage of the MOS like device), (ii) a slow relaxation of G after the system was perturbed and put out of equilibrium by an abrupt change of e.g. V_{gate} or the temperature T, (iii) memory effects and a simple ageing law. Reports of the same kind of phenomena have been given in thin metal films [2–4]. Our purpose is to conduct the same studies in another kind of disordered insulator in order to test the generally of the phenomena and gain a complementary insight into their mechanisms. We chose to study granular metals which can be monitored through a metal to insulator transition but have a rather different microstructure than InO_x films, whose microscopic parameters (grain size, inter-grain tunnel barriers) can in principle be controlled.

2   Experimental

The samples were prepared as described previously [5] by e-beam evaporation on sapphire substrates. Each sample consists of a MOS-like device obtained by successively depositing the aluminium gate, the Al_2O_3 insulator and the granular Al *channel*, as well as a granular Al *strip* directly deposited on the substrate simultaneously to the MOS *channel*. The granular Al films (200 Å thick) consists of metallic grains covered with insulating Al_2O_3 layers. Preliminary STM images have shown the granular films to consist of a rather compact ensemble of grains with a broad size distribution (30–300 Å). They can be made metallic or insulating depending on the evaporation conditions (evaporation rate and O_2 pressure). The effects this paper deals with were only observed in insulating samples (in contrast to the effects reported in [6], and we will concentrate here on samples with resistance ratios R(4 K)/R(300 K) in the range 10^4–10^5 and R (4 K)~100 MΩ to a fewGΩs. The R(T) curves cannot be fitted to a simple law in the whole temperature range (2 K–300 K), but a simple activation law is observed below T~30 K in all our samples (with activation energies of a few meV), which is quite unexpected since the Efros-Schlovskii variable range hopping type of dependence is almost always observed in granular metals and

---

[*] Corresponding author: e-mail: thierry.grenet@grenoble.cnrs.fr, Phone: +00 476 887 461, Fax: +00 476 887 988

cermets. All the measurements shown below were performed in the ohmic regime, at 4 K unless otherwise stated.

3   Field effect and slow relaxation: results and discussion

We have already shown [5] that the granular aluminium *channels* exhibit a symmetrical field effect "dip" which is very similar to the one observed in highly doped InO$_x$. In Fig. 1 we show its temperature dependence: as T is increased the "dip" amplitude vanishes and its FWHM increases.

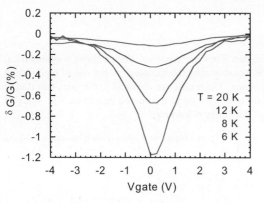

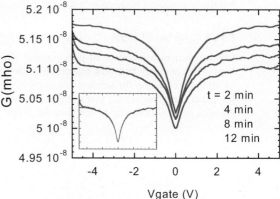

Fig. 1 Field effect curves measured at different temperatures (decreasing T from top to bottom curve)

Fig. 2 Dependence of the field effect curves on the measurement time (increasing V$_{gate}$ scan duration from top to bottom curve), measured at 4 K. Inset: same curves after amplitude normalisation.

An important feature is that these measurements are performed out of equilibrium. The amplitude of the field effect "dip" and the value of the conductance outside it are smaller for longer measurement times (slower V$_{gate}$ scans), as is illustrated in Fig. 2. However the shapes of the curves remain unchanged and are perfectly superposed when their amplitudes are normalised, as is shown in the inset of Fig. 2. Thus, although the amplitudes depend on the measurement conditions, the broadening only depends on the temperature.

Non-equilibrium and slow relaxation effects are best illustrated by the "two dip" experiment [7]. A sample is first equilibrated at V$_{gate}$ = 0. The gate voltage is then changed and maintained at its new value for a waiting time t$_w$, after which it is set back to zero. As is shown in Fig. 3, during t$_w$ a second "dip" appears centred on the waiting gate voltage, at the expense of the V$_{gate}$ = 0 "dip". When V$_{gate}$ is set back to zero, the second "dip" vanishes logarithmically with time, and the initial one recovers its amplitude. These processes can be very long and involve hours or days. The effect of the measurement speed in Fig. 2 can then be understood, as a "second dip" starts to form during the measurement and follows the V$_{gate}$ scan, its amplitude being larger for slower measurements. It was also shown in [5] that, as in InO$_x$, the longer the time taken to write the second dip, the longer it took to erase it. More precisely, a simple ageing law was observed: the erasing time being simply proportional to the "writing" time.

Another way to perturb the sample and to put it out of equilibrium is by abruptly changing the temperature. In an experiment analogous to the "two dip" one, the sample, first equilibrated at 4 K (V$_{gate}$ = 0 throughout the experiment), was heated up and maintained for a time t$_w$ at the higher temperature, and then quench cooled to 4 K by plunging in the liquid helium. Long after the sample was at 4 K, its conductance continued to relax logarithmically with time down to its equilibrium value, as is illustrated in Fig. 4. For these measurements, the bath temperature is measured precisely (less than 1 mK sensitivity) to correct for the effect of small temperature drifts arising for small helium pressure variations.

The slow relaxation of the conductance after the sample was cooled is related to a memory of the "dip" thermal broadening it experienced. This can be shown directly by measuring the whole G(V$_{gate}$)

curve evolution after cooling the sample. The result of such an experiment is shown in Fig. 5. The sample was cooled from room temperature down to T = 4 K within t ~ 15 min, and then $G(V_{gate})$ curves were regularly measured. It is seen in Fig. 5 that the shape of the curves slowly evolves, in particular their FWHM diminishes with time to the equilibrium value. The top curve clearly does not have an equilibrium shape, as the sample experienced a continuously decreasing temperature. Were it in equilibrium, its FWHM would correspond to the T = 6 K thermal broadening of the "dip". However, the sample has its T = 4 K conductance (the T = 6 K value would be ten times larger). This unambiguously shows that, although the sample temperature is indeed 4 K, it keeps a memory of the thermal broadening it previously experienced. Again, the "dip" depth increases logarithmically with time.

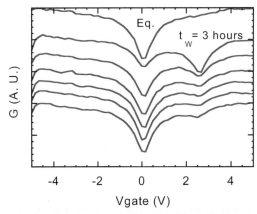

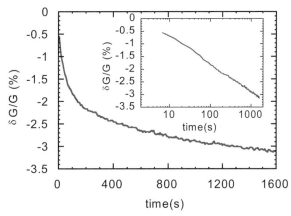

Fig. 3 "Two dip" experiment. The sample is first equilibrated with V_{gate} = 0 (top curve). V_{gate} is set to 2.5 V for 3 hours, which results in a second dip (second top curve). After V_{gate} is set back to zero, the second "dip" vanishes (five lower curves, measured each 24 minutes). The curves are shifted for clarity (T = 4 K).

Fig. 4 Slow relaxation of the conductance of a granular Al *channel* which, after equilibration at 4 K, was maintained at T = 12.5 K for one hour and then quench-cooled at 4 K at time t = 0. The inset shows the logarithmic time relaxation. V_{gate} = 0.

All the results presented above show that insulating granular aluminium samples can be driven out of equilibrium at low temperature by changing either T or the gate voltage, and that slow logarithmic relaxation of the conductance then follow. All the phenomena we have described are strikingly similar to the ones reported in InO_x insulating films, showing that the same physics is involved in both systems. The mechanism of this glassy behaviour remains unclear. Very similar memory effects had been reported in the capacitance versus voltage curves of "tunnel capacitors", which are capacitors containing small metallic islands in the dielectric close to one of the plates [8]. They were interpreted as arising from Coulomb blockade by the islands and slow polarisation processes of the dielectric. Adkins et al. [2] also interpreted their data qualitatively in a similar manner, invoking a slow polarisation of the substrate. We checked that the slow conductance relaxation following a temperature change (as in Fig. 4) is also observed in granular aluminium *strips*, with the same slope as the nearby *channel*. This indicates that the substrate probably does not play a role, since one could have expected a difference between the disordered evaporated Al_2O_3 and the single crystalline sapphire substrate. The effect of carrier concentration on the dynamics in InO_x shows that the phenomena are intrinsic to the samples, and that electron-electron interactions may be important in the glassy behaviour [7], which may be due to the realisation of a Coulomb glass.

In the case of granular aluminium, we want to know what role the granularity plays in the phenomena, and what is the behaviour of small samples where the averaging over large numbers of grains is less effective. In a preliminary attempt, we made a "small" sample of dimensions 20µm × 20µm. We show its field effect behaviour in Fig. 6. The prominent feature is the presence of reproducible conductance oscillations which are superposed to the central "dip". When the sample is put out of equilibrium by imposing a variable gate voltage, the central "dip" is erased and only the oscillations are present (upper curve). Once V_{gate} is set to zero, the central "dip" slowly appears as is expected, while the conductance oscillations remain essentially unchanged.

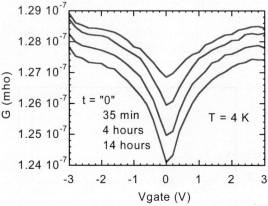

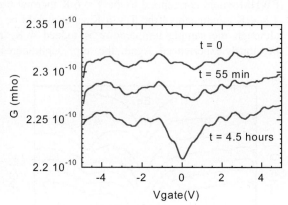

Fig. 5 Slow evolution of the $G(V_{gate})$ "dip" after the sample was cooled from room temperature to 4 K. Increasing time from top to bottom curves.

Fig. 6 Field effect curves for a "small" sample. Conductance oscillations are present. In the top curve, the central "dip" was erased by applying a variable gate voltage for one day. At time $t = 0$, it was set to zero, and the central "dip" appeared, while the oscillations remained unchanged (the curves are shifted for clarity).

Actually, the oscillation pattern is seen to change on longer time scales (days) or when the sample is heated. It will be interesting to determine whether the dynamics of the central "dip" and of the oscillation pattern are independent or not. The oscillations may be related to Coulomb blockade effects on single grains, in which case they should be influenced by the potential disorder or polarisation surrounding the grains, and thus be a good sensor of any slow dynamics. Further work is needed on this topic. In order to help elucidating the origin of the glassy behaviour, it will be of course also be interesting to control the microstructure of the films and study its influence on the phenomena.

Acknowledgements   We thank A. Barbara for the STM examination of the samples and F. Gay for his technical help for the field effect measurements.

References

[1] for a review see Z. Ovadyahu, this Conference and the references therein.
[2] C. J. Adkins, J. D. Benjamin, J. M. D. Thomas, J. W. Gardner and A. J. Mc Geown, J. Phys. C: Solid State Phys. **17**, 4633 (1984).
[3] G. Martinez-Arizala, D. E. Grupp, C. Christiansen, A. M. Mack, N. Markovic, Y. Seguchi, and A. M. Goldman, Phys. Rev. Lett. **78**, 1130 (1997).
[4] N. Markovic, C. Christiansen, G. Martinez-Arizala, and A. M. Goldman, Phys. Rev. B **65**, 012501 (2001).
[5] T. Grenet, Eur. Phys. J. B **32**, 275 (2003).
[6] E. Bielejec and W. Wu, Phys. Rev. **87**, 256601 (2001).
[7] A. Vaknin, Z. Ovadyahu, and M. Pollak, Phys. Rev. Lett. **81**, 669 (1998).
[8] J. Lambe and R. C. Jaklevic, Phys. Rev. Lett. **22**, 1371 (1969).

phys. stat. sol. (c) **1**, No. 1, 13–16 (2004) / **DOI** 10.1002/pssc.200303621

Field-induced space charge limited current flow in disordered ultrathin films

L. M. Hernandez, A. Bhattacharya, Kevin A. Parendo, and **A. M. Goldman**[*]

School of Physics and Astronomy, University of Minnesota, 116 Church St. SE, Minneapolis, MN 55455, USA

Received 1 September 2003, revised 10 September 2003, accepted 17 September 2003
Published online 28 November 2003

PACS 72.20.Ee, 73.50.–h, 73.61.Ng

Slow, nonexponential relaxation of electrical transport accompanied by memory effects can be induced in quench-condensed ultrathin amorphous Bi films by the application of a parallel magnetic field. This behavior, which appears to be a form of space charge limited current flow, is found in extremely thin films on the insulating side of the superconductor–insulator transition.

1 Introduction

Electronic systems such as doped semiconductors and disordered metals, which exhibit electrical versions of glass behavior, are often described as Coulomb or charge glasses. There is a substantial literature describing hysteretic, slow non-exponential relaxation, and memory effects in these systems, some of which is devoted to magnetic-field-induced effects [1–5]. Here we describe a parallel magnetic field induced glass-like phenomenon in quench-condensed ultrathin films of amorphous Bi (a-Bi) [6], which manifests itself as a field-induced space-charge-limited flow of current [7]. The effects were similar to those reported for granular Al films in Ref. 5. However in the present case the sheet resistances of the films were never less than 10^5 Ω at 300 mK, a full decade greater than those of Ref. 5, and superconductivity was not involved.

2 Experimental

The fabrication and subsequent study of a-Bi films were carried out using a system that combined an Oxford Instruments dilution refrigerator (Kelvinox 400) with a "bottom loading" sample transfer/film growth system [8]. The epi-polished $SrTiO_3$ substrates were first cooled to helium temperatures and then pre-coated with a thin layer of a-Ge, before the Bi growth was initiated. Films prepared in this manner are believed to be disordered on microscopic length scales [9]. The current leads were 100 Å thick pre-evaporated Pt films and the films themselves were 2.5 mm long and 0.5 mm in width, the dimensions defined by a shadow mask. Four-terminal DC measurements were carried out using a current of 7×10^{-11} A.

In zero magnetic field, a voltage, as measured across two electrical leads, which were 0.5 mm apart and centered between the current leads, developed in response to this current in a time on the order of the RC time constant of the complete measuring circuit. At temperatures above about 100 mK and in zero magnetic field, the four-terminal resistance could be described by a variable range hopping form [10], but with an exponent slightly greater than 0.5.

---

[*] Corresponding author: e-mail: goldman@physics.umn.edu

In parallel fields above a certain temperature-dependent threshold, at times later than the RC time constant, the measured voltage continued to change in a manner that was not exponential in time. This phenomenon was first noticed in an 11.38 Å thick film at 50 mK. The change in voltage with time was slower, the higher the magnetic field, as shown in Fig. 1. In fields less than 0.1 T, after the initial transient, there was no slow relaxation. In a 12 T field at 50 mK and at long times the field grew logarithmically with time. After about 10^4 seconds the fraction of the measured voltage drop changing with time

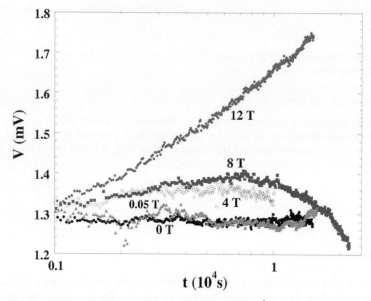

Fig. 1   Voltage vs. time at T = 50 mK for an 11.38 Å thick film in zero magnetic field and in fields of 0.02, 0.05, 0.1, 1.0, 4, 8, and 12 T. In zero magnetic field this film's sheet resistance was greater than 10^7 Ω at 50 mK.

was about 1/3 of the total measured voltage. These glass-like effects vanished above 200mK even in a magnetic field of 12 T.

From Fig. 1 it appears that the drift curves obtained in intermediate magnetic fields do not exhibit simple logarithmic dependence on time, and in some instances do not change monotonically with time. Simple logarithmic forms were found only in the largest fields at very long times. The observation of a nonmonotonic temporal evolution of the voltage was originally believed to be a signature of the voltage compliance limit of the current source being reached. It was subsequently realized that this was not the case, and that the behavior was nearly identical to that found in space-charge-limited current flow in the presence of trapping sites [7, 11]. In this circumstance charge enters the film from the current leads at its ends. A charge front forms and migrates away from the leads. The charge is eventually uniformly spread through the film. A simple picture is to consider the current electrodes as plates of a planar capacitor. The space-charge regions that extend away from the electrodes into the film are effectively extensions of these plates. For electrons the charge front would progress from the negative electrode, and for holes from the positive. As the front moves into the film, the electric field increases as the effective spacing of the plates decreases. When the space charge is spread out over the film the electric field and the measured voltage fall. A schematic of this process is shown in Fig. 2. The rise and fall of the electric field is quite apparent in the 4 T and 8 T curves of Fig. 1. In models of the transient behavior of space charge limited current flow, the time of the maximum in the voltage response is associated with the transit time for charge across the film.

The fraction of the current responding in a glass-like manner decreased with increasing film thickness and the effect vanished completely when the film thickness reached 11.95 Å. This is shown in Fig. 3 with data taken at a temperature of 50 mK and in a 12 T magnetic field. The extraordinarily slow response of the glassy regime precluded any truly systematic examination of the dependence of the resistance on current, thickness, temperature, and magnetic field. As a consequence the nature of the onset of this slow relaxing state and its phase diagram, are not known.

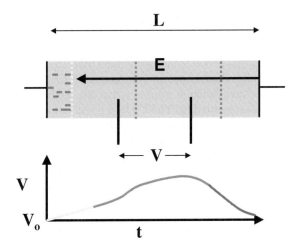

Fig. 2 Schematic of four-terminal arrangement showing current and voltage leads on film of length **L** (upper). Plot of measured voltage drop vs. time (lower). The space charge front (vertical dashed line) migrates into film from left so that electric field **E**, associated with current source is applied over shorter and shorter distances. This results in the voltage difference **V**, initially increasing with time **t** as shown. At larger times, the electric field is screened by space charge, causing it to decrease. The time variation of **V** is color-coded (shaded) to correspond to the positions of the charge front.

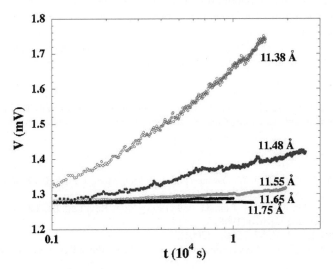

Fig. 3 Voltage vs. time at T = 50 mK and H = 12 T for films of different thicknesses. The vertical axis has been adjusted to account for small offsets due to thermal emfs.

There were memory effects in the response of these films. For example, if the current source was switched off for a short time, the measured voltage was observed to fall to a lower value (actually negative) in a time given by the circuit time constant. When the current source was restored, the voltage was observed to rise within the circuit time constant to the value it exhibited before the current source was disconnected, at which point it resumed its upward drift. The negative voltage found with the current source off suggests that the film was losing stored space charge during that time. A second memory effect involved temperature. Starting at a temperature of 50 mK and a field of 12 T, the current source was turned off and the temperature rose to 100 mK. The current was then restored and the voltage was seen to first grow to its previous value at 50 mK, and then to decay to a value associated with 100 mK, after which time it continued to drift upward, but at a different rate.

3 Discussion

One can rule out all explanations of this behavior that derive from superconductivity, as the films did not exhibit either superconductivity or superconducting fluctuations in zero magnetic field. One can also rule out effects associated with reduction of the wave function overlap transverse to the field direction, which would decrease the hopping rate. This was suggested by Yu [12] as a possible cause of glass-like behavior in a magnetic field found in capacitive charging experiments on In_2O_3 films. A field of 0.1 T would suggest a length of ~ 50 Å which is much greater than the hopping length (~ 5 Å) and is therefore not large enough to reduce the wave function overlap in any significant way as required by this picture.

Positive magnetoresistance [13] in disordered films is known to follow from the polarization of the spins of the electrons involved in hopping conduction [14, 15]. Channels that involve singlet configurations of electrons on hopping cites are eliminated by polarization of the charge carriers. The effect saturates when the Zeeman energy is the order of k_BT. The absence of saturation in the change of the slow relaxation with increasing magnetic field suggests that there must be an energy scale greater than k_BT that determines the onset of the glass-like state.

4 Conclusions

We speculate that the observed effects result from a collective state occurring in a disordered, spin-polarized, two-dimensional electronic system. A possible candidate for this state would be the disordered Wigner glass described by Chakravarty and collaborators [16]. The application of an electric field to such a system could result in the glass sliding. The dynamics of this sliding might resemble our observations. Unfortunately we do not know the carrier density in these systems as the Hall resistance could not be measured, possibly as consequence of a very low carrier mobility or equal numbers of electrons and holes. Thus it is difficult at this time to achieve a precise understanding of the cause of the observed glass-like behavior.

Acknowledgements   We acknowledge very useful discussions with F. Zhou, and L. I. Glazman. This work was supported by the National Science Foundation Condensed Matter Physics Program under grant Grant NSF-0138209.

References

[1] C. J. Adkins, J. D. Benjamin, J. M. D. Thomas, J. W. Gardner, and A. J. McGeown, J. Phys. C **17**, 4633 (1984).
[2] Zvi Ovadyahu and Michael Pollak, Phys. Rev. Lett. **79**, 459 (1997).
[3] A. Vaknin, Z. Ovadyahu, and M. Pollak, Phys. Rev. Lett. **81**, 669 (1998); Phys. Rev. Lett. **84**, 3402 (2000).
[4] G. Martinez-Arizala *et al.*, Phys. Rev. B **57**, R670 (1998).
[5] E. Bielejec and Wenhao Wu, Phys. Rev. Lett. **87**, 256601 (2001).
[6] see: L. M. Hernandez, A. Bhattacharya, Kevin A. Parendo, and A. M. Goldman, Phys. Rev. Lett., in press, for a more detailed account.
[7] Albert Rose, "Concepts in Photoconductivity and Allied Problems," Interscience Publishers, New York (1963).
[8] L. M. Hernandez and A. M. Goldman, Rev. Sci. Instrum. **73**, 162 (2002).
[9] M. Strongin, R. S. Thompson, O. F. Kammerer, and J. E. Crow, Phys. Rev. B **1**, 1078 (1970).
[10] A. L. Efros and B. I. Shklovskii, J. Phys. C **8**, L49 (1975).
[11] A. Many and G. Rakavy, Phys. Rev. **126**, 1980 (1962).
[12] Clare C. Yu, Phys. Rev. Lett. **82**, 4074 (1999).
[13] K. M. Mertes *et al.*, Phys. Rev. B **60**, R5093 (1999).
[14] A. Kurobe and H. Kamimura, J. Phys. Soc. Jpn. **51**, 1904 (1982).
[15] K. Matveev, L. I. Glazman, Penny Clarke, D. Ephron, and M. R. Beasley, Phys. Rev. B **52**, 5289 (1995).
[16] S. Chakravarty, S. A. Kivelson, C. Nayak, and K. Voelker, Philos. Mag. B **79**, 859 (1999).

phys. stat. sol. (c) **1**, No. 1, 17–20 (2004) / **DOI** 10.1002/pssc.200303641

Quantum mode-coupling theory for vibrational excitations of glasses

W. Schirmacher[*, 1], **E. Maurer**[1], and **M. Pöhlmann**[1]

Physik-Department E13, Technische Universität München, 85747 Garching, Germany

Received 1 September 2003, revised 25 September 2003, accepted 25 September 2003
Published online 28 November 2003

PACS 63.50.+x

We exploit the fact that the mean-field theory for the vibrational anomalies of disordered solids, derived recently by the authors, is equivalent to a quantum mode-coupling theory. For small anharmonicities this theory predicts a continuous transition to a different glassy state if the degree of disorder is increased ("glass-to-glass transition"). It is argued that the vibrational glassy anomalies (boson peak, anomalous temperature dependence of the specific heat) are precursor phenomena of this transition.

1 Introduction The anomalous low-frequency vibrational properties of glasses and other disordered solids and their associated low-temperature thermal properties keep being a matter of controverse debate both among experimentalists and theorists. Such anomalies are the deviation of the vibrational density of states (DOS) from Debye's law at low frequencies ("boson peak") and the almost-linear temperature variation of the specific heat at very low temperatures, followed by a boson-peak-like anomaly at higher temperatures. Progress in understanding the nature of the boson peak has been achieved by noting that models with spatially fluctuating elastic constants [1–4] exhibit the boson-peak anomaly as observed in the experimental data. These model ideas are corroborated by a recent nuclear-inelastic scattering investigation [5] of several glassy materials which shows that the boson-peak excitations are of collective character. This rules out models [6–8] in which the boson peak is due to localized excitations.

The present authors have developed recently a field-theoretical treatment of a disordered solid which allows for the inclusion of anharmonic interactions into the considerations [9]. The mean-field theory derived within this formalism boils down to the self-consisten Born approximation (SCBA) in the absence of the anharmonicity. The SCBA is equivalent to the coherent-potential approximation (CPA) [1, 3] for small degree of disorder. Within the SCBA experimentally measured boson peak data [5] (i.e. the reduced DOS $g(\omega)/\omega^2$) with a characteristic maximum in the meV regime) can be explained quantitatively with only two adjustable parameters, namely, the mean sound velocity c_0 and its mean-square fluctuations ("degree of disorder") $\gamma \propto \langle (\Delta c)^2 \rangle$. It was mentioned in Ref. [9] that in the presence of the anharmonic coupling the mean-field equations derived from the field-theoretical treatment are mathematically equivalent to the quantum-mechanical analogon of Götze's [10, 11] mode-coupling equations for the dynamic susceptibility of glass-forming viscous liquids. Here we exploit this idea further and show that (i) there is a continuous transition to another non-ergodic state if γ exceeds a critical value γ_c, and that (ii) the boson peak and the anomalous temperature variation of the specific heat can be considered as precursor phenomena of this transition.

---

[*] Corresponding author: e-mail: wschirma@ph.tum.de, Phone: +49 089 289 12455, Fax: +49 089 289 12473

2 Mean-Field equations As in Ref. [9] we deal with a model solid with longitudinal excitations only. We consider elastic disorder, characterized by the fluctuation parameter γ and an anharmonic coupling, featuring the Grüneisen parameter g. The mean-field equations derived from an effective field theory by means of a saddle-point approximation are expressed in terms of the local dynamic susceptibility $Q(z) = i \int_0^\infty dt \exp\{izt\} Q(t)$ with

$$Q(t) = \lim_{r \to r'} \frac{\tilde{m}_0}{6} \overline{\langle [\nabla \cdot \boldsymbol{u}(r, t + t_0), \nabla \cdot \boldsymbol{u}(r', t_0)] \rangle}. \tag{1}$$

Here m_0 is the mass density [12], $\boldsymbol{u}(r, t)$ are the displacement fields, and $z = \omega + i0$. The self-consistent mean-field equations are

$$Q(z) = \frac{1}{2} \sum_{|k| < 1} \frac{k^2}{-z^2 + k^2[1 - \Sigma(z)]} = \frac{1/2}{(1 - \Sigma(z))} [(1 + z^2 G(z)], \tag{2}$$

with the self energy

$$\Sigma(z) = \gamma Q(z) + \tilde{g} Q_2(z). \tag{3}$$

The mode-coupling parameter is given by $\tilde{g} = 4\gamma g^2/m_0$, where g is the Grüneisen coupling constant. $Q_2(z)$ is the convoluted dynamical susceptibility. Its imaginary part $Q_2''(\omega)$ can be related to the Fourier transform $S(\omega)$ of the van-Hove correlation function $S(t) = \lim_{r \to r'} \frac{\tilde{m}_0}{6} \overline{\langle \nabla \cdot \boldsymbol{u}(r, t + t_0) \nabla \cdot \boldsymbol{u}(r', t_0) \rangle}$ by $Q''(\omega) = \frac{1}{2}(1 - \exp\{-\beta\omega\}) S(\omega)$ with $\beta = [k_B T]^{-1}$. A similar relation holds for Q_2 and S_2, and we have

$$S_2(t) = S(t)^2. \tag{4}$$

In Eq. (2) $G(z)$ is the Green's function

$$G(z) = \sum_{|k| < 1} \frac{1}{-z^2 + k^2[1 - \Sigma(z)]}, \tag{5}$$

from which the DOS $g(\omega)$ can be obtained as

$$g(\omega) = \frac{2\omega}{\pi} \Im m\{G(z)\}. \tag{6}$$

We can rewrite the mean-field equations in terms of Kubo's relaxation function $\Phi(t) = \int_0^\beta d\lambda S(t - i\lambda)$, which is related to the susceptibility by $Q(t) = i \frac{d}{dt} \Phi(t)$, which implies $Q(z) = Q_0 + z\Phi(z)$ with $Q_0 = Q(z{=}0) = \Phi(t{=}0)$. In the same fashion we define $Q_2(z) = Q_2(z{=}0) + z\Phi_2(z)$ and $\Sigma(z) = \Sigma(z{=}0) + zM(z)$, where $M(z)$ is called memory-function. Introducing normalized quantities $\phi(z) = \Phi(z)/Q_0$, $\phi_2(z) = \Phi_2(z)/Q_0^2$, and $m(z) = 2Q_0 M(z)$ we obtain a self-consistency equation for $\phi(z)$

$$m(z) = \lambda_1 \phi(z) + \frac{1}{T} \lambda_2 \phi_2(z) = \frac{\phi(z) - zG(z)}{1 + z\phi(z)} \tag{7}$$

where we have introduced the two parameters $\lambda_1 = 2\gamma Q_0^2$ and $\lambda_2 = T\tilde{g}Q_0^3 = T\gamma g^2 4Q_0^3/m_0$. In the classical limit $\omega \ll T$ we have $\Phi(t) = S(t)/T$, from which follows $\Phi_2(t) = T[\Phi(t)]^2$ In this limit (7) agrees mathematically with the Mode-coupling equation for the so-called λ_1–λ_2 model [10] (or F_{12} model [11]) of W. Götze This equation is a schematic representation of the mode-coupling theory for the liquid-glass transition of viscous liquids [11], because it shows the same critical phenomena as the full theory.

3 Non-Ergodicity transition It has been noted in the literature dealing with disordered harmonic models [1, 3] that if the disorder-induced fluctuations of the elastic constants are too strong, an instability occurs, which manifests itself by the appearance of negative values of ω^2 in the spectrum, implying that $G(z)$ is no more a positive analytic function. The unstable situation arises because in the harmonic theory with too strong force constant fluctuations some atoms are sitting on potential hills instead in potential wells. Such an instability occurs also in the harmonic version of the present theory, namely in the SCBA (Eqs. (2) and (3) with $\tilde{g} = 0$) for $\gamma > \gamma_c = 0.5$. This unphysical situation is, of course, an artifact due to the absence of anharmonic interactions. Within our anharmonic mean-field theory (quantum mode-coupling theory) there appears also an instability for $\gamma > \gamma_c$. But now the spectrum remains positive. Instead of the non-physical unstable eigenvalues a phase transition to a non-ergodic state occurs. The order parameter for such a transition (non-ergodicity parameter) is

$$f = \lim_{t \to \infty} \phi(t) = \lim_{z \to 0} (-z\phi(z)) \tag{8}$$

from Eq. (7) we obtain the following self-consistency equation for f

$$\frac{f}{1-f} = \lambda_1 f + \lambda_2 f^2 \tag{9}$$

Because this is a homogeneous equation for f, there is always the solution $f = 0$. However, it has been shown [11] that the stable solution of the mode-coupling equations is always that with the largest f if a nontrivial solution $f > 0$ exists. The regions in the (λ_1, λ_2) plane where such nonergodic solutions exists are depicted in Fig. 1. For $\lambda_2 < 1$ the transition is continuous, otherwise it is discontinuous. In our case, we are interested in the situation at low temperatures and small anharmonicity which means $\lambda_2 \ll 1$. Therefore we expect a continuous transition for λ_1 exceeding $\lambda_{1,c} = 1$. Because for this value $Q_0 = 1$, this is equivalent to γ exceeding $\gamma_c = 0.5$.

How can we interpret such a transition? The non-ergodicity parameter can be shown [11] to play the role of a Debye-Waller factor, which, in the case of the glassy freezing of a liquid, characterizes the structure of the frozen material. As we have started in our investigation with an already frozen material, f has to be interpreted as a shift of the Debye-Waller factor which occurs, because the disorder has become so large, that a global rearrangement of the structure takes place. We call this transition a glass-to-glass transition.

From the previous investigation of harmonic models [1, 3–5, 9] it has been shown that the boson peak (maximum in $g(\omega)/\omega^2$) becomes more strongly pronounced (and shifts to smaller frequencies) as γ approaches γ_c. Therefore the boson-peak anomaly can be interpreted as a precursor phenomenon of the glass-to-glass transition.

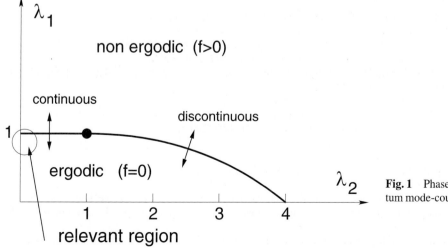

Fig. 1 Phase diagram of the quantum mode-coupling theory.

From the investigation of the dynamics of a liquid near the glass transition by means of the mode-coupling theory it is known [11] that strongly pronounced dynamical critical phenomena in the low-frequency asymptotics of the Kubo function appear. It can be shown [10, 11] that for $\phi''(\omega)$ near $\lambda_1 = 1$ the following asymptotic scaling behaviour holds

$$\phi''(\omega) = A(\lambda_2) \begin{cases} \omega_\epsilon^{a-1} & \omega < \omega_\epsilon \\ \omega^{a-1} & \omega > \omega_\epsilon \end{cases} \tag{10}$$

where $0 < a < 0.5$ and $a \to 0.5$ as $\lambda_2 \to 0$, and $A(\lambda_2)$ is a constant which has to be determined numerically. The cross-over frequency $\omega_\epsilon = [1 - \lambda_2]^{1/a}$ can be estimated to ly near the boson-peak frequency which (in units of the Debye frequency) is roughly given by $\omega_{\mathrm{boson}} \approx 1 - 2\gamma$. In the regime $\omega < \omega_\epsilon$ we have a constant Kubo spectral function $\phi''(\omega) = \phi''(0) = A(\lambda_2)/[1 - \lambda_2]^{\frac{1}{a}-1}$ ("white-noise spectrum" [11]) This means that the dynamical response function $Q''(\omega)$ varies linearly with frequency, and therefore the broadening of the Brillouin line (inverse of the mean free path) quadratically with ω, as it is observed frequently in experiments.

Such a white-noise spectrum of $\phi''(\omega)$ implies also a specific heat which varies linearly with temperature in the following way [13]:

From the effective action in saddle-point approximation one can show that the averaged internal energy $\bar{H}$ contains an anomalous contribution $\Delta\bar{H}$, which is essentially the thermal average of the function $Q''(\omega) = \omega\phi''(\omega)$ and therefore its temperature dependence reflects the low-ω asymptotics of $\phi''(\omega)$:

$$\Delta\bar{H} \propto \int\limits_0^\infty d\omega \, \frac{1}{\exp\{\beta\omega\} - 1} \, \omega\phi''(\omega) \propto \begin{cases} T^2\omega_\epsilon^{a-1} & \omega < \omega_\epsilon \\ T^{a+1} & \omega > \omega_\epsilon \end{cases}. \tag{11}$$

This implies a T-linear specific heat for $T < \omega_\epsilon$.

In conclusion we have shown that in a model with static disorder and an anharmonic coupling, which is described by a quantum mode-coupling theory, a glass-to-glass transition occurs, which means that with increasing disorder a state with a different ergodic structure is reached. Within this scenario the boson peak anomaly and the T-linear specific heat are precursor phenomena of this transition.

Acknowledgements   It is a pleasure to thank D. Belitz and W. Götze for numerous helpful and encouraging discussions. This research was supported by the Deutsche Forschungsgemeinschaft under Schi/6-1.

References

[1] W. Schirmacher, G. Diezemann, and C. Ganter, Phys. Rev. Lett. **81**, 136 (1998).

[2] J. W. Kantelhardt, S. Russ, and A. Bunde, Phys. Rev. B **63**, 064302 (2001).

[3] S. N. Taraskin, Y. L. Loh, G. Natarajan, S. R. Elliott, Phys. Rev. Lett. **86**, 1255 (2001).

[4] T. S. Grigera, V. Martin-Mayor, G. Parisi, P. Veroccio, Nature **422**, 289 (2003).

[5] A. I. Chumakov, I. Sergueev, U. van Bürck, W. Schirmacher, T. Asthalter, R. Rüffer, O. Leupold, and W. Petry, to be published in Science.

[6] V. G. Karpov, M. I. Klinger, F. N. Ignatiev, Sov. Phys. JETP **57**, 493 (1983).

[7] M. I. Klinger, A. M. Kosevich, Phys. Lett. A **280**, 365 (2001).

[8] V. L. Gurevich, D. A. Parshin, H. Schober, JETP Letters **76**, 650 (2002).

[9] W. Schirmacher, M. Pöhlmann, and E. Maurer, phys. stat. sol. (b) **230**, 31 (2002).

[10] W. Götze, Z. Phys. **56**, 139 (1984).

[11] W. Götze in: Liquids, Freezing and Glass Transition, J. P. Hansen, D. Levesque and J. Zinn-Justin, Eds., North-Holland, Amsterdam, 1991, p. 287.

[12] We use units in which lengths are measured in Debye lengths k_D^{-1}, times in inverse Debye frequencies $\omega_D^{-1} = [c_0 k_D]^{-1}$ and actions in units of $\hbar$. In these units $c_0 = 1$ and $m_0 = \bar{m}_0 c_0 \hbar^{-1} k_D^{-1}$, where $\bar{m}_0$ is the mass density in the usual SI units.

[13] Note that this mechanism of the temperature-linear specific heat is different from that anticipated in Ref. [9].

phys. stat. sol. (c) **1**, No. 1, 21–24 (2004) / **DOI** 10.1002/pssc.200303644

Non-equilibrium transport in arrays of type-II Ge/Si quantum dots

N.P. Stepina[*], **A.I. Yakimov**, **A.V. Dvurechenskii**, **A.V. Nenashev**, and **A.I. Nikiforov**

Institute of Semiconductor Physics, prospekt Lavrenteva 13, 630090 Novosibirsk, Russia

Received 1 September 2003, revised 18 September 2003, accepted 18 September 2003
Published online 28 November 2003

PACS 72.20.Ee, 73.50.Pz, 73.63.Kv

We study the effect of interband light excitation on hopping conductivity in dense arrays of Ge/Si quantum dots. Both negative and positive photoeffects depending on dot occupations with holes were observed. Long-time conductivity dynamics (typically, $10^2 - 10^4$ sec at T=4.2 K) has been revealed as well as after switch on and switch off the illumination, displaying a sluggish temporal dependence. Our observation is explained by spatial separation of electrons and holes due to the presence of potential barriers created by positively charged Ge quantum dots. The time-dependent equalization of barrier heights as a result of hole trapping into the dots was proposed as an additional effect explaining persistent photoconductivity.

1 Introduction   Ge/Si (001) quantum dots (QDs) structures exhibit a type-II band lineup. The large (~0.7 eV) valence-band offset characteristic of this heterojunction leads to an effective localization of holes in Ge regions. It has been shown [1, 2] that at low temperatures (<20K) charge transport through such a system is dominated by hole hopping between the dots. The Hartree potential of holes induces a quantum well for free electrons of Si at Ge/Si interface. Thus a fundamental feature of these dots is that trapping potential for electrons would be greatly enhanced with charging the dots with holes. The spatial separation of electron and hole in Ge/Si QDs causes many exciting phenomena, such as negative interband photoconductance [3], anomalous quantum-confined Stark effect [4], blue shift of the interband transition with formation of exciton complexes [5]. In this paper we describe experimental results that demonstrate anomalous dynamics of hopping photoconductivity (PC) created by exposure to interband illumination.

2 Experimental details   The samples were grown on a (001) oriented Si substrate with a resistivity of 1000 Ω cm by molecular-beam epitaxy in the Stranskii-Krastanov growth mode. To supply holes on the dots, a boron δ-doping Si layer inserted 5 nm below the Ge QD layer was grown. Si cap and buffer layers were boron doped at a level of ~10^{16} cm^{-3}. Details of sample preparation have been described elsewhere [1]. To avoid influence of surface effects on photoconductivity measurements, Ohmic contacts were fabricated by Al evaporation into the preliminary etched pits followed by annealing at 400 °C in N$_2$ atmosphere. The etching depth was such as the electrical contacts to buried Ge QD layer are formed. To separate response from the dots, the test samples were grown under conditions similar to the dot samples, except that no Ge was deposited. Photoconductivity experiments have been carried out using a GaAs light-emitting diode (LED) with an emission maximum at a wavelength of 0.9 μm for excitation.

---

[*] Corresponding author: e-mail: stepina@isp.nsc.ru, Phone: +7 3832 33 26 24, Fax: +7 3832 33 26 24

3 Experimental results   Hopping transport of holes along two-dimensional layers of Ge quantum dots in Si have been investigated in [1, 2]. From the temperature dependence of conductivity the mechanism of charge transfer was identified as being variable-range hopping in a density of state determined by long-range Coulomb interaction between the dots.   Interband illumination of such a system results in a complicated transient behavior of the photoconductivity during reaching illuminated or dark steady state. Our experiments involve the following procedure. The samples are cooled to the measuring temperature with a zero drain voltage, and are allowed to equilibrate for several hours. The time dependence of conductance under the condition of LED illumination (on) and stopping illumination (off) was taken several times serially. Figure 1 a shows typical conductance transient traces for two structures with different δ-boron doping. In contrast to the test Si samples, in which the PC is always positive, both positive and negative photoeffects depending on dot filling factor are observed in dot samples. In both cases, kinetics of the recovery as well as of the decay are extremely slow. When the LED illuminates on the sample, the resistivity changes rapidly at the beginning and slowly at the end, while the residual photoconductance can persist for several hours after the light switch off. This phenomenon characterizes a typical effect of persistent photoconductivity (PPC). The PPC level is about 90% of the initial PC value after more than 5000 seconds of decay. Long-time relaxation process was not observed in test samples containing no dots (Fig. 1b).

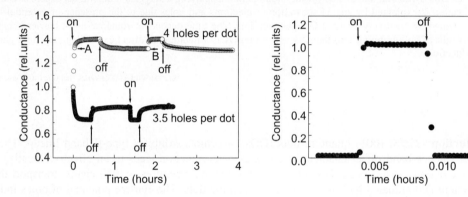

Fig. 1 a) Conductivity traces for the samples with different average numbers of holes per one Ge dot. The solid line is the fitting to the experimental data by a logarithmic law. b) Photoconductivity for the Si sample without Ge quantum dots.

Transient curves for the samples with positive photoconductance after the first light switch on are shown in Figure 2 a for different light intensities. All curves were normalized to conductivity values in saturation (G_{max}). The negative photoconductivity behaviour is similar to sign that for the positive one. One can see that the rate of PC changing at the initial part of the curves increases with intensity of illumination. Figure 3 shows the conductance decay traces at different temperatures and demonstrates the quenching of the PPC effect at high temperatures. Experimental data have been normalized to unity at $t = 0$ (at a moment when the illumination is terminated) according to $G_{PPC}(t) = (G(t) - G_d)/(G(0) - G_d)$. Here, $G(0)$ is the conductance level immediately after the termination of the excitation source, G_d is the initial conductance in the dark.

4 Discussion   To explain the experimental observations we propose the following model. Quantum dots occupied with holes induce a band bending which corresponds to a potential barrier for free holes and a potential well for electrons. When an electron-hole pair is photoexcited, electron is trapped by a dot while the hole cannot recombine with this electron because of the QD's repulsive potential. Recombination of electrons with equilibrium holes in Ge nanoclusters reduces the potential barrier for hole capture. Thus, under illumination, the hole trapping into QDs occurs in a condition of permanent decreasing of the potential barrier height. Changing of hole concentration in the dots can be described by

$$\frac{dp}{dt} = -J_e - C_{em} + C_{cap}^0 \exp\left(-\frac{V}{kT}\right),$$ (1)

where J_e is the electron flux into the dots, V is the barrier height, C_{em} and C_{cap} are the emission and capture rates of holes in the dots, correspondently. Since the equilibrium state, in which the hole emission and capture rates are equal, cannot be reached over the time-window of experiment, one may neglect the emission term in this equation. If concentration of non-equilibrium holes Δp in QDs is small

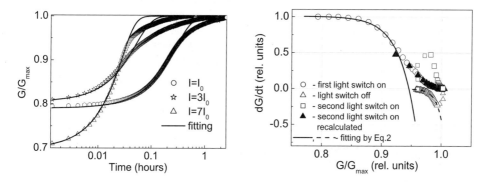

Fig. 2 a) Light intensity dependence of photoconductivity buildup. Solid lines are the fit to the experimental data by numerical solution of Eq. (2). I is the light intensity. b) Derivative of experimental relaxation curve as a function of conductivity.

enough, then the change in the barrier height and conductivity would be proportional to Δp. In this case, expression for conductivity takes the form

$$\frac{dG}{dt} = AI + B\exp(-\gamma G),$$ (2)

where I is the light intensity, A, B and γ are the constants. Oscillating behaviour of conductivity with dot occupancies was found in [1]. Then the different sign of photoconductivity observed in our experiment can be due to the change of hole number in dots under interband illumination. Solid lines in Fig. 2 a depict the fit to the experimental data by a numerical solution of Eq. (2). The initial part of the curves is described satisfactorily by Eq. (2), while the further conductivity growth becomes slower than the calculated one. In Fig. 2 b, we show derivative of the experimental relaxation curve dG/dt as a function of G. The exponential behavior characterized by Eq. (2) (solid line) is really violated when the conductivity reaches the value $G \approx 0.93 G_{max}$. Discrepancy between the predicted and experimental data was proposed to be due to the alignment of barrier heights which, in turn, is a result of photohole capture into the dots under illumination. Nonuniformity of dot sizes yields the dispersion of the barrier heights around them; the effective barrier height (EBH) corresponds to a barrier with minimum value. During the capture process, holes enter the dots with predominately smallest positive charge, being resulted in equalization of the dot occupation. Obviously, for the uniform hole distribution through the dots, the EBH is in excess of that for the equilibrium state. Increasing of the effective barrier height causes the conductivity to rise slowly.

After the light switch off, holes continue to be captured by QDs and the barrier height steadily rises. Then the conductivity obeys the following equation:

$$\frac{dG}{dt} = B\exp(-\gamma G).$$ (3)

It has an analytical solution $G(t) = G_0 + C \ln(t - t_0)$, where G_0, C and t_0 are the constants. One can see in Fig. 1 a that the experimental data are well described by logarithmic law (solid line in Fig. 1 a) that corresponds to exponential dependence of dG/dt on G depicted by open triangles in Fig. 2 b. Dash line in Fig. 2 b indicates fitting to the experimental data by Eq. (3).

New conductive state resulting from the relaxation of the system after light termination differs from the initial one, because the PPC is realized at low temperature when the hole trapping into the dots is strongly limited. Moreover, this state as well as the stationary state under illumination is characterized by uniform distribution of holes through the dots. The second switching on of light takes place when the barrier heights are equal with their absolute value being lower then after the first illumination. Thus the point A in Fig. 1 corresponds to the barrier height larger than that at point B. If it is the case, we should get a coincidence of derivative curves depicted in Fig. 2 b for the first and second light switching on by extending the latter curve toward the values characterized by lower barrier height. It can be seen that after such a transformation (filled triangles in Fig. 2 b) these curves are really superposed.

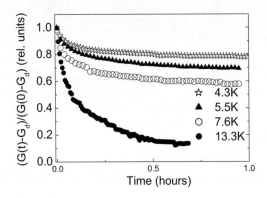

Fig. 3   PC decay at different temperatures.

Increase of the temperature accelerates the hole capture by dots results in the quenching of PPC effect (Fig. 3), that give further evidence to support the proposed model.

5 Summary   Non-exponential slow kinetics of photoconductivity excitation and effect of persistent photoconductivity have been observed in p-type Ge/Si heterostructures containing Ge quantum dots. The essential points for the explanation of the experimental data are proposed to be (i) dependence of the hole trapping into the quantum dots on QD charge state, and (ii) equalization of the number of holes per each dot in a QD ensemble during the photoconductivity excitation and relaxation.

Acknowledgements   This work was supported by the Russian Foundation for Basic Research (Grant № 01-02-17329) and by the Program "Surface Atomic Structure" (Grant № 40.012.1.1.1153).

References

[1] A. I. Yakimov, A. V. Dvurechenskii, A. I. Nikiforov, and A. A. Bloshkin, JETP Lett. **77**, 445 (2003).
[2] A. I. Yakimov, A. V. Dvurechenskii, V. V. Kirienko, Yu. I. Yakovlev, A. I. Nikiforov, and C. J. Adkins. Phys.Rev. B **61**, 10868 (2000).
[3] A. I. Yakimov, A. V. Dvurechenskii, A. I. Nikiforov, O. P. Pchelyakov, and A. V. Nenashev. Phys. Rev. B **62**, 16283 (2000).
[4] A. I. Yakimov, A. V. Dvurechenskii, A. I. Nikiforov, V. V. Ulyanov, A. G. Milekhin, and A. O. Govorov, S.Schulze, and D.R.T. Zahn, Phys. Rev. B **67**, 125318 (2003).
[5] A. I. Yakimov, N. P. Stepina, A. V. Dvurechenskii, A. I. Nikiforov, and A. V. Nenashev. Semicond. Sci. Technol. **15**, 1125 (2000).

phys. stat. sol. (c) **1**, No. 1, 25–28 (2004) / **DOI** 10.1002/pssc.200303649

Why study 1/*f* noise in Coulomb glasses

Clare C. Yu[*]

Department of Physics and Astronomy, University of California, Irvine, CA 92697, USA

Received 1 September 2003, accepted 4 September 2003
Published online 28 November 2003

PACS 71.23.Cq, 72.70.+m, 72.20.Ee, 72.80.Ng, 72.80.Sk

We briefly review 1/f noise in Coulomb glasses. We then argue that measurements of the second spectrum of the noise in Coulomb glasses could help to determine if electron fluctuations are correlated.

An electron glass is an amorphous insulator in which the electrons see a random potential and are localized at random positions. If the electrons interact with one another via long range Coulomb interactions, then we have a special case of an electron glass known as a Coulomb glass.

In a broad sense there have been two theoretical points of view of electron hopping in Coulomb glasses. The single particle approach models the electronic excitations as quasiparticles. Transport is described by variable range hopping which depends on the single particle density of states. The Coulomb interactions between localized electrons produce a Coulomb gap in the single particle density of states that is centered at the Fermi energy. The Coulomb gap makes the ground state stable with respect to single electron hops [1–3]. This approach has been very successful when compared to experiment. It finds that the conductivity goes as $\exp\left[-(T_o/T)^{1/2}\right]$ which has been seen experimentally. Tunneling experiments have also seen the Coulomb gap [4].

In the other approach the interactions between electrons produces many body excitations [5]. As a result transport is accomplished by correlated electron hopping, e.g., sequential hops as well as simultaneous or collective hopping. This approach is more difficult to deal with theoretically and is harder to verify experimentally. In this paper we argue that noise measurements could help to determine if correlated electron motion is involved in transport in Coulomb glasses.

To an experimentalist, noise is a nuisance at best and a serious problem hindering measurements at worst. However noise comes from the fluctuations of microscopic entities and it can act as a probe of what is happening physically at the microscopic scale. Let us set up our notation and define what we mean by noise. Let $\delta I(t)$ be a fluctuation in some quantity I at time t. If the processes producing the fluctuations are stationary in time, i.e., translationally invariant in time, then the autocorrelation function of the fluctuations $\langle \delta I(t_2)\,\delta I(t_1)\rangle$ will be a function $\psi_I(t_2 - t_1)$ of the time difference. In this case the Wiener–Khintchine theorem can be used to relate the noise spectral density $S_I(f)$ to the Fourier transform $\psi_I(f)$ of the autocorrelation function [6]: $S_I(f) = 2\psi_I(f)$ where f is the frequency. 1/f noise, which is ubiquitous and dominates at low frequencies, corresponds to $S_I(f) \sim 1/f$.

In Coulomb glasses electron hopping can occur on very long time scales which can produce low frequency noise. Experimental studies on doped silicon inversion layers have shown that low frequency 1/f noise is produced by hopping conduction [7]. Shklovskii had suggested that 1/f noise is caused by fluctuations in the number of electrons in an infinite percolating cluster [8]. These fluctuations are caused by the slow exchange of electrons between the percolating conducting cluster and

[*] e-mail: cyu@uci.edu, Phone: 1-949-824-6216, Fax: 1-949-824-2174

small isolated donor clusters. Subsequent theoretical [9–13] and experimental [14–16] work found $1/f$ noise at low temperatures and low frequencies in Coulomb glasses.

However, it is still unclear what the temperature dependence of the noise amplitude is. Experiments on doped silicon (Si:B) at temperatures above 1.5 K with a fixed bias current of 4.5 mA found that the noise increased with increasing temperature [14]. On the other hand measurements on ion implanted silicon (Si:B:P) at low temperatures ($T < 0.5$ K) found that the noise decreased with increasing temperature [15]. Other measurements on doped silicon done between 2 K and 20 K find that the noise is independent of temperature at lower temperatures and then decreases with temperature at higher temperatures [16]. The crossover temperature depends on the current bias. This suggests that the temperature dependence is sensitive to the amount of current bias as well as the temperature [16].

Theoretically, the picture is equally murky. Starting from the model that noise comes from fluctuations δN_P in the number of electrons in the percolating cluster due to electron exchange with small isolated clusters [9], Shtengel and Yu [13] found that the noise increases with increasing temperature due to the increase in the thermally activated electron hopping. However Shklovskii [12] argued that the noise decreases with increasing temperature because the probability of finding a isolated cluster decreases with increasing temperature. (A cluster is isolated if it has no neighbors to which it can hop within a certain amount of time.) Another temperature dependent factor is the normalization of the noise amplitude by the size of the percolating cluster N_P:

$$\frac{S_I(f)}{I^2} = \frac{2\langle \delta N_P(t_2) \, \delta N_P(t_1)\rangle_f}{N_P^2} . \tag{1}$$

where I is the average DC current, and $\langle \ldots \rangle_f$ is the Fourier transform of the autocorrelation function. The size of the percolating cluster increases with increasing temperature. To understand this, note that two sites have a bond if electrons can hop between them faster than a certain rate which includes thermally activated hopping. Preliminary calculations [17] on noninteracting electrons with a flat density of states indicates that $N_P \sim T^3$. If this is included in the calculations of Shtengel and Yu [13] where the isolated clusters were single sites, the noise amplitude decreases with increasing temperature. In short, the temperature dependence of the noise is still an open question. Setting this issue of temperature dependence aside, measurements of the 1/f noise spectrum in Coulomb glasses are not really able to tell if the fluctuations were single particle or collective. A more relevant measurement would be the so–called second spectrum of the noise. To understand the second spectrum, consider the following. Suppose we take a time series on Monday and calculate the first spectrum, i.e., the noise spectrum $S(f)$. Then we do the same thing on Tuesday, Wednesday, etc. (In practice one would want to take sequential spectra as close together in time as possible.) So now we have a set of noise spectra $S(f)$ taken at different times t_2. The second spectrum is the power spectrum of the fluctuations of $S(f)$ with time, i.e., the Fourier transform of the autocorrelation function of the time series of $S(f)$ [18–20]. To calculate the second spectrum, we can divide a first spectrum into octaves. An octave is a range of frequencies from f_L to f_H where typically $f_H = 2f_L$. We can discretize the first spectrum by associating each octave with the total noise power in that octave. We do this for each data set. For each octave this gives us a set of numbers with one number from each data set labeled by t_2. Now we can calculate the fluctuations in the noise power in a given octave labeled by frequency f. Then we can calculate the autocorrelation function of these fluctuations, Fourier transform it and obtain the noise power $S_2(f_2, f)$ which is the second spectrum.

The second spectrum looks for correlations in the fluctuations that produce the first spectrum. So the second spectrum can tell us if the fluctuators are correlated or independent. If the second spectrum is white (independent of f_2) and equal to the square of the first spectrum [19–21], the fluctuators are not correlated. Such noise is called Gaussian. If the fluctuators are correlated, then the noise is non-Gaussian, and $S_2 \sim 1/f_2^{1-\beta}$ where the exponent $(1 - \beta) > 0$. It is often useful to plot the second spectrum $S_2(f_2, f)$ as a function of the ratio f_2/f because over a given time interval the high frequency fluctuations get averaged more than the low frequency fluctuations.

The second spectrum has been used to differentiate between the hierarchical model and the droplet model of spin glasses [19, 20] because these two models assume different correlations between the fluctuators. These models were originally developed for short range spin glasses. We will take the liberty of adopting their qualitative features for the case of electron glasses. In the droplet model, clusters or droplets of rearranging electrons produce fluctuations [22–24]. There are fewer large droplets than small droplets, and the big droplets rearrange more slowly than the small droplets. So the large clusters contribute to the low frequency noise and the small fast clusters contribute to the high frequency noise. In the simplest case, the droplets are noninteracting and produce a white second spectrum. A more sophisticated version has interacting droplets. Large droplets are more likely to interact with other droplets than are small droplets. So non-Gaussian noise and the second spectrum will be larger at lower frequencies f_1.

In the hierarchical model [19, 20, 25–30] the states (or electron arrangements) of the electron glass lie at the endpoints of a bifurcating hierarchical tree (which looks more like the roots of a tree). The Hamming distance D between two states is the fraction of electrons that must rearrange to convert one state into another. It turns out that D corresponds to the minimum height to which one must go in the tree in order to get from one endpoint (state) to another. The farther apart 2 states are, the longer the time to go between them. The tree structure is self similar. As a result, the hierarchical model predicts that S_2 will be scale invariant and will only depend on f_2/f and not on the frequency f, while the interacting droplet model predicts that for fixed f_2/f, S_2 will be a decreasing function of f [19, 20]. Because of this, the second spectrum can differentiate between the droplet model and the hierarchical model. Measurements of resistance fluctuations in the spin glass CuMn find that its behavior is consistent with the hierarchical model [19, 20].

Measurements of the second spectrum of the noise in silicon inversion layers in MOSFETs in the vicinity of the metal–insulator transition have found that the exponent $(1 - \beta)$ changes from being approximately zero in the metallic phase to a finite value of order unity [31] in the glassy or insulating phase. Similar results were found for doped silicon crystals Si:P(B) where $(1 - \beta)$ was small in the metallic phase and became greater than 1 in the insulating phase [16]. The frequency dependence of S_2 indicates that the electronic fluctuations are correlated. Furthermore in the experiments on the Si MOSFETs, second spectra plots versus f_2/f for different values of f fall along one curve, implying that the hierarchical picture is better suited to describing the insulating phase [31].

Returning to our original question, we see that the measurements of the second spectra imply that the correlated electron fluctuations are important. We then need to how to reconcile this with the single quasiparticle picture that has been so successful. One possibility is that correlated electron motion produces conductance and tunneling characteristics similar to those predicted by the single quasiparticle picture. Another possibility is the dynamical current redistribution (DCR) model [32] in which non-Gaussian noise statistics can result from statistically independent fluctuators. As a simple example of this, consider two fluctuating resistors in parallel. The amplitude of the fluctuation of the total resistance due to one resistor depends on the state of the other resistor. The DCR model is most effective near the percolation threshold where a small number of large fluctuators produce frequency dependence in S_2 over a range of 1 or 2 decades. This bandwidth is determined by the frequency of the independent fluctuators. It could be that single quasiparticle hopping produces non-Gaussian noise according to the DCR model if one allows for a large number of quasiparticles with a broad distribution of hopping rates. A third possibility is that the observed non-Gaussian noise arises from nonequilibrium aging which would result in the first spectrum $S(f)$ deviating from $1/f$. For example, if the fluctuations had a drift that increased linearly in time, then $S(f)$ would go as $1/f^2$. Even though the Si MOSFET experiments do see a first spectrum that approaches $1/f^2$ with decreasing electron density, the experimentalists were careful to rule out aging [33].

In any event it is clear that more work needs to be done both theoretically and experimentally to understand the implications of the second spectrum. Noise measurements should be done on the insulating side further away from the metal-insulator transition. Theoretical modeling and simulations could shed some light on the implications of the second spectra measurements.

Acknowledgements We thank Michael Weissman and Michael Pollak for helpful discussions. This work was supported in part by DOE grant DE-FG03-00ER45843 and ONR grant N00014-00-1-0005.

References

[1] M. Pollak, Discuss. Faraday Soc. **50**, 13 (1970).
[2] A. L. Efros and B. I. Shklovskiĭ, J. Phys. C **8**, L49 (1975).
[3] B. I. Shklovskiĭ and A. L. Efros, Electronic Properties of Doped Semiconductors (Spinger-Verlag, Berlin, 1984).
[4] B. Sandow et al., Phys. Rev. Lett. **86**, 1845 (2001).
[5] M. Pollak, phys. stat. sol. (b) **205**, 35 (1998).
[6] S. Kogan, Electronic Noise and Fluctuations in Solids (Cambridge University Press, Cambridge, 1996).
[7] R. F. Voss, J. Phys. C **11**, L923 (1978).
[8] B. Shklovskiĭ, Solid State Commun. **33**, 273 (1980).
[9] S. M. Kogan and B. I. Shklovksiĭ, Sov. Phys. Semicond. **15**, 605 (1981).
[10] V. I. Kozub, Solid State Commun. **97**, 843 (1996).
[11] S. Kogan, Phys. Rev. B **57**, 9736 (1998).
[12] B. I. Shklovskii, Phys. Rev. B **67**, 045201 (2003).
[13] K. Shtengel and C. C. Yu, Phys. Rev. B **67**, 165106 (2003).
[14] J. G. Massey and M. Lee, Phys. Rev. Lett. **79**, 3986 (1997).
[15] D. M. et al., phys. stat. sol. (b) **230**, 197 (2002).
[16] S. Kar, A. K. Raychaudhuri, and A. Ghosh, (2002), cond-mat/0212165.
[17] C. C. Yu and M. Pollak, unpublished.
[18] P. J. Restle, R. J. Hamilton, M. B. Weissman, and M. S. Love, Phys. Rev. B **31**, 2254 (1985).
[19] M. B. Weissman, N. E. Israeloff, and G. B. Alers, J. Magn. Magn. Mater. **114**, 87 (1992).
[20] M. B. Weissman, Rev. Mod. Phys. **65**, 829 (1993), and references therein.
[21] M. B. Weissman, Rev. Mod. Phys. **60**, 537 (1988).
[22] A. J. Bray and M. A. Moore, Phys. Rev. Lett. **58**, 57 (1987).
[23] D. S. Fisher and D. A. Huse, Phys. Rev. B **38**, 373 (1988).
[24] D. S. Fisher and D. A. Huse, Phys. Rev. B **38**, 386 (1988).
[25] A. T. Ogielski and D. L. Stein, Phys. Rev. Lett. **55**, 1634 (1985).
[26] G. Paladin, M. Mézard, and C. D. Dominicis, J. Phys. (Paris) Lett. **46**, L985 (1985).
[27] M. Schreckenberg, Z. Phys. B **60**, 483 (1985).
[28] C. Bachas and B. A. Huberman, Phys. Rev. Lett. **57**, 1965, 2877 (1986).
[29] A. Maritan and A. L. Stella, J. Phys. A **19**, L269 (1986).
[30] P. Sibani, Phys. Rev. B **35**, 8572 (1987).
[31] J. Jaroszyński, D. Popović, and T. M. Klapwijk, Phys. Rev. Lett. **89**, 276401 (2002).
[32] G. T. Seidler, S. A. Solin, and A. C. Marley, Phys. Rev. Lett. **76**, 3049 (1996).
[33] D. Popović, private communication.

phys. stat. sol. (c) **1**, No. 1, 29–32 (2004) / **DOI** 10.1002/pssc.200303614

Hopping conductivity in carbynes. Magnetoresistance and Hall effect

S. V. Demishev [*, 1, 2], **A. A. Pronin**[1], **N. E. Sluchanko**[1], **N. A. Samarin**[1], **V. V. Glushkov**[1, 2], **A. G. Lyapin**[3] , **T. D. Varfolomeeva**[3], **S. V. Popova**[3], and **V.V. Brazhkin**[3]

[1] General Physics Institute of RAS, Vavilov street, 38, 119991 Moscow, Russia
[2] Moscow Institute of Physics and Technology, Dolgoprudnii, 141700 Moscow Region, Russia
[3] Institute of High Pressure Physics of RAS, Troitsk, 142190 Moscow Region, Russia

Received 1 September 2003, revised 16 September 2003, accepted 16 September 2003
Published online 28 November 2003

PACS 72.20.Ee, 72.20.My, 72.30.+q, 72.80.Cw

In 2D and 3D carbyne samples synthesized under high pressure magnetoresistance and Hall effect have been studied in the region of variable range hopping conductivity $T < 40$ K. For both 2D and 3D cases magnetoresistance is positive, but analysis shows that shrinkage of the wave function in magnetic field does not allow describing experimental data quantitatively and anomalous contribution probably coming from quantum interference effects may be important. In 3D case Hall coefficient follows Mott's law similar to that observed in the temperature dependence of resistivity.

1 Introduction

Carbyne is an allotropic carbon form based on sp-type bonds and possessing a pronounced linear-chain structure [1–3]. Contrary to other allotropic modifications with sp^2 (graphite) or sp^3 (diamond) bonds, carbyne cannot be synthesized as a perfect crystal, because its chains contain "built-in" disorder, probably due to the instability of large linear carbon clusters [3]. Although up to now there is no commonly accepted model structure of carbyne, it is customarily believed that linear sp segments of the polymeric carbon molecule in carbyne alternate with sp^2-hybridized carbon atoms [1–3]. The sp^2 centers give rise to chain kinks, and the dangling bonds appearing at the kinks may attach impurity atoms or form interchain links (in the absence of the sp^2 defects, the carbon chains are bonded by weak van der Waals forces). As a result, the carbon chains may form complex globular structures.

The fraction of sp^2 bonds in carbyne and related sample resistivity ρ can be varied smoothly by applying high pressure and temperature [1]. An increase in the synthesis temperature T_{syn} above 600 °C under pressure of 7 GPa allows to observe low temperature variable range hopping (VRH) transport for $T < 40$ K [1, 4–6]. It was found [4–6] that hopping exponent n in the Mott's law

$$\rho = \rho_0 \cdot \exp((T_0 / T)^n) \tag{1}$$

strongly depends on the synthesis conditions and decreases with T_{syn} from $n = 1/2$ first to $n = 1/3$ and than to $n = 1/4$. In the absence of long range correlations index n is given [4,7] by

$$n = \frac{1}{d+1}, \tag{2}$$

where d is space dimension (the discussion of applicability of Eq. (2) to 1D case is given in [4]), and the variation of n may reflect changes in effective dimension of the Miller-Abrahams network. At the same

---

[*] Corresponding author: e-mail: demis@lt.gpi.ru, Phone: +7-095-1328253, Fax: +7-095-1358129

time structural studies [1,6] show that increase of T_{syn} induces the $sp \rightarrow sp^2$ transition, i.e. in carbynes modified under high pressure the formation of a disordered 2D graphite-like network from the structure dominated by 1D chains may be expected. This opens an intriguing opportunity of creation of an experimental system with variable dimensionality based on carbynes [4].

The detailed study of DC and AC conductivity and thermopower in hopping regime have been carried in [4–6] and confirmed above supposition. It was found [4–6] that quasi 1D hopping conductivity persists up to $T_{syn} \sim 700°C$ where the appearance of new sp^2 centers likely leads to the disorder and bending of individual chains in this range of synthesis temperatures. At $T_{syn} > 700°C$, the increase in the fraction of the sp^2 bonds induces cross-linking between the chains, rendering the conduction two-dimensional. On a further increase in T_{syn} and a rise in the concentration of sp^2 centers in the carbyne matrix, the topology of quasi-two-dimensional carbon layers becomes more complicated and they start to interact with each other, leading to the 3D character of conduction. It is worth to note that observed value $n=1/2$ in 1D case does not related with Coulomb correlations and agrees with Eq. (2) [4].

In the present work we report results of the study of magnetoresistance in the hopping region for carbyne samples with $n=1/3$ (2D) and $n=1/4$ (3D). In the latter case we have also carried out first measurements of the hopping Hall effect. As starting material carbyne with chains ...=C=C=C=C=... of the cumulene type was used. Samples were prepared by the method used in [1]. Synthesis was carried out at a pressure of 7.7 GPa. In coordinates $\log(\rho) = f(T^{-n})$ linear segments characteristic to VRH have been observed at $T < T^* \sim 40$ K down to $T \sim 1.8$ K (Fig. 1).

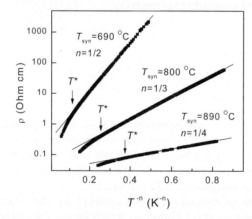

Fig. 1 Temperature dependences of resistivity in Mott's coordinates for carbyne samples obtained at various T_{syn}.

2 Magnetoresistance

The typical field dependences of magnetoresistance at various temperatures for carbyne samples are presented in Fig. 2. Interesting that "visible" negative magnetoresistance typical for various hopping systems in low magnetic fields is missing in carbyne samples. The positive magnetoresistance at first glace may originate from the shrinkage of the wave function [8, 9]. As long as values of T_0 are known from the temperature dependences of resistivity (Fig. 1) the above assumption have allowed to calculate magnitude of effective localization radius a_{eff} from equation $a_{eff} = \left(c^2\hbar^2 A(T)/e^2 t_d\right)^{1/4} \left(T/T_0\right)^{3n/4}$ [8, 9], where parameter $A(T) = \partial \ln \rho(H,T)/\partial H^2$ have been determined from the low field square part of each $\rho(H)$ curve (Fig. 2) and t_d is numerical coefficient calculated in [8, 9] and slightly dependent of the space dimension. The calculated empirical localization radius demonstrate pronounced temperature dependence and increase 1.5–2 times when temperature is lowered from $T \sim 20–30$ K to $T=1.8$ K (see inset in fig. 2). However, if shrinkage mechanism dominates, the localization radius should not depend on temperature: $a_{eff}(T) \approx const$. The latter condition in our experiments was valid only for 3D sample with $n=1/4$ in the range $T \leq 3$ K (inset in Fig. 2).

The observed anomalous behavior of magnetoresistance indicates that another contribution to magnetoresistance different from shrinkage effects may be important. A candidate possibly may be the quantum interferential effects. Note that interferential effects as a rule dominate in weak magnetic fields and gives rise to a negative magnetoresistance [9].

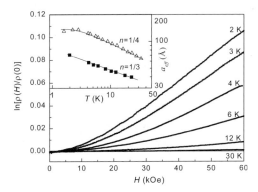

Fig. 2 Magnetoresistanse of carbyne sample with $n=1/3$. Inset shows temperature dependence of effective localization radius (see text).

We suppose that in the case of carbynes the character of the quantum interference is considerably changed and the related contribution becomes a square function of magnetic field in the same diapason as for the part originating from shrinkage. Following this assumption it is possible to write

$$\ln \frac{\rho(H)}{\rho(0)} = \frac{t_d e^2}{c^2 \hbar^2} \cdot \left[a^4 \left(\frac{T_0}{T} \right)^{3n} + f(T) \right] \cdot H^2 , \tag{3}$$

where first term in brackets represents a standard expression for the shrinkage mechanism [8, 9], parameter a denotes true value of localization radius and function $f(T)$ accounts quantum interference. From Eq. (3) we get

$$a_{eff} = \left[a^4 + f(T) \left(\frac{T}{T_0} \right)^{3n} \right]^{1/4} . \tag{4}$$

Than data in fig. 2 including saturation of a_{eff} at $T < 3$ K may be explained assuming that $f(T)T^{3n} \to 0$ at $T \to 0$ and $f(T) < 0$. Negative $f(T)$ corresponds to negative magnetoresistance as in convenient hopping systems, but in carbynes en essential modification of the quantum interference term may be expected. The checking of this supposition is a subject of future work.

3 Hopping Hall effect

Studying of the hopping Hall effect belongs to one of the most difficult experimental tasks as long as in this case a measurement of the small Hall voltage from a sample with high resistivity is required. We find that for the sample with $n=1/4$ in the hopping region $T < 40$ K Hall coefficient $R(T)$ follows the law $\ln R \sim (T_0^*/T)^{1/4}$, similar to the temperature dependence of resistivity (Fig. 3). It is worth to note that the value of T_0^* for the Hall coefficient is different from the value T_0 for the resistivity (Eq. (1)). From the data in Fig. 3 we find the ratio $(T_0^*/T_0)^{1/4} \approx 0.8$.

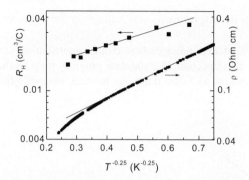

Fig. 3   Hopping Hall coefficient and resistivity in carbyne sample with $n=1/4$.

Mott's type temperature dependence of Hall coefficient with $T_0^* < T_0$ have been predicted in [10,11]. However the estimated value of $(T_0^*/T_0)^{1/4}$ is 5/8=0.625 [10,11] and somewhat smaller than observed experimentally. Nevertheless the qualitative agreement of experimental results with the calculations [10,11] indicates that observed discrepancy may be probably understood in the new theoretical studies of the hopping Hall effect.

4   Conclusion

In conclusion we have studied magnetoresistance and Hall effect in the region of variable range hopping conductivity $T < 40$ K for 2D and 3D carbyne samples synthesized under high pressure. In both 2D and 3D cases magnetoresistance is positive, but analysis shows that shrinkage of the wave function in magnetic field does not allow describing experimental data quantitatively and anomalous contribution probably coming from quantum interference effects may be important. In 3D case Hall coefficient follows Mott's law similar to that observed in the temperature dependence of resistivity.

Acknowledgements   The support from RFBR grant 00-02-16403, INTAS grant 00-807, programs of RAS "Low dimensional quantum structures" and "Strongly correlated electrons in semiconductors, metals, superconductors and magnetic materials" is acknowledged. Authors are grateful for support from program "Integration" and grant PD02-1.2-336 of the Russian Ministry of Education. AAP is supported by grant MK-2188.2003.02 of the President of Russian federation. SVD and VVG are grantees of Russian Science Support Foundation.

References

[1] A. G. Lyapin, V. V. Brazhkin, S. G. Lyapin et al., phys. stat. sol. (b) **211**, 401 (1999).
[2] Yu. P. Kudryavtsev, S. E. Evsyukov, M. V. Guseva et al., Izv. Akad. Nauk, Ser. Khim., No. 3, 450 (1993).
[3] B. M. Bulychev and I. A. Udod, Ross. Khim. Zh. **39**, 9 (1995).
[4] S. V. Demishev, A. A. Pronin, N. E. Sluchanko et al., JETP Lett. **72**, 381 (2000).
[5] S.V. Demishev, A.A. Pronin, N.E. Sluchanko et al., Phys. Solid State **44** (4), 607 (2002).
[6] S. V. Demishev, A. A. Pronin, V. V. Glushkov et al., JETP **95**, 123 (2002).
[7] A. Hunt, Solid State Commun. **86**, 765 (1993).
[8] B. I. Shklovskii and A. L. Efros, Electronic Properties of Doped Semiconductors (Springer, New York, 1984), chap. 9.
[9] M.E. Raikh, J. Czingon, Qiu-yi Ye et al., Phys. Rev. B **45** , 6015 (1992).
[10] M. Grüenewald, H. Müeller, P. Thomas, and D. Würtz, Solid State Commun. **38**, 1011 (1981).
[11] M. Grüenewald, H. Müeller, and D. Würtz, Solid State Commun. **43**, 419 (1982).

phys. stat. sol. (c) **1**, No. 1, 33–36 (2004) / **DOI** 10.1002/pssc.200303619

Magneto-mechanical effects in nano-magnets

A. Cohen and **A. Frydman**[*]

Minerva Center, Department of Physics, Bar Ilan University, Ramat Gan 52900, Israel

Received 1 September 2003, revised 19 September 2003, accepted 22 September 2003
Published online 28 November 2003

PACS 73.23.-b, 73.40.Rw, 75.50.Cc, 75.75.+a, 75.80.+q

The magnetoresistance curves of quench-condensed granular Ni very close to the electric percolation threshold exhibit sharp reproducible resistance jumps. These have been interpreted as signs for magneto-mechanical effects which occur in the magnetic particles due to interactions with the surrounding grains. In this paper we present additional experimental results that are consistent with a model in which the grains that act as bottlenecks in the resistance network exhibit buckling-like magneto-mechanical distortions leading to a resistance increase.

Granular metals are a mixture of metal and non-metal materials. When the metal concentration is small the structure takes the form of isolated metallic islands imbedded in an insulating matrix. In this configuration the system is an insulator exhibiting hopping conductivity at low temperature. Like a strongly localized system, a granular insulator is highly non-homogeneous. A natural way to treat such a system is to model it by a resistor network [1, 2], each resistor representing a pair of grains. A percolation approach to such a network leads to the understanding that the transport is governed by critical resistors. The conductivity of the critical resistors determines the conductivity of the entire system [3].

Recently we reported on studies performed on dilute granular metals, which are close to the electric percolation threshold [4]. These samples exhibit a number of features that demonstrate that the transport flows through a very small number of grains, presumably due to the fact that the percolation network is governed by one or few critical resistors. In particular, when the grains are ferromagnetic, their magnetoresistance (MR) curve shows sharp reproducible resistance jumps. These were interpreted as signs of magneto-mechanical distortions which occur in single grains that act as bottlenecks in the dilute percolation network. In this paper we present further experimental results on such systems. We suggest that the findings are consistent with a model in which the grains experience a buckling-like deformation due to the external magnetic field and interactions with the surrounding grains.

The granular systems in this study were prepared by quench-condensation [5–8] i.e. evaporation on a cryogenically cold substrate (T < 10K) and in UHV environment. Quench condensing Ni on a SiO substrate gives rise to a random array of separated Ni grains which have a diameter of ~100Å when the system reaches electric continuity [9–11]. As more Ni is quench-condensed, the average distance between the grains decreases and the resistance drops. A significant advantage of this method is the very sensitive control on the sample growth process which allows one to terminate the evaporation at any desired stage of the material deposition and "freeze" the morphological configuration. In particular, it enables one to stop the film growth at a thickness at which measurable conductivity first appears across the sample. We name these systems "dilute systems" as opposed to the conventional 2D systems.

Figure 1 compares the MR properties of a conventional 2D quench-condensed granular Ni sample and a dilute quench-condensed granular Ni sample. The field in these experiments was applied perpendicular to the layers. Both samples exhibit a negative MR at small fields. Such behavior is typical for all granu-

---

[*] Corresponding author: e-mail: frydman@mail.biu.ac.il, Phone: +972-3-5318102, Fax: +972-3-5357678

lar ferromagnets whether prepared by quench condensation or by other methods such as co-sputtering of magnetic and non-magnetic materials [9–17]. This feature occurs as the moments of the grains align with an external field because a hop between grains with opposite spin orientation is a more resistive process than a hop between grains with the same spin orientation. The unusual feature which appears primarily in quench-condensed systems is a positive MR feature at high magnetic fields. In the conventional samples a smooth positive MR background is observed, extending, in many cases, to fields larger than 5 T (Fig. 1a). In the dilute samples this manifests itself by sharp resistance jumps at specific magnetic fields after which the MR trace saturates (Fig. 1b).

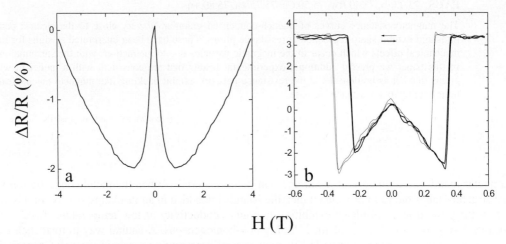

H (T)

Fig. 1 Magnetoresistance of two quench condensed Ni samples at T=4 K. Frame a shows the measurement on a conventional 2D sample. Frame b shows the measurement on a dilute sample in which the evaporation process was stopped at the point at which conductivity was first detectable. Heavy lines are a sweep from left to right and light lines are a sweep from right to left. Note that there are two sweeps for each direction.

The positive MR feature is rather puzzling since the application of an external magnetic field acts to reduce the magnetic disorder by orienting the magnetic moments of the grains thus reducing the resistance. Even more baffling is the fact that the resistance at high field reaches a value which is higher than that at zero field. This led us to speculate that the positive MR features stem from slight mechanical distortions due to magnetic repulsion between each grain and it's surrounding. The tunneling probability thus increases giving rise to a high resistance state. In the following paragraphs we present a set of experimental findings that demonstrate that the sharp resistance switches originate from a physical mechanism different than that of the negative MR associated with alignment of grains.

Figure 1b includes several MR curves of a dilute system while sweeping the field back and forth a number of times. It is seen that while the curves retrace each other when sweeping in a specific direction, there is a clear hysteresis when the field is swept in opposite directions. Interestingly, the hysteresis is observed only with regard to the sharp resistance jumps and not with regard to the negative MR at low fields. Hysteresis in the MR is usually attributed to orientation of the magnetization moments of the ferromagnetic grains. Indeed, systems of larger grains display a hysteretic magnetoresistance peak [9–11]. In our samples this is clearly not the case since the low-field magnetic peak occurs at H=0, indicating that the grains are superparamagnetic and no hysteresis is associated with the magnetization. The hysteretic nature of the sharp jumps has to be attributed to a different mechanism.

There are a number of other parameters, other than the sweep direction, to which the sharp resistance jumps are sensitive while the negative MR centered at H = 0 is robust. Some examples are demonstrated in Fig. 2. Figure 2a shows the MR of a dilute sample as a function of the sweep rate. It is seen that the resistance jump shifts as a function of rate until, for fast enough sweeps, the jump disappears altogether and the high resistance state is observed throughout the entire sweep. At the same time, the low field negative MR is not affected by the sweep rate (as long as the resistance jump actually occurs). Figure 2b compares the MR traces of a zero-field cooled sample versus a 2 T field cooled sample. The resistance jump in the field-cooled samples is clearly shifted towards larger fields while the low-field negative MR is not affected by the cooling field. Figure 2c shows a similar effect as the temperature is raised and Figure 2d illustrates this tendency as the resistance of the sample is lowered by annealing the sample at T > 30 K.

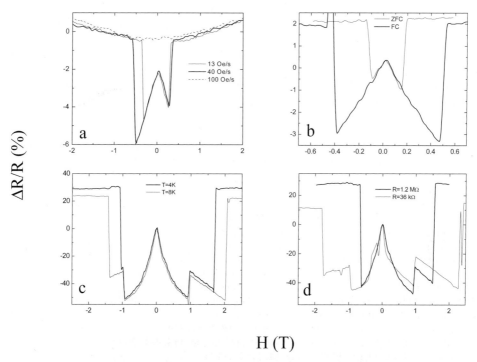

Fig. 2 a) Magnetoresistance measurements for different magnetic field sweep rates (field swept from right to left). b) Magnetoresistance curves for a dilute system measured under zero field cooling (light line) and after cooling the sample from 30 K at a field of 2 T (heavy line). Measurements were performed at T=4 K. c) MR traces at T=4 K (heavy line) and T=8 K (light line). d) MR curves as a function of sample resistance which is varied by annealing the sample at temperatures above 35 K. Heavy lines are for R=1.2 MΩ and light lines are for R=36 kΩ.

The above results lead to the conclusion that the sharp resistance jumps stem from a mechanism different than that leading to the negative MR. It should be noted, however, that we observed sharp reproducible resistance features only as a function of magnetic field and not of other parameters such as temperature or electric field. Hence, it is the magnetic field that causes the resistance to sharply increase above the original resistance value. This is consistent with our postulation that the magnetic field causes morphologic changes in the grain geometries causing a resistance rise. The sharpness of the resistance change leads us to suggest that it is caused by a buckling-like instability. We envision that as the field is in-

© 2004 WILEY-VCH Verlag GmbH & Co. KGaA, Weinheim

creases, the grains align parallel to each other, all perpendicular to the substrate. The dipole-dipole repulsion between a grain and its surrounding neighbors leads to stress which, at a critical point, causes a mechanical distortion thus increasing the tunneling resistance. Such a process is hysteretic in nature. Hence the hysteresis observed in Fig. 1b is due to the instability point rather than to the magnetization properties of the grains.

This model can account for the findings of Fig. 2. Since the buckling instability is mechanically hysteretic, the dependence of the jumps on sweep rate is natural, while the negative MR feature is not likely to be sensitive to the rate of magnetic field sweep (Fig. 2a). Raising the temperature of the sample increases the entropy. Hence, one can expect that the buckling point will require larger magnetic energy and the resistance jump will shift towards larger fields. This will have only a limited influence on the magnetization of the grains (Fig. 2c). As for the cooling field and the annealing effect (Figs. 2b and 2d), though these finding are far from being understood, we assume that they are associated with slight changes in the geometric configurations of the grains. This will have a much larger effect on the magneto-distortion than on the shape of the MR due to alignment of the grains.

Finally we note that magnetostriction (MS) effects in bulk ferromagnets have been widely studied. Nickel, for example, shows a negative MS coefficient of about 10^{-5} [18]. These effects are much less studied in nanoparticles since the detection of minute geometrical variations provides a large experimental challenge. Tunneling conductivity, which depends exponentially on the distance between particles, is a natural candidate for the investigation of nano-distortions. Studies such as the present one have the potential of shedding light not only on the physics of hopping in a granular system but also on the mechanism of magneto-mechanical effects in magnetic nanostructures.

Acknowledgements We gratefully acknowledge illuminating discussions with R. Berkovits, R.C. Dynes, Z. Ovadyahu and M. Pollak. This research was supported by the Israel Science Foundation.

References

[1] A. Miller and E. Abrahams, Phys. Rev. **120**, 745 (1960).
[2] C.J. Adkins, J. Phys. C: Solid State Phys. **15**, 7143 (1982).
[3] V. Ambegeokar, B.I. Halperin, and .S. Langer, Phys. Rev. B **4**, 2162 (1971).
[4] A. Cohen, A. Frydman, and R. Berkovits, to be published; *cond-mat 0212627*.
[5] M. Strongin, R. Thompson, O. Kammerer, and J. Crow, Phys. Rev. B **1**, 1078 (1970).
[6] R.C. Dynes, J.P. Garno and J.M. Rowell, Phys. Rev Lett. **40**, 479 (1978).
[7] H.M. Jaeger, D.B. Haviland, B.G. Orr, and A.M. Goldman, Phys. Rev B **40**, 182 (1989).
[8] R.P. Barber and R.E. Glover III, Phys. Rev. B **42**, 6754.
[9] A. Frydman and R.C. Dynes, Solid State Commun. **110**, 485 (1999).
[10] A. Frydman, T.K. Kirk, and R.C. Dynes, Solid State Commun. **114**, 481 (2000).
[11] A. Frydman and R.C. Dynes, Philos. Mag. **81**, 1153 (2001).
[12] J.I. Gittelman, Y. Goldstein and S. Bozowski, Phys. Rev. B **5**, 3609 (1972).
[13] A. Milner, A. Gerber, B. Groisman, M. Karpovsky, and A. Gladkikh, Phys. Rev. Lett. **76**, 475 (1996).
[14] W. Yang, Z.S. Jiang, W.N. Wang, and Y.W. Du, Solid State Commun. **104**, 479 (1997).
[15] S. Honda, T. Okada, M. Nawate and M. Tokumoto, Phys. Rev. B **56** 14566 (1997).
[16] S. Sankar and A.E. Berkowitz, App. Phys. Lett. **73**, 535 (1998).
[17] J.S. Helman and B. Abeles, Phys. Rev. Lett. **37**, 1429 (1976).
[18] E.W. Lee, Rep. Prog. Phys. **18**, 184 (1955).

phys. stat. sol. (c) **1**, No. 1, 37–41 (2004) / **DOI** 10.1002/pssc.200303635

Positive magnetoresistance in a weak magnetic field in germanium with impurity concentration near the metal-insulator transition

R. Rentzsch[*, 1] and **A. N. Ionov**[2]

[1] Institut für Experimentalphysik, Freie Universität Berlin, 14195 Berlin, Germany
[2] Ioffe Physico-Technical Institute, St. Petersburg 194021, Russia

Received 1 September 2003, revised 22 September 2003, accepted 2 October 2003
Published online 28 November 2003

PACS 71.30.+h, 72.20.My, 72.80.Cw

On the metallic side of the metal-insulator transition (MIT) the theory of quantum corrections to the conductivity provides an accurate description of the temperature dependence of conductivity and magneto-conductivity in a wide variety of disordered conductors. However, we found that there exists a concentration region $1 \leq N/N_c < 1.4$ (here N_c is the critical concentration of the MIT) where the quantum interference corrections do not manifest themselves in the low-temperature resistivity and magnetoresistance experiments. Experiments have been done on isotopically engineered bulk n-type Ge with a high homogeneity of the impurity distribution, which was obtained by neutron-transmutation doping on the metallic side of the MIT close to N_c. We observed an abnormal positive magnetoresistance at relatively small magnetic fields and low temperatures (down to 30 mK) which we consider to be due to the Maki-Thomson correction to the metallic conductivity [1] and a second contribution which we attribute to be due to the tunneling between neighboring electron lakes localized in long range potential fluctuations.

1 Introduction

On the metallic side of the metal-insulator transition (MIT), the theory of quantum interference corrections to residual (Drude) conductivity such as enhanced electron-electron interaction and weak localization provides an accurate description of the temperature dependence of conductivity and magnetoresistance in a wide variety of disordered conductors with different dimensionality [1]. It is well known that at sufficiently low temperatures the temperature dependence of conductivity for three-dimensional systems obeys the law $\sigma(T) = \sigma(0) + mT^{1/2} + b(D(T) \tau_\varphi(T))^{-1/2}$, where $\sigma(0)$ is the value of the conductivity at $T = 0$, and m and b are constants; the second term is due to the electron – electron interaction and the third term is due to weak localization, where D is the diffusion coefficient and τ_φ is the phase-breaking time for inelastic collisions of electrons. At relatively low magnetic fields the weak localization correction is suppressed and in experiments it results in a negative magnetoresistance [1]. Close to the metal-insulator transition where the impurity concentration $N \approx N_c$ (N_c is the concentration of the MIT), the disorder increases. In this case, because of the statistical arrangement of impurities, there can be a situation in which for some part of the impurities the Hubbard repulsion energy will exceed the Fermi energy. As a result, singly occupied states with unpaired spins appear. Thus, in the direct proximity of the metal-insulator transition there is a coexistence of localized and delocalized states [2, 3]. In this case, the phase-breaking time is modified by an additional term τ_s due to the scattering time at localized spins. At $\tau_s \ll \tau_\varphi(T)$ the modified phase relaxation time is completely defined by τ_s which is assumed to have no tem-

---

[*] Corresponding author: e-mail: rentzsch@physik.fu-berlin.de

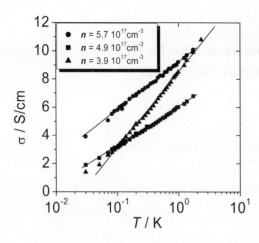

Fig. 1 Temperature dependence of conductivity of n-Ge at different electron concentrations.

perature dependence. This condition is simple to fulfill at low temperatures because of the common condition that $\tau_\varphi \propto T^{-s}$, with $s \cong 1$. Therefore, the temperature dependence of the conductivity in this case should be totally determined by the interaction effect. Consequently at $N \approx N_c$ a dependence $\sigma(T) \sim T^{1/3}$ has been predicted [4] and is indeed experimentally observed in disordered systems with impurity correlation [5].

However, in disordered systems without impurity correlation such as for example n-Ge , for barely metallic samples in the vicinity of the metal-insulator transition the temperature dependence of the conductivity obeys a logarithmic-like behavior within the experimental range of temperatures from 2 K down to 30 mK (Fig. 1) [6]. At the moment there are two explanations for the logarithmic T behavior. This dependence has been explained in [7] by different kinds of magnetic scattering. For example, if there are localized paramagnetic centers they can scatter the conduction electrons. In a magnetic field these spins of impurities are oriented along the magnetic field direction and as a result spin-flip scattering is suppressed. In a weak magnetic field one should observe a negative magneto-resistance depending quadratically on the magnetic induction [8].

Recently a new theoretical model was presented to explain the logarithmic behavior of the conductivity in granular metals [9]. It was demonstrated that for any dimension of the grains in the case of large tunneling conductance the temperature dependence is logarithmic. The model of the granular metal may be useful also to describe a disordered electron system at low electron concentration. In such a system electrons can spend a considerable time in traps which originated from strong fluctuations of the long range disorder potential. In this case, the potential wells where many electrons could be trapped ("electron lakes") correspond to the metallic grains as proposed in the model of the granular metal [9]. The tunneling takes place in this case between neighboring electron lakes.

2   Results and discussion

Experiments have been done on isotopically-engineered bulk n-type Ge with high homogeneity of the impurity distribution and with a compensation degree $K = N_A/N_D$ of about 12%, which was obtained by neutron-transmutation doping on the metallic side of the MIT very close to N_c. The temperature dependence of the conductivity obeys a logarithmic-like behavior [6] (Fig. 1). We observed at $T < 100$ mK an abnormal positive magnetoresistance at relatively small magnetic fields (Figs. 2a–d). This result contradicts the model of a simple magnetic scattering [7] because no negative magnetoresistance was obtained. The abnormal positive magnetoresistance depends on the sweep speed of the magnetic field and has properties which are – apart from the positive magnetoresistance – connected with the common shrinkage effect of the impurity wave function in a magnetic field as obtained on the dielectric side of the MIT. In a control experiment on NTD-n-type GaAs with metallic conductivity we found a negative magnetoresistance (Fig. 3) as predicted by the theory of quantum corrections to the diffusive transport [1] (see

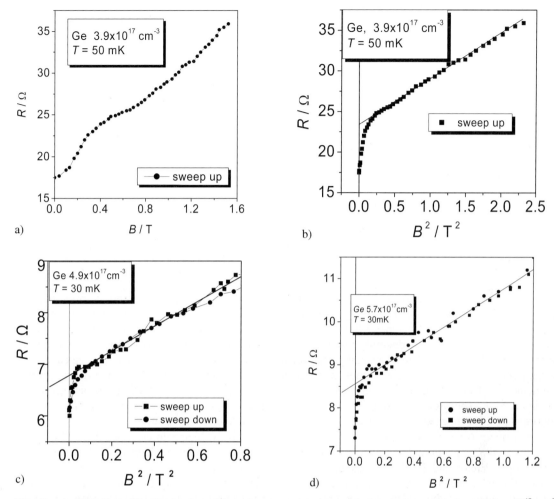

Fig. 2 a) NTD-Ge with abnormal positive magnetoresistance et electron concentration $n = 3.9 \times 10^{17}$ cm^{-3}, b) NTD-Ge with abnormal positive magnetoresistance at $n = 3.9 \times 10^{17}$ cm^{-3}, c) abnormal positive magnetoresistance at $n = 4.9 \times 10^{17}$ cm^{-3}, d) abnormal positive magnetoresistance at $n = 5.3 \times 10^{17}$ cm^{-3}.

also in [5]). However, even in the AIII-BV semiconducting compounds with n-type conductivity (n-GaAs, n-InP, n-InAs), it was shown that above N_c the absolute value of the negative magnetoresistance effect is much less than given simply by taking into account the weak localization correction to the conductivity [10, 11]. Even in weak magnetic fields the positive magnetoresistance due to scattering of the charge carriers on superconducting fluctuations has a remarkable influence. This is the so called Maki-Thomson correction which must be taken into account even in non-superconducting material where the Coulomb interaction is purely repulsive [12]; it serves as major contribution suppressing the negative magnetoresistance in low magnetic fields [10, 11]. The quantum correction to the diffusive transport at low temperatures and in low magnetic fields can be expressed as $\Delta\sigma(H) = \Delta\sigma_{wl}(H) + \Delta\sigma_{MT}(H) = (1 - \beta) \Delta\sigma_{wl}(H)$, where $\Delta\sigma_{wl}$ is the correction due to the weak localization and $\Delta\sigma_{MT}$ is the correction due to the scattering on superconducting fluctuations (Maki-Thomson correction). It was observed that at $k_F l \geq 1$, where k_F is the Fermi wave vector and l is the mean free path of the charge carriers, $\beta \approx 0.1$, and steeply increases at $k_F l < 1$. At $k_F l \approx 0.4$ a maximum value of $\beta \approx 0.7$ was found. In [13, 14] it was found that in Ge and Si with n-type conductivity due to the complicated structure of the conduction band $\Delta\sigma_{MT}$ is even enlarged in comparison with the AIII-BV semiconducting compounds with the simple conduction band structure.

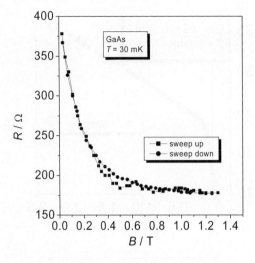

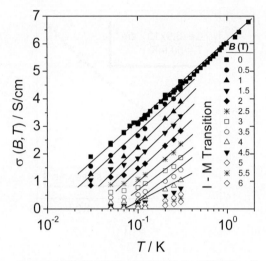

Fig. 3   Negative magnetoresistance in n-GaAs.   **Fig. 4**   Magnetoconductivity at $n = 4.9 \times 10^{17}$ cm^{-3}.

Because of the closeness to the MIT ($k_F l < 1$) in our experiments on n-Ge it cannot be ruled out that the Maki-Thomson correction completely compensates the weak localization effect resulting in a positive magnetoresistance at small magnetic fields $B \le 0.5$ T (Fig. 2a–d). We assume that the majority of the electrons takes part in the conductivity resulting in a tunneling conductance between nearest fluctuations of the disorder potential, whereas a small part of the electrons with energies just above the typical energy of the fluctuations due to electron – phonon interaction forms superconducting fluctuations, from which the tunneling electron scatters. The magnetic field suppresses these superconducting fluctuations resulting in the Maki-Thomson correction to the conductivity [12] which can serve as the main reason for the abnormal positive magnetoresistance at low magnetic fields and low temperatures.

Above about $B \approx 0.5$ T we observed a second process of positive magnetoresistance depending quadratically on the magnetic induction. At present, there is no theory which explains this effect. However, the theory of Efetov and Tschersich has not been expanded to predict the sign of the magnetoresistance. It was only mentioned that the logarithmic temperature dependence of conductivity holds independent of the dimensionality of the system and the presence of a magnetic field which is in accordance with the finding in our experiment [15, 16] (Fig. 4).

Acknowledgements   We wish to thank Dr. P. Fozooni and Prof. M. Lea (Royal Holloway University of London, Egham, U.K.) for their help in the experiments and for valuable discussions. One of us (A. I.) thanks for the financial support by the fond RFBR, grant No 03-02-17516 and the Freie Universität Berlin for its hospitality.

References

[1] B. L. Altshuler and A. G. Aronov, in Electron–Electron Interaction in Disordered Systems, edited by A. L. Efros and M. Pollak (North-Holland, Amsterdam, 1985).

[2] A. N. Ionov and I. S. Shlimak, Pis'ma Zh. Eksp. Teor. Fiz. **39**, 208 (1984) [JETP Lett. **39**, 247 (1984)].

[3] H. von Löhneisen, Phil. Trans. Royal Society London A **356**, 139 (1998).

[4] B. L. Altshuler and A. G. Aronov, Zh. Eksp. Teor. Fiz. Pisma **37**, 349 (1983) [Eng. Transl. JETP Lett. **37**, 410 (1983)].

[5] K. J. Friedland, A. N. Ionov, R. Rentzsch, Ch. Gladun, and H. Vinzelberg, J. Phys.: Condensed Matter **2**, 3759 (1990).

[6] A. N. Ionov, R. Rentzsch, P. Fozooni, and M. J. Lea, phys. stat. sol. (b) **205**, 257 (1998); Ann. Phys. **8**, 507 (1999).

[7] H. Fukuyma, Technical Report of ISSP, Ser **A**, No. 1445 (1981).

[8] K. Yosida, Phys. Rev. **107**, 396 (1957).

[9] K. B. Efetov and A. Tschersich, cond-mat/0302257 (2003).

[10] T. I. Voronina, O.V. Emeljanenko, T. S. Lagunova, Z. I. Tschugujeva, and Z. Sch. Janovitzkaya, Fiz. Tech. Poluprov. **18**(4), 743 (1984).

[11] T. I. Voronina, O. V. Emeljanenko, T. S. Lagunova, and D. D. Nedoglo, Fiz. Tech. Poluprov. **20**(6), 1025 (1986).

[12] A.T. Larkin, Zh. Eksp. Teor. Fiz. Pisma **31**, 239 (1980) [JETP Lett. **31**, (1980)].

[13] T. Chui, P. Lindenfeld, W. L. McLean, and K. Mui, Phys. Rev. Lett. **47**, 1617 (1981).

[14] T. A. Poljanskaya and I.I. Saidaschev, Pis'ma Zh. Eksp. Teor. Fiz. **34**, 378 (1981) [JETP Lett. **34**, (1981)].

[15] R. Rentzsch, A. N. Ionov, P. Fozooni, and M. J. Lea, Ann. Physik (Leipzig) **8**, SI225 (1999).

[16] R. Rentzsch, A. N. Ionov, P. Fozooni, and M. J. Lea, Physica B **284–288**, 1185 (2000).

phys. stat. sol. (c) **1**, No. 1, 42–45 (2004) / **DOI** 10.1002/pssc.200303632

Variable range hopping in the Coulomb gap

A. M. Somoza[1], **M. Ortuño**[1], and **M. Pollak**[*, 2]

[1] Departamento de Física, Universidad de Murcia, Murcia 30071, Spain
[2] Department of Physics, University of California, Riverside, CA, USA

Received 1 September 2003, revised 5 October 2003, accepted 6 October 2003
Published online 28 November 2003

PACS 72.20.Dp, 72.20.Ee

We studied by computer simulation hopping conduction in the Coulomb glass in the strongly localized limit. Full account was taken of dynamic many-body effects by considering the configurations of the entire system. Many-electron transitions were seen to be important at very low temperatures.

1 Introduction

Coulomb glass and electron glass are the customary names used for Anderson insulators in the presence of inter-site Coulomb interactions. In the absence of interactions, hopping conduction deep in the Anderson insulator is rather well established theoretically as well as experimentally. Mott [1] showed that the interplay between the difficulties of hopping long distances and those of hopping to high energies lead to what he termed variable range hopping – the typical hopping length decreases with increasing temperature. He showed that when the density of states is constant near the Fermi level, the conductivity is

$$\sigma \propto \exp\left[-(T_0/T)^{1/(d+1)}\right] \tag{1}$$

where d is the dimensionality, T_0 is a constant that depends on the concentration and on the localization radius. Efros and Shklovskii [2] (ES) found that the one-particle density of states in the presence of Coulomb interactions behaves as E^{d-1}, where E is measured from the Fermi level and d is the dimensionality of the system. Extending Mott's argument to such densities of states, they obtained for the conductivity

$$\sigma \propto \exp[-(T_0/T)^{1/2}], \qquad T_0 = \beta e^2/\varepsilon k a \tag{2}$$

where only the value of β depends on d: $\beta = 2.8$ for $d = 3$ and $\beta = 6.2$ for $d = 2$.

When Coulomb interactions are important, many-body effects can be expected to play a role in determining the physical properties. In particular, correlated motion of electrons, not considered in the derivation of (2), may affect the conductivity. Two such effects have been proposed for the strongly localized regime – collective transitions of several electrons and sequential correlations, in which the order of transitions matters. The method here intrinsically accounts for both of these effects. We study primarily the role of collective correlations. To fully account for the many-body effects the method in this work considers the states of the entire system rather than just one-particle states. Thus the usual percolation theory in real space used for hopping conduction with no interaction is replaced by a percolation theory in configuration space. A similar, much more restricted study was reported by our group earlier [3].

---

[*] Corresponding author: e-mail: mpollak@earthlink.net

2 The model

This work relates to two-dimensional strongly localized Coulomb glasses where the resonance energy t is much smaller than all other relevant energies and can be neglected in the Hamiltonian. Consistent with this regime is the assumption that on-site interaction energies are very large and can be accounted for by disallowing double occupation of a site, i.e. limiting site occupation to 0 or 1. Finally, spin interactions can be neglected because inter-site exchange energies are proportional to t^2. The Hamiltonian then includes a random energy and an interaction energy:

$$H = \sum_i \varepsilon_i n_i + \sum_{i<j} \frac{n_i n_j}{r_{ij}} , \tag{3}$$

where i,j labels sites, ε is the random site energy and r_{ij} is the distance between two sites.

The Hamiltonian is the energy of a state (configuration) of a system, determined by the occupation n_i of each site. For transport properties it is necessary to know transition rates between the states of the system. Any two configurations I,J are related by a transition of a number n of electrons, so we need the transition rates for many-electron transitions. These were shown [4] to be

$$w_{IJ} = w_0 \gamma^{n-1} \exp\left(-\frac{E_m}{kT}\right) \cdot \exp\left(-\frac{2\sum r_{ij}}{a}\right) . \tag{4}$$

E_m is the larger of the configuration energies E_I, E_J, the sum is the minimized (from among all exchange variants) sum over the hopping distances of the n electrons in the transition, and γ, is a complicated expression, but is a measure of the relative importance of an interaction energy as compared to random energies. It vanishes in the absence of interactions and thus its main function is to eliminate many-electron transitions for distant electrons (with $\gamma = 0$ eq. (4) is non-vanishing only for $n = 1$). However, this is also accomplished by the exponential hopping energy factor in (4). Unless interaction reduces E_m below that of single electron hops, the exponential distance factor will always favour one-particle transitions. The inclusion of γ is therefore not important. The resistances R_{IJ} used in percolation theory are related to the transition rates by $R_{IJ} \propto 1/w_{IJ}$. Similarly to the familiar percolation by 1-electron hops, only the exponential factors in (4) are of concern. We shall refer to this full exponent as $-\ln(R_{IJ})$.

Clearly the procedure accounts for collective correlations by including many-electron transitions. It also accounts for successive correlations because the intermediate configurations in a succession of transitions are different for different sequences and the percolation path can choose the easier intermediate state.

3 The algorithm

We take the number of electrons to be half the number of sites. The sites are arranged at random, but with a minimum separation between them, which we choose to be $0.05r$. Cyclic boundary conditions are applied in both directions. We make $r=n^{1/2}$ the unit of length and e^2/r the unit of energy (n is the total number of sites). We also assume $k_b = 1$. The random energies ε_i are distributed uniformly over an energy range equal to e^2/r. The localization length was fixed to $a=2r$.

We used a numerical algorithm to obtain the ground state and the lowest energy many-particle configurations of the system . The algorithm consists of the two stages –. in the first we find a "back- bone" of the set of low-energy states formed by metastable states, in the second we complete this set by exploring the neighborhood of the states found. This is an extension of that developed by A. Diaz [5].

In the first stage, we repeatedly start from states chosen at random and relax each sample by means of a local search procedure. In an iterative process, we look for neighbors of lower energies and always accept the first such state found. The procedure stops when no lower energy neighboring states exist and ensures stability with respect to all one-electron jumps and compact two-electron jumps. We then consider a set of metastable states found by the process just described and look for the sites which maintain the same occupation in all of them. These sites are assumed to be passive -, since. they are not allowed to change occupation the relaxation algorithm is applied onlyto the other, active sites. The whole procedure is repeated until no new passive sites are found with the set of metastable states considered. In the second stage, we complete the set of low-energy configurations by generating all the states that differ by one- or two-electron transitions from any configuration stored. In order to speed up this process, which is very CPU time consuming, we again assume passive and active sites and at first look for neighboring configurations by changing the occupation of active sites only. We later relax this restriction in the final filling stage. The completeness of the final set of low lying states was checked by comparison with the configurations obtained with a modified Monte Carlo procedure (which consumes much more computer time).

4 Results and discussion

The procedure yields percolation paths for each sample at different temperatures. From this we can extract the maximal resistance R in each sample and thus obtain sets $\{R(T,L)\}$ of values for each T and each L, with the number in each set corresponding to the number ν of samples of a given size $L = n^{-1/2}$. Specifically our sets here are $\nu = 24$, $n = 2500$; $\nu = 52$, $n = 1600$; $\nu = 92$, $n = 900$; $\nu = 54$, $n = 400$ and $\nu = 52$, $n = 225$. The number of lowest energy configurations obtained for each sample is 150000. Because of the exponential dependence of the Rs on random variables, we expect, and observe, a mesoscopic scatter within each set. To obtain a definitive value $R_m(T,L)$, we imagine a tile-work of samples with a distribution of Rs obtainable from the distribution within $\{R(T,L)\}$. Arguing that interactions between the "tiles" must be much weaker than the interaction within a tile, we are justified to use simple percolation theory for the tile system. This procedure then yields $R_m(T,L)$. It turns out that $R_m(T,L)$ has an appreciable size dependence, particularly at low T, as can be seen in Fig. 1. We finally take the value of $R_m(T)$ for a macroscopic system to be the extrapolated value of $R(T,L)$ to $1/L = 0$. The result for $R(T)$ is plotted in figure 2 where it is also compared with the predictions of ES.

To obtain a quantitative measure of the importance of collective hopping, we follow the above procedure, but when only one electron hops are allowed in the percolation. We then compare these values of $R(T)$ with those obtained for percolation including multi-electron transitions. The results, as shown in Figs. 1 and 2, indicate that collective correlations are very important at very low temperatures, but their importance is rather diminished above $T = 0.01$. In fact we find that around $T = 0.05$ nearly half of the samples R is the same for many-electron percolation as for one-electron percolation. Above this temperature then collective hopping should cease altogether to be important. We observe a deviation with respect to ES at low temperature, both for one- and for many-electron percolation but our data in this region have large statistical errors. These errors are due to the reduced number of samples studied and to larger size effects at the lowest temperatures. The calculations presented here constitute a huge numerical effort, which exhausts our present computational capabilities. It is desirable to extend this study in the future to larger samples in order to fully resolve the disagreement with ES predictions.

It is of interest to follow the change of the percolation path with temperature. One can generate "movies" of the changing path by producing frames of the path for different T. Studies of the changing path throws some light on the dynamical behaviour of the Coulomb glass. For example, one can observe the usual variable range hopping, i.e. typical hops becoming shorter at higher temperatures. Another interesting process one can observe is a change from collective hops to sequentially correlated hops or uncorrelated hops with increasing T (a process sometimes referred to as variable number hopping).

In conclusion, we improved on a previous algorithm, capable to fully study many-body effects in interacting systems at low temperatures. We applied it to an extensive study of conduction in the two dimensional Coulomb glass. We found that dynamical many-body effects increase monotonically with decreasing temperature and with increasing size. The first is clear from Fig. 1 by comparing the intersects at $1/L=0$ from left and from right, the second from the larger steepness of the lines on the left of the figure. The conductivity at higher temperatures is in reasonable agreement with ES.

The method is an excellent tool to study the dynamical processes of the Coulomb glass. Understanding details of the dynamics is important for understanding, in addition to transport, details of noise and of the glassy behavior of the Coulomb glass.

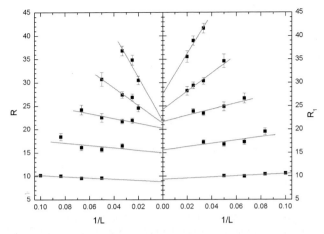

Fig. 1 The maximal resistance $R_m(T,L)$ as a function of size and temperature. The right side of the figure is for one-electron hopping, the left side for multi-electron hopping. T from top to bottom are 0.003, 0.005, 0.008, 0.015 and 0.04.

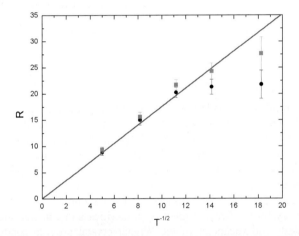

Fig. 2 The resistance R_m as a function of temperature. Squares represent 1-e hopping, circles represent multi-e hopping. The straight line correspond to the ES prediction.

Acknowledgement We acknowledge partial support by Dirección General de Investigación Científica y Técnica.

References

[1] N. F. Mott, J. Non-Crys. Solids **1**, 1 (1968).
[2] A.L. Efros and B.I. Shklovskii, J. Phys. C **8**, L49 (1975).
[3] A. Pérez-Garrido et al., Phys. Rev. B **55**, R8630 (1997).
[4] M. Pollak, J. Phys. C **14**, 2977 (1981).
[5] A. Díaz-Sánchez, Ph.D. thesis, Universidad de Murcia (1998).

phys. stat. sol. (c) **1**, No. 1, 46–50 (2004) / **DOI** 10.1002/pssc.200303642

Coulomb gap and variable range hopping in a pinned Wigner crystal

B. I. Shklovskii[*]

William I. Fine Theoretical Physics Institute, University of Minnesota, Minneapolis, Minnesota 55455, USA

Received 1 September 2003, revised 16 September 2003, accepted 16 September 2003
Published online 28 November 2003

PACS 72.20.Ee

It is shown that pinning of the electron Wigner crystal by a small concentration of charged impurities creates the finite density of charged localized states near the Fermi level. In the case of residual impurities in the spacer this density of states is related to nonlinear screening of a close acceptor by a Wigner crystal vacancy. On the other hand, intentional doping by a remote layer of donors is a source of a long range potential, which generates dislocations in Wigner crystal. Dislocations in turn create charged localized states near the Fermi level. In both cases Coulomb interaction of localized states leads to the soft Coulomb gap and ES variable range hopping at low enough temperatures.

The growth of GaAs-heterostructures mobilities has made possible to study collective properties of the hole gas with a very small concentration. The average distance between holes r_s, calculated in units of the hole Bohr radius, has reached 80 and exceeded the theoretically predicted freezing point of the hole liquid, $r_s = 38$, where it should become the Wigner crystal. The transport of such low density systems with and without magnetic field was studied recently. If a strong magnetic field applied along the hole plane the low temperature conductivity of the Wigner-crystal-like phase seems to be related to the variable range hopping (VRH) [1]. Thus, it seems timely to discuss ohmic transport of the Wigner crystal. Of course, any Wigner crystal is always pinned by impurities and can not slide as a whole in a small electric field. Interstitials and vacancies of the Wigner crystal provide conductivity but they have large activation energies and freeze out at low temperatures. On the other hand, the VRH conductivity requires nonzero density of states of charge excitations near the Fermi level, both in the case of Mott law or Efros–Shklovskii (ES) law. If pinning centers produce such "bare" density of states then the long range Coulomb interaction creates the Coulomb gap around the Fermi level and leads to the ES variable range hopping. In this paper, I study the origin of the bare density of states of the pinned Wigner crystal, the Coulomb gap on its background and the resulting VRH conductivity at low temperatures.

Let me start from a clean electron Wigner crystal on uniform positive background. Consider an additional electron added to the perfect Wigner crystal and let the lattice relax locally. This addition costs the interstitial energy $0.14e^2/\kappa a$, where a is the lattice constant or the optimal triangular lattice, κ is the dielectric constant of GaAs [2]. On the other hand, the energy cost of the extraction of an electron and the corresponding relaxation around the vacancy is equal to the vacancy energy $0.26e^2/\kappa a$. Thus, the density of states of relaxed excitations (electronic polarons [3]), consists of two delta-like peaks at energies $-0.26e^2/\kappa a$ and $0.14e^2/\kappa a$. The lower peak corresponds to occupied states, the upper one contains empty states. Between them there is the hard (completely empty) gap of

---

[*] e-mail: shklovs@physics.umn.edu; shklo001@umn.edu

the width $0.40e^2/\kappa a$. The Fermi level of electronic polarons (the zero of energy) is situated in this gap, therefore the density of states at the Fermi level is zero.

In typical samples the Wigner crystal is pinned by charged impurities: donor and acceptors. Even if there is no intentional doping and the two-dimensional electrons gas (2DEG) is created in a hetero-structure by a distant metallic gate, there is always a small concentration of residual donors and acceptors. Many high mobility heterostructures are intentionally doped by a donor layer, which is separated from 2DEG by the spacer with a large width $s \gg a$. This is an additional source of disorder. In the beginning of this paper we neglect this source and deal with residual impurities only. We return to the role of doped layer in the second part of this paper.

Let me argue that close residual donors and acceptors create localized states at the Fermi level. I use theory of pinning of the Wigner crystal by charge impurities developed in Ref. [4]. As in that paper I talk about Wigner crystal of electrons and concentrate on close residual acceptors which according to Ref. [4] produce stronger pinning than donors. (For the hole crystal the role of acceptors is of course played by donors.) Because the spacer width, s, and the lattice constant of the Wigner crystal, a, are much larger than the lattice constant of AlGaAs the authors of Ref. [4] assumed that acceptors are randomly distributed in space with three-dimensional concentration N. Each acceptor is negative because it has captured one electron from 2DEG.

It was found [4] that effect of an acceptor dramatically depends on its distance, d, from the plane of the Wigner crystal. If d is larger than $0.68a$ the electron crystal adjusts to acceptor in such a way that in the ground state an interstitial position of the crystal is exactly above the acceptor. In this case, interaction of the acceptor with the Wigner crystal can be calculated in the elastic approximation. On the other hand, when d is smaller than $0.68a$ acceptor creates a vacancy in the crystal, which positions itself right above the acceptor. This means that the close acceptor effectively builds in the crystal cite pushing away to infinity its electron. We can consider the latter acceptor as an empty localized state for electron, while the former acceptor can be considered as a localized state occupied by an electron. Their energies are equal at $d = 0.68a$, so that all acceptors with $d > 0.68a$ are occupied and all accep-tor with $d < 0.68a$ are empty. Thus, acceptors with $d = 0.68a$ are at the Fermi level and the bare density of states near the Fermi level

$$g_B \sim \frac{Na}{e^2/\kappa a}, \tag{1}$$

where Na is an estimate of the two-dimensional concentration of acceptors located within distance a (only they provide substantial pinning) and $e^2/\kappa a$ is the characteristic energy of their pinning. The non-zero density of states g_B makes possible VRH conductivity of the pinned Wigner crystal. For example, an electron from an acceptor with $d = 0.68a + 0$ can hop to a distant one with $d = 0.68a - 0$.

The long range Coulomb interaction between localized states leads to the soft Coulomb gap at the Fermi level [5]. Shape of the Coulomb gap depends on the interaction, $U(r)$, of the hopping electron in the final state of the hop and the empty place it has left behind (excitonic term) [3, 5]. If the interaction can be described by the standard Coulomb law $U_C(r) = -e^2/\kappa r$, where r is the two-dimen-sional distance between initial and final state, then the Coulomb gap has the standard for two-dimen-sional case form

$$g(E) = \frac{2}{\pi} \frac{|E| \kappa^2}{e^4}. \tag{2}$$

The attraction potential $U(r)$ generally speaking is different from $U_C(r)$ due to screening of the Cou-lomb interaction by elastic deformations of the Wigner crystal or in other words by its polarization. In order to calculate Fourier image $U(q)$ of $U(r)$ we have to introduce the Larkin length L at which acceptors destroy the long range order of the crystal. In the two-dimensional Wigner crystal pinning by close acceptors is so strong that this length is close to the average distance between them, $L \sim (Na)^{-1/2}$. Using L for $qa \ll$ we can write

$$U(q) = 2\pi/q\kappa\epsilon(q), \tag{3}$$

where

$$\epsilon(q) = 1 + \frac{2}{qr_D} \frac{q^2}{q^2 - L^{-2}}. \tag{4}$$

is given by interpolation between large and small q cases. Here $r_D \sim -0.32a$ is the linear Debye screening radius of the Wigner crystal. (The asymptotics of Eq. (4) at large and small q are similar to those of the dielectric constant of a two-dimensional electron gas in a strong magnetic field [6–8], where the Larmour radius plays the role of L.)

In the real space the Fourier image of Eq. (4) results in the Coulomb potential $U(r) = U_C(r)$ only at $r \gg L^2/a = 1/Na^2$. One can say that by these distances all electric lines of a probe charge located in the reference point in the plane of Wigner crystal leave the plane. This leads to the standard Coulomb gap Eq. (2), but only at energies $|E| < e^2Na^2/\kappa$. At smaller distances $r \ll 1/Na^2$ potential $U(r)$ is weaker than $U_C(r)$ and grows only logarithmically with decreasing r. But this does not lead to faster (exponential) growth of the density of states at $|E| \sim e^2Na^2/\kappa$ as it would if g_B were very large. The reason is that at energy $|E| \sim e^2Na^2/\kappa$ the Coulomb gap density of states, Eq. (2), already reaches the bare density of states, g_B, given by Eq. (1). The Coulomb gap is just a depletion of the density of states on the background of the bare density. Therefore, the density of states in the Coulomb gap can not be larger than g_B [3]. This means that at $|E| \gg e^2Na^2/\kappa$ the density of states saturates at the level of Eq. (1).

The role of distance $L^2/a = 1/Na^2$ can be interpreted in another way. When we add an electron to the clean Wigner crystal it makes an interstitial but its charge spreads to infinity. In the a pinned Wigner crystal an added charge is smeared in the disc of the finite radius R. We can find R minimizing the sum of the Coulomb energy $e^2/\kappa r$ of the disc of radius r and the shear energy $\mu(u/L)^2r^2$ necessary for the disc dilatation by the area a^2. Here $\mu \sim e^2/a^3$ is the shear modulus and $u \sim a^2/r$ is the necessary displacement. It is important that the characteristic distance of the variation of the shear displacement (which percolates between strongly pinning acceptors) is the average distance between close acceptors $L \sim (Na)^{-1/2}$. Minimizing this sum we find that $R = L^2/a = 1/Na^2$. This again tells us that at distances larger than $L^2/a = 1/Na^2$ we deal with the unscreened Coulomb potential.

At low enough temperatures the Coulomb gap leads to Efros–Shklovskii(ES) law of the temperature dependence of the variable range hopping [3, 5]

$$\sigma = \sigma_0 \exp\left[-(T_{ES}/T)^{1/2}\right]. \tag{5}$$

Here σ_0 is a prefactor, which has only an algebraic T-dependence and

$$T_{ES} = Ce^2/\kappa\xi, \tag{6}$$

where C is a numerical coefficient close to 6 and ξ is the localization length for electron tunnelling with energy close to the Fermi level in the Wigner crystal. We will discuss the value of ξ in the end of the paper.

When with increasing temperature the width of the band of energies used for ES hopping $(TT_{ES})^{1/2}$ reaches e^2Na^2/κ ES law is replaced by the Mott conductivity

$$\sigma = \sigma_0 \exp\left[-(T_M/T)^{1/3}\right], \tag{7}$$

where $T_M = C_M/(g_B\xi^2)$. Transition from the ES law to the Mott law happens at $T \sim T_{ES}^3/T_M^2 \sim T_{ES}(Na^2\xi)^2$. Thus, while ES law does not depend on the acceptor concentration N the range of ES law shrinks when concentration of acceptors decreases.

This example emphasizes universality of the Coulomb gap and ES law. The Coulomb gap was derived for lightly doped semiconductors, where disorder is as strong as interactions [5]. Later Efros [9] suggested a model, where electrons are located on sites of a square lattice with +1/2 charges, the number of sites being twice larger than number of electrons. Random energies of sites are uniformly distributed in the band A (they are measured in units of Coulomb interaction of two electrons on nearest sites). In this model the Coulomb gap does not survive in the small disorder case $A \ll 1$

because in this case in spite of small disorder positive charges of empty sites and negative charges of occupied sites alternate in the perfect NaCl-like order. One could, therefore, say that the Coulomb gap is the property of strong disorder only. The above example of the Wigner crystal on the uniform background pinned by rare strong impurities however shows that such impression may be misleading. Even in the case when concentration of acceptors N is small (one can call it the weak disorder case) the Coulomb gap and ES law survive.

One can interpret what happens with the Wigner crystal in the terms of the Efros model. At strong disorder when $Na^3 \sim 1$ our model is close to the Efros model with $A \sim 1$. When parameter Na^3 becomes much smaller than unity and the disorder becomes weak, the energy scatter of states created by impurities stays at the level of $e^2/\kappa a$, while the long Coulomb interaction between these states becomes much smaller, of the order of $e^2(Na)^{1/2}\kappa$. Introducing a two-dimensional random lattice with cites occupied by acceptors we can come to a renormalized Efros model with $A \sim 1/(Na^3)^{1/2} \gg 1$. This brings us again to the Coulomb gap at small energies.

Until now we talked about the role of residual acceptors. Let us qualitatively discuss intentionally doped heterojunctions where donors are situated in the narrow layer parallel to the plane of the 2DEG (delta-doping) at a large distance $s \gg a$ from 2DEG. Random distribution of donors in this layer creates fluctuations of their potential of all wavelengths, but only harmonics of the random potential with wavelengths larger than s reach 2DEG. Let us assume that the two-dimensional concentration of charged donors in the doped layer is N_2. In principle N_2 can be different from the two-dimensional concentration of electrons in the Wigner crystal n (2DEG can be compensated by acceptors or created by illumination). If $N_2 \ll n$ the Wigner crystal screens external potential by small, purely elastic deformations. Due to these deformations energies of an interstitial and a vacancy depend on a coordinates. Therefore, both delta-like peaks of the density of states are somewhat smeared. Their width, however, is much smaller than the hard gap between them. Thus, the hard gap is preserved and no states at the Fermi level appear. When the concentration of donors, N_2, grows, the amplitude u of displacements of electrons in the Wigner crystal with wavelength of the order s reaches the lattice constant of the crystal a. Such large deformations resolve themselves by creation of dislocations [10, 11], because the energy price of the dislocation core becomes smaller than an elastic energy, which is eliminated by the dislocation. This happens when the donor concentration, N_2, is of the order of electron concentration, n. Indeed, if we cover the layer of donors by squares with the side s the typical fluctuation of number of charges in a square is equal $(N_2 s^2)^{1/2}$. Potential of the charge of this fluctuation reaches 2DEG practically without compensation by oppositely charged fluctuations in neighboring squares. As a result at $n = N_2$ 2DEG has to provide $(N_2 s^2)^{1/2} = s/a$ new electrons to screen random potential in a square. This may be done by an additional raw of electrons in the square or, in other words, by two dislocations. Thus approximately one dislocation appears in a square with a side s. This picture is actually a simple visualization of the Larkin domain. Below we concentrate on the case $N_2 \sim n$. In this case, the length s plays the role of the Larkin length L.

An isolated dislocation brings the electronic polaron state right to the Fermi level. Indeed, if we add an electron from the Fermi level to the end of an additional raw terminated by a dislocation and let the rest of electrons relax, the dislocation just moves along this additional raw by one lattice constant a and the energy remains unchanged. Similarly, if we extract an electron from the end of the raw to the Fermi level, the dislocation moves in the opposite direction and the energy does not change. This means that if we neglect the interaction of dislocation with long range potential of donors and interaction between dislocations they create a delta peak of the density of states right at the Fermi level. This peak is normalized on the concentration of dislocations. The Fermi level is pinned in the middle of this peak.

In the long range fluctuating potential of donors the peak of the density of dislocation states is smeared, because dislocations strongly interact with the gradient of potential. The peak is smeared also due to the interaction between the dislocations. When an additional electron is absorbed by a dislocation and, as a result, the dislocation moves by one lattice site, the logarithmic interaction with other dislocation changes.

To understand the role of Coulomb interaction in the density of states of the pinned Wigner crystal one has to concentrate on the fate of the charge of an added electron. In the clean Wigner crystal, if we have a single dislocation and move it by one lattice constant adding a new electron, the charge of this electron spreads to infinity, leaving the dislocation neutral.

If we have a gas of dislocations in the positions fixed by disorder and their interaction, a single charge can only spread to the finite distance, R. It can be estimated in the way we did for residual randomly distributed acceptors. If we assume that dislocation are fixed in space by fluctuating donor potential the extra electron charge spreads optimizing the sum of its Coulomb energy and the energy of the shear deformation. Optimization leads to $R \sim s$. Thus, the Coulomb potential of charges inside the pinned Wigner crystal is valid at distances in the plane $r \gg s$.

This system is clearly similar to the Coulomb glass and therefore has the Coulomb gap. Indeed, as we mentioned above the derivation of the Coulomb gap is based upon the observation that when a localized electron is transferred to another localized state one should take into account its $1/r$ Coulomb attraction with the hole it has left (the excitonic effect). We claimed above that that the Coulomb interaction is valid if $r \gg s$. Therefore, the linear in energy Coulomb gap appears at the Fermi level. The width of this gap is $e^2/\kappa R = e^2/\kappa s$. At $s \gg a$ the Coulomb gap occupies only small fraction of the energy range between interstitial and vacancy peaks. Away from the Coulomb gap the density of states is almost constant. Thus, in the case of intentional δ-doping by remote donors we again arrive to the Coulomb gap of density of states and correspondingly to ES law at low temperatures.

Let us discuss the value of localization length ξ in Eq. (6). In classical Wigner crystal $\xi \sim a/r_s^{1/2}$. This is a quite small value which leads to too large T_0 and very large resistances in the range of ES law. However, close to the melting point of the Wigner crystal ξ can be much larger, making an observation of ES law in pinned Wigner crystal more realistic. Magnetoresistance, observed in Ref. [1] may be related to ES variable range hopping conductivity.

In conclusion, I emphasize again that I am dealing with the Wigner crystal which in the absence of impurities slides on the positive background. Impurities pin this Wigner crystal and lead to ES VRH conductivity due to hops between its pieces. This situation is similar to what happens in a system of many quantum dots situated in random electrostatic potential of stray charges or in a system of many electron puddles with random positive charge of their background. Similar physics was recently theoretically studied in quasi-one-dimensional systems [12].

Acknowledgements I am grateful to M. M. Fogler, A. L. Efros and S. Teber for many important discussions. This paper is supported by NSF DMR-9985785.

References

[1] H. Noh, Jongsoo Yoon, D. C. Tsui, and M. Shayegan, Phys. Rev. B **64**, 081309 (2001).

[2] D. S. Fisher, B. I. Halperin, and R. Morf, Phys. Rev. B **20**, 4692 (1979).

[3] B. I. Shklovskii and A. L. Efros, Electronic Properties of Doped Semiconductors, (Springer, New York, 1984).

[4] I. M. Rouzin, S. Marianer, and B. I. Shklovskii, Phys. Rev. B **46**, 3999 (1992).

[5] A. L. Efros and B. I. Shklovskii, J. Phys. C **8**, L49 (1975).

[6] I. V. Kukushkin, S. V. Meshkov, and V. B. Timofeev, Usp. Fiz. Nauk **155**, 219 (1988) [Sov. Phys. Usp. **31**, 511 (1988)].

[7] I. L. Aleiner and L. I. Glazman, Phys. Rev. B **52**, 11296 (1995).

[8] M. M. Fogler, A. A. Koulakov, and B. I. Shklovskii, Phys. Rev. B **54**, 1853 (1996).

[9] A. L. Efros, J. Phys. C **9**, 2021 (1976).

[10] D. Fisher, M. P. A. Fisher, and D. A. Huse, Phys. Rev. B **43**, 130 (1991).

[11] Min-Chul Cha and H. A. Fertig, Phys. Rev. B **50**, 14388 (1994).

[12] M. M. Fogler, S. Teber, and B. I. Shklovskii, cond-mat/0307299.

phys. stat. sol. (c) **1**, No. 1, 51–54 (2004) / **DOI** 10.1002/pssc.200303648

Two-dimensional phononless VRH conduction in arrays of Ge/Si quantum dots

A. I. Yakimov[*1], **A. V. Dvurechenskii**[1], **A. V. Nenashev**[1], **A. I. Nikiforov**[1], **A. A. Bloshkin**[2], and **M. N. Timonova**[2]

[1] Institute of Semiconductor Physics, prospekt Lavrenteva 13, 630090 Novosibirsk, Russia
[2] Novosibirsk State University, Pirogova 2, 630090 Novosivirsk, Russia

Received 1 September 2003, accepted 3 September 2003
Published online 28 November 2003

PACS 72.20.Ee, 73.21.La, 73.40.Gk, 73.63.Kv

We report measurements of a two-dimensional variable-range hopping conductance in delta-doped Ge/Si heterostructures with a layer of Ge nanometer-scale quantum dots. We found that the conductance σ vs. temperature T follows the Efros-Shklovskii behavior $\sigma = \sigma_0 \exp[-(T_0/T)^{1/2}]$ with the temperature-independent prefactor $\sigma_0 \sim e^2/h$. A strong reduction of the measured value of T_0 from that calculated for single-particle hopping was observed. All these results provide a manifestation of interaction-driven many-electron correlated hopping in dense arrays of quantum dots.

1 Introduction In general, the temperature dependence of the conductivity for variable-rangle hopping (VRH) is given by $\sigma(T) = \sigma_0(T) \exp[-(T_0/T)^x]$. VRH conductivity in the presence of long-range Hartree interaction between localized single-particle excitations obeys the Efros-Shklovskii (ES) law [1] $\sigma(T) = \sigma_0 \exp[-(T_0/T)^{1/2}]$, where $k_B T_0 = Ce^2/\kappa\xi$ is the characteristic interaction energy scale, C is a numerical coefficient that depends on dimensionality, k_B is the Boltzmann constant, κ is the relative permittivity of the host lattice, ξ is the localization length of electrons. Within the mechanism of phonon-assisted VRH, the prefactor σ_0 takes the form $\sigma_0 = \gamma/T^m$, where γ is a temperature-independent parameter and $m \sim 1$ [2]. The theoretical value of the constant C for single-particle hopping in two dimensions (2D) is $C \simeq 6$ [2, 3].

Several authors have argued that under certain conditions dc VRH conduction can be dominated by many-particle Coulomb correlations between electronic transitions [4, 5]. Sequential correlations appear when the hops of an electron facilitates the hopping probability of another electron due to rearrangement of the local potentials and/or site occupations in the vicinity of the initial and final states for tunneling process. There can be also interaction-driven simultaneous hopping of several electrons resulting in a lowering of the energy configuration of the system. Because formation of such dressed polaron state provides partial screening of Coulomb interaction at large distances, the characteristic interaction energy in correlated hopping is reduced relative to its single-particle value [6, 7]. Pérez-Garrido et al. [4] showed that in a regime of many-particle excitations, $\sigma(T)$ dependence has the ES form with numerical constant $C = 0.6 \pm 0.2$, i.e., the parameter T_0 turns out to be about one order of magnitude smaller than Efros and Shklovskii's prediction for single-electron hoping. A reduction of hopping constants from the single-particle value in gated GaAs/AlGaAs heterostructures has been observed in [8]. Kozub, Baranovskii and Shlimak, assuming that interaction-assisted fluctuations of energies of hopping sites have spectral density

* Corresponding author: e-mail: yakimov@isp.nsc.ru, Phone: +07 3832 332 624, Fax: +07 3832 332 466

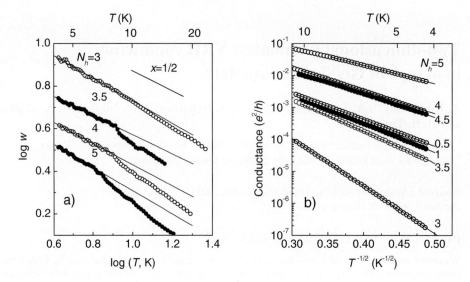

Fig. 1 a) Temperature dependence of the logarithmic derivative $w(T) = \partial \ln \sigma(T)/\partial \ln T$ for samples with different average numbers of holes per one Ge QD. The solid lines are least-square fits to linear dependence. b) The conductance $\sigma(T)$ vs $T^{-1/2}$ for different dot occupation. Symbols correspond to experimental data, solid lines are approximations of the experimental data using equation $\sigma(T) = \gamma/T^m \exp[-(T_0/T)^{1/2}]$. T_0, γ, and m are variable parameters.

$1/f$, demonstrated that sequential Coulomb correlations in a Coulomb glass can result in a phononless VRH with a temperature-independent universal prefactor $\sigma_0 \approx e^2/h$ [5].

Usually, gated disordered semiconductors are exploited to look for the many-particle correlations in 2D VRH. To drive conductivity of the system, one is obliged to change carrier concentration thereby approaching inevitably the metal-insulator transition (MIT). However, since localization degrades the screening of electron-electron interaction, correlation effects should be particularly important on the insulator side far from the MIT. From this point of view, we believe that it is more reasonable to use dense arrays of quantum dots (QDs) to study correlated hopping because one can fix QDs density and change only carrier wavefunctions by varying the dot filling factor, being deep in the insulator phase.

Previously, using an artificial screening provided by a metallic plane, parallel to a layer of Ge QDs in Si, we have proved that the VRH transport in arrays of self-assembled Ge/Si(001) QDs is strongly affected by long-range interactions [9]. In this paper, we examine in detail measurements of 2D variable-range hopping conductance in delta-doped Ge/Si heterostructures with a layer of Ge nanometer-scale QDs grown by molecular-beam epitaxy in the Stranskii-Krastanov growth mode. The average size of the dot base length is around 10 nm, the height is ~ 1 nm. The areal density of the dots is $n_{QD} = 4 \times 10^{11}$ cm^{-2}. To supply holes on the dots, a boron δ-doping Si layer inserted 5 nm below the Ge QD layer was grown. After spatial transfer, the average number of holes per dot was varied from $N_h = 1/2$ to $N_h = 5$ by varying the doping. The sample preparation and data analysis are described in detail elsewhere [10].

2 Temperature dependence of conductivity   In order to obtain detailed information on the functional dependence $\sigma(T)$, we used the differential method for an analysis of the temperature dependence of the reduced activation energy [11] $w(T) = \partial \ln \sigma(T)/\partial \ln T = m + x(T_0/T)^x$. In this approach, if $m \ll x(T_0/T)^x$, then $\log w(T) = A - x \log T$, and $A = x \log T_0 + \log x$. Plotting $\log w$ as a function of $\log T$, one can find the hopping exponent x from the slope of the straight line. The parameter A can be found by the intersection point of the straight line with the ordinate axis, which gives the characteristic temperature $T_0 = (10^A/x)^{1/x}$. Typical plots of $\log w(T)$ versus $\log T$ for several samples are given in Fig. 1 a. At

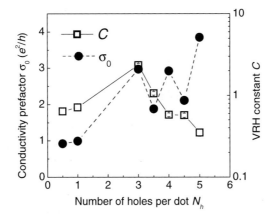

Fig. 2 Dependence of the conductivity prefactor and numerical parameter C associated with variable-range hopping in arrays of Ge/Si quantum dots on the average number of holes in QDs.

$T < 10$ K, a linear relationship is observed between $\log w(T)$ and $\log T$, implying that $m \ll x(T_0/T)^x$ at these temperatures. From the slope angle of the approximating straight lines (solid lines in Fig. 1 a), we found that the exponent x takes approximately the same value $x = 0.51 \pm 0.05$ for all samples.

Because it was already established that $x \simeq 0.5$, the method of non-linear regression can be used for further determining the exponent m in the region of low temperatures. With this aim, the experimental data $\sigma(T)$ at $T < 10$ K were approximated by the equation $\sigma(T) = \gamma/T^m \exp[-(T_0/T)^{1/2}]$, and the parameters γ, m and T_0 were varied to obtain the best fit. Figure 1 b shows the conductivity in units of e^2/h, the quantum of conductance, of samples with different QD occupation plotted versus $T^{-1/2}$; the symbols are the experimental points and the solid lines are the least-squares fits to the ES equation. We found that m lies in the region 0.16 ± 0.09. This means that the conductivity prefactor σ_0 virtually does not depend on temperature at low T and signals *against* the conventional phonon-assisted hopping mechanism.

The dependence $\sigma_0(N_h)$ is presented in Fig. 2. An impressive feature is that the prefactor having a value of order e^2/h is not constant but quantized in units of the conductance quantum. Although, currently, there is no preconceived explanation of the oscillating behavior of σ_0, we consider universality of the prefactor as a manifestation of the 2D VRH conduction stimulated by the sequential Coulomb correlations [5].

To obtain further evidence for correlated VRH in arrays of Ge/Si QDs, it is necessary to measure the ES characteristic temperature T_0 and compare it with the theoretical predictions. In fact, since T_0 depends on the localization length ξ and hence on the electronic configuration of occupied hole state in the dots, it is more convenient to find the universal constant C whose value is inverse proportional to intensity of the many-particle effects. It may be done taking into account the values of T_0 and ξ.

3 Analysis of the parameters associated with VRH in Ge/Si QDs Asymptotic values of the hole localization length resulting from hole tunneling between coupled Ge/Si quantum dots were obtained by computer modeling. The simulation was performed on a square lattice of 15×15 sites with the lattice constant $n_{QD}^{-1/2} + \delta r$, where δr is a random value with a Gaussian distribution. Only overlapping between nearest neighbours were included. We use the Hamiltonian

$$\hat{H} = \sum_{i,\alpha} E_{i,\alpha} \hat{a}_{i,\alpha}^+ \hat{a}_{i,\alpha} + \sum_{i,j,\alpha,\beta} J_{i,j,\alpha,\beta} \hat{a}_{j,\beta}^+ \hat{a}_{i,\alpha}, \tag{1}$$

where index i counts the dots, index α denotes the hole bound state number in QD (we consider only nine bound states in each dot); $\hat{a}_{i,\alpha}^+$ ($\hat{a}_{i,\alpha}$) the creation (annihilation) operator for a hole in state α of ith QD, $E_{i,\alpha}$ is the hole energy in this state, $J_{i,j,\alpha,\beta}$ is the integral of overlapping between the αth state in ith

QD and βth state in jth QD. The random hole energies $E_{i,\alpha}$ were taken as the size-quantization energies in quantum dots whose dimensions are characterized by a Gaussian distribution with the mean square deviation of 20%. Dependence of the energy levels on dot size has been calculated previously using the sp^3 tight binding model with inclusion of spin-orbit interaction and deformation effects [12].

The following procedure was used to determine the overlap integrals. We calculated energies of hole states in a model structure containing Ge quantum dot inside a Si box. Periodic boundary conditions, $\psi(-d/2, y, z) = \psi(d/2, y, z)$ or $\psi(-d/2, y, z) = -\psi(d/2, y, z)$, where d is the box size in x direction, $\psi(x, y, z)$ the hole wave function, were considered. The same boundary conditions were used for the y and z directions. Overlap integral in plane of QD array in x direction was defined as $J(d) = |E_+ - E_-|/4$, where E_+ and E_- are the hole energies corresponding to the boundary conditions given above. The obtained dependence $J(d)$ can be rewritten in the form $J(d) = A_\alpha \exp(-B_\alpha d)$, where coefficients A_α and B_α depend on the energy level number α, and B_α equals to the inverse localization length of a hole in α state of isolated QD. Integrals $J_{i,j,\alpha,\beta}$ were determined as geometrical mean of overlap integrals between α and β states: $J_{i,j,\alpha,\beta} = \sqrt{J_{i,j,\alpha,\alpha}J_{i,j,\beta,\beta}} = \sqrt{A_\alpha A_\beta}\exp[-(B_\alpha + B_\beta)d_{ij}/2]$, where $d_{i,j}$ is the distance between QDs. Simulation was carried out using 5000 random realizations of QD array with the filling factor 1/2 for the ground s-state or for the first excited p-state in the dots. For each realization, we calculated probability p_i of hole to occupy corresponding s- or p-state in each dot. The probability value was then approximated by the equation $p_i = a \exp(-2d_i/\xi_r)$, where d_i is the distance between ith QD and the dot with maximum local hole wavefunction amplitude. Localization length ξ was obtained by averaging ξ_r through all array realizations. We found $\xi = 2.3$ nm for $N_h \leq 2$ and $\xi = 2.8$ nm for $2 < N_h \leq 6$.

It now remains to determine the magnitude of C using equation $C = k_B T_0/(e^2/\kappa\xi)$ and experimental values of T_0. These results are presented in Fig. 2. The value of hopping constant turns out to be $C = 0.9 \pm 0.4$. This implies that characteristic interaction temperature T_0 is considerably smaller than the value from simulation of single-electron transport that provides an additional argument in favor of correlated hopping in 2D arrays of Ge/Si quantum dots.

4 Summary   In summary, we have investigated the variable-range hopping transport of holes in Ge/Si self-assembled quantum dots. We find the universal temperature-independent conductivity prefactor $\sigma_0 \sim e^2/h$ and demonstrate the hopping constant C to be much smaller that the single-particle value. We believe that our results provide experimental evidence for many-electron correlated hopping in two-dimensional arrays of quantum dots.

Acknowledgements   This work was supported by the Russian Foundation for Basic Research (Grant No. 03-02-16526) and by the Program of President of Russian Federation (Grant No. MD-28.2003.02).

References

[1] A. L. Efros and B. I. Shklovskii, J. Phys. C **8**, L49 (1975).
[2] D. N. Tsigankov and A. L. Efros, Phys. Rev. Lett. **88**, 176602 (2002).
[3] V. L. Nguen, Fiz. Tekn. Poluprov. **18**, 335 (1984) (Soviet Phys. Semicond. **18**, 207 (1984)).
[4] A. Pérez-Garrido, M. Ortuño, E. Cuevas, and J. Ruiz, Phys. Rev. B **55**, R8630 (1997).
[5] V. I. Kozub, S. D. Baranovskii, and I. S. Shlimak, Solid State Commun. **113**, 587 (2000).
[6] A. G. Zabrodskii and A. G. Andreev, Zh. Eksper. Teor. Fiz., Pisma **58**, 809 (1993) (JETP Lett. **58**, 758 (1993)).
[7] M. Lee and J. G. Massey, phys. stat. sol. (b) **205**, 25 (1998).
[8] F. W. Van Keuls and A. J. Dahm, phys. stat. sol. (b) **205**, 21 (1998).
[9] A. I. Yakimov, A. V. Dvurechenskii, V. V. Kirienko, Yu. I. Yakovlev, A. I. Nikiforov, and C. J. Adkins, Phys. Rev. B **61**, 10868 (2000).
[10] A. I. Yakimov, A. V. Dvurechenskii, A. I. Nikiforov, and A. A. Bloshkin, Zh. Eksper. Teor. Fiz., Pisma **77**, 445 (2003) (JETP Lett. **77**, 376 (2003)).
[11] A. G. Zabrodskii and K. N. Zinoveva, Zh. Eksper. Teor. Fiz. **86**, 727 (1984) (Soviet Phys.-JETP **59**, 425 (1984)).
[12] A. V. Dvurechenskii, A. V. Nenashev, and A. I. Yakimov, Nanotechnology **13**, 75 (2002).

phys. stat. sol. (c) **1**, No. 1, 55–58 (2004) / **DOI** 10.1002/pssc.200303650

Disorder dependence of phase transitions in a Coulomb glass

Michael H. Overlin[1], **Lee A. Wong**[1], and **Clare C. Yu**[*,1]

Department of Physics and Astronomy, University of California, Irvine, CA 92697, USA

Received 1 September 2003, accepted 4 September 2003
Published online 28 November 2003

PACS 64.60.Fr, 64.70.Pf, 65.60.+a, 71.23.Cq, 71.23.An

We have performed a Monte Carlo study of a three dimensional system of classical electrons with Coulomb interactions at half filling. We systematically increase the positional disorder by starting from a completely ordered system and gradually transitioning to a Coulomb glass. The phase transition as a function of temperature is second order for all values of disorder. We use finite size scaling to determine the transition temperature T_C and the critical exponent ν. We find that T_C decreases and that ν increases with increasing disorder.

Electrons with long range Coulomb interactions in three dimensions display a rich and complex behavior. If there is translational invariance and a background of compensating positive charge, the system forms a Wigner crystal at low densities where the potential energy dominates the kinetic energy [1, 2]. In the presence of quenched disorder the competition between interactions and disorder produces a Coulomb glass. Comparing these two extremes reveals similarities and differences. For example both undergo a phase transition when the temperature is lowered. In one case an ordered arrangement of electrons is formed while in the case of the Coulomb glass a highly disordered arrangement is frozen into place. Yet both low temperature phases have a gap in their single particle density of states.

In this paper we study the effect of gradually introducing disorder into a three dimensional system of electrons with long range Coulomb interactions. The system is discrete in the sense that the electrons sit on half of the available sites. In the ordered case the sites form a cubic lattice. The disorder is introduced in the positions of the sites and their deviation from a cubic lattice. The Hamiltonian is

$$H = \sum_{i>j} \frac{(n_i - K)(n_j - K)}{r_{ij}} \tag{1}$$

where we set the charge $e = 1$, n_i is the number operator for site i, $r_{ij} = |\mathbf{r}_i - \mathbf{r}_j|$, and K is a compensating background charge making the whole system charge neutral. $n_i = 1$ (-1) for an occupied (unoccupied) site. We consider half–filling with $K = 1/2$.

We have simulated three dimensional systems of linear size $L = 4$, 6, and 8. We place $N = L^3$ sites in the system. We have only considered the case of half filling in order to take advantage of the particle-hole symmetry. For the ordered case the sites form a cubic lattice. In the ground state, every other site is occupied; the occupied sites form a face centered cubic (FCC) lattice. We can gradually introduce disorder by allowing the deviation of a site from its position in a cubic lattice to be chosen from a Gaussian distribution with a standard deviation of σ. This gives the radial distance from the cubic lattice site. The angular coordinates of the site are chosen randomly using a uniform distribution. The ordered case corresponds to $\sigma = 0$. $\sigma = 1$ corresponds to a very disordered case with a standard deviation equal to the cubic lattice constant a. For all values of the disorder, the system

---
[*] Corresponding author: e-mail: cyu@uci.edu, Phone: 1-949-824-6216, Fax: 1-949-824-2174

undergoes a second order phase transition as the temperature is lowered. (In the ordered case, constraining the electrons to sit on lattice sites rather than allowing them to have continuous translational degrees of freedom results in a second order phase transition. This is consistent with our observations of the lack of coexistence of the ordered and disordered phases at T_C and with the absence of hysteresis.) We study the effects on the thermodynamics of this phase transition as a function of the disorder.

We use infinite periodic boundary conditions in which the simulation box is infinitely replicated in all directions to form a lattice. We use an Ewald summation technique [3] so that an electron on a given site interacts with the other electrons and all their images via the Coulomb interaction.

We used a Monte Carlo heat bath algorithm. We keep a table of the potential energy at each site. Each electron is looked at sequentially and moved to one of the available $N/2 + 1$ sites (its own site or one of the available $N/2$ unoccupied sites), chosen with a Boltzmann probability. If the site chosen is the electron's originial location, the potential energies are unchanged; if the electron hops to a new site, we update all the potential energies. If the electron chooses its initial site, which it does with high probability at low temperatures, we do not have to recompute the potential energies. This speeds up the simulation considerably, partially compensating for the much longer equilibration times needed at low temperatures. Our longest run (for $L = 4$ at $T = 0.01$) had 3×10^6 Monte Carlo steps per electron. Depending on the system size and temperature, the sample averages involved between 5 and 190 disorder configurations.

Let $S_i = 2(n_i - K)$ be an effective spin associated with the occupation of site i so that $S_i = 1\ (-1)$ for an occupied (unoccupied) site. The Edwards–Anderson order parameter is defined as $q \equiv [\langle S_i \rangle^2]$; we will denote thermal averages by $\langle \ldots \rangle$ and disorder averages by $[\ldots]$. We use the moments of the overlap to define Binder's g [4, 5]:

$$g = \frac{1}{2}\left(3 - \frac{[\langle q^4 \rangle]}{[\langle q^2 \rangle]^2}\right) \tag{2}$$

Binder's g provides a way to monitor the phase transition. Since g is dimensionless, we expect that it should satisfy a scaling form

$$g(L, T) = \hat{g}(L^{1/\nu}(T - T_C)). \tag{3}$$

Thus at the critical temperature, $g(L, T_C)$ should have the same value independent of the system size L (as long as L is sufficiently large for finite size scaling to apply) [4, 5]. We have determined the critical exponent ν and the transition temperature T_C as a function of the disorder σ through the finite size scaling of $g(L, T)$ [5, 6]. In Fig. 1 we plot $g(L = 8, T)$ versus T for various values of σ.

Notice that the transition region moves to lower temperatures with increasing disorder. This reflects the decrease in T_C with increasing σ. The transition temperature corresponds to the temperature where the curves of $g(L, T)$ versus T for all sizes cross.

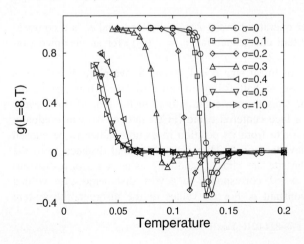

Fig. 1 $g(L = 8, T)$ vs. T for $\sigma = 0$ (45 runs), $\sigma = 0.1$ (10 runs), $\sigma = 0.2$ (5 runs), $\sigma = 0.3$ (15 runs), $\sigma = 0.4$ (115 runs), $\sigma = 0.5$ (45 runs), and $\sigma = 1$ (108 runs). The number of runs in parentheses is the number of runs that were averaged to obtain the data.

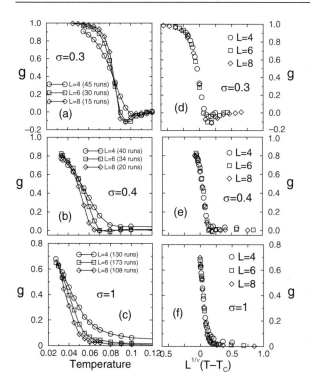

Fig. 2 a)–c) $g(L, T)$ versus T for $\sigma = 0.3$, 0.4, and 1.0 at $L = 4$, 6, and 8. The solid lines are guides to the eye. ($g(L = 8, T)$ vs. T is virtually identical.) The number of runs in parentheses is the number of runs that were averaged to obtain the data. d) $g(L, T)$ for $\sigma = 0.3$ scaled using $\hat{g}\left(L^{1/\nu}(T - T_C)\right)$ with $T_C = 0.085 \pm 0.002$ and $\nu = 0.071 \pm 0.02$. e) $g(L, T)$ for $\sigma = 0.4$ scaled using $\hat{g}\left(L^{1/\nu}(T - T_C)\right)$ with $T_C = 0.045 \pm 0.001$ and $\nu = 1.05 \pm 0.05$. f) $g(L, T)$ for $\sigma = 1$ scaled using $\hat{g}\left(L^{1/\nu}(T - T_C)\right)$ with $T_C = 0.028 \pm 0.001$ and $\nu = 1.30 \pm 0.1$.

To more accurately determine T_C, we use the scaling hypothesis to collapse the data for a given value of σ onto a single curve as shown in Fig. 2. T_C and ν are used as adjustable parameters to collapse the data. We can estimate the errors in the critical temperature and the critical exponent ν by how well the curves can be made to collapse. The values of ν and T_C at various values of σ are given in table 1. In Fig. 3 we plot T_C and ν versus σ.

We can see that the transition temperature decreases from $T_C = 0.128 \pm 0.001$ at $\sigma = 0$ to $T_C = 0.028 \pm 0.001$ at $\sigma = 1$. ν increases from $\nu = 0.55 \pm 0.05$ at $\sigma = 0$ to $\nu = 1.30 \pm 0.10$ at $\sigma = 1$.

It is interesting that T_C is much lower than the characteristic energies of the system which are of order unity. This is especially true for large values of the disorder. The reason for this was given by Grannan and Yu [6] and is as follows. At the temperatures of our simulations, nearby pairs of sites will with high probability consist of an occupied and an unoccupied site. Since these strongly coupled pairs of sites are close together, they are guaranteed to have small dipole moments. Therefore, they will interact weakly with the rest of the system, remaining active down to temperatures much lower than the bare interaction energy.

To summarize, we have performed a Monte Carlo study of a classical three dimensional Coulomb system of electrons at half filling. We systematically increase the positional disorder by introducing deviations from positions in a cubic lattice. We start from a completely ordered system and gradually transition to a Coulomb glass. The phase transition as a function of temperature is second order for all values of disorder. We use finite size scaling to determine the transition temperature T_C and the critical exponent ν. We find that T_C decreases and that ν increases with increasing disorder.

Table 1 The values of T_C and ν for different valuse of σ.

σ	T_C	ν
0.0	0.128 ± 0.001	0.55 ± 0.05
0.1	0.123 ± 0.001	0.57 ± 0.05
0.2	0.110 ± 0.001	0.61 ± 0.05
0.3	0.085 ± 0.002	0.71 ± 0.02
0.4	0.045 ± 0.001	1.05 ± 0.05
0.5	0.030 ± 0.001	1.35 ± 0.05
1.0	0.028 ± 0.001	1.30 ± 0.10

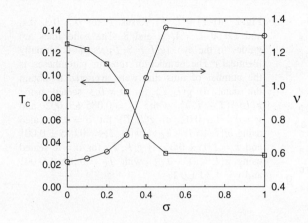

Fig. 3 The transition temperature T_C (□) and the critical exponent ν (○) versus the disorder σ.

Acknowledgements We thank Peter Young, Robijn Bruinsma, Alina Ciach, and Sylvain Grollau for helpful discussions. This work was supported by DOE grant DE-FG03-00ER45843.

References

[1] E. P. Wigner, Phys. Rev. **46**, 1002 (1934).
[2] E. P. Wigner, Trans. Faraday Soc. **34**, 678 (1938).
[3] S. W. de Leeuw, J. W. Perram, and E. R. Smith, Proc. R. Soc. Lond. A **373**, 27 (1980). (We omitted the dipole term in the Ewald summation. We have done some runs with the dipole term and find no qualitative and only a slight quantitative difference.)
[4] K. Binder, Z. Phys. B **43**, 119 (1981).
[5] R. N. Bhatt and A. P. Young, Phys. Rev. B **37**, 5606 (1988).
[6] E. R. Grannan and C. C. Yu, Phys. Rev. Lett. **71**, 3335 (1993).

phys. stat. sol. (c) **1**, No. 1, 59–62 (2004) / **DOI** 10.1002/pssc.200303607

Electron localization in an external electric field

O. Bleibaum[*,1] and **D. Belitz**[2]

[1] Institut für Theoretische Physik, Otto-von-Guericke Universität Magdeburg, 39016 Magdeburg,
PF 4120, Germany
[2] Department of Physics and Materials Science Institute, University of Oregon, Eugene OR 97403, USA

Received 1 September 2003, revised 16 September 2003, accepted 16 September 2003
Published online 28 November 2003

PACS 72.15.Rn

The impact of a weak electric field on the weak-localization corrections is studied within the framework of a nonlinear σ-model. Two scaling regimes are obtained. In one, the scaling is dominated by temperature; in the other, by the electric field. An explicit expression is derived for the crossover temperature between the two regimes.

1 Introduction   The physics of weakly disordered systems has been the subject of considerable interest over the past years. If interactions effects can be ignored, it is known that quantum interference effects lead to the localization of all electronic states in two-dimensions, even for arbitrarily weak disorder [1]. That is, the ohmic conductivity of such systems vanishes. It is well understood how the localization is affected by external fields that destroy the phase coherence underlying the quantum interference effects, e.g., magnetic fields [1]. However, there still are open questions about the impact of an electric field, which does not break time reversal invariance, and thus has no direct impact on the interference effects. This is an important question, since transport experiments invariably involve electric fields. In the theoretical literature, one can find arguments that a weak electric field has no impact on the localization corrections at all [2], to arguments that predict a strong impact on the localization corrections already at very small fields [3], to arguments that predict immediate delocalization [4], to arguments that predict delocalization beyond a critical field strength [5]. Here we revisit this question. In agreement with Ref. [4], we find that an arbitrarily weak field destroys the localization. However, for weak disorder this effect is not observable at realistic temperatures.

2 The model   We consider a Hamilton operator

$$H = \frac{\hat{p}^2}{2m} + \boldsymbol{F} \cdot \boldsymbol{x} + V(\boldsymbol{x}) \,. \tag{1}$$

Here $\hat{\boldsymbol{p}}$ is the momentum operator, m is the electron mass, $\boldsymbol{x} = (x, y)$ is the position in real space, $\boldsymbol{F} = (F, 0)$ is the electric field, and $V(\boldsymbol{x})$ is a random potential. For the latter we assume a Gaussian distribution with zero mean and a second moment given by

$$\langle V(\boldsymbol{x}) \, V(\boldsymbol{y}) \rangle = \frac{\hbar}{2\pi\nu\tau} \, \delta(\boldsymbol{x} - \boldsymbol{y}) \,, \tag{2}$$

where ν is the density of states and τ is the relaxation time in the self-consistent Born approximation.

---

[*] Corresponding author: e-mail: olaf.bleibaum@physik.uni-magdeburg.de, Phone: +49 391 67 12474, Fax: +49 391 67 11217

To investigate the impact of the field on the electron localization, we focus on the density relaxation, as described by the integral equation

$$n(\boldsymbol{x}, E \mid \Omega) = \int d\boldsymbol{y} \ P(\boldsymbol{x}, \boldsymbol{y} \mid E, \Omega) \ n_0(\boldsymbol{y}, E). \tag{3}$$

Here $n_0(\boldsymbol{x}, E)$ is the initial distribution of the number of particles with energy E at position $\boldsymbol{x}$, the propagator P describes the evolution of this distribution, and $n(\boldsymbol{x}, E \mid \Omega)$ is the Laplace transform of the time dependent particle number density $n(\boldsymbol{x}, E \mid t)$.

To calculate the density propagator P we generalize the nonlinear σ-model of Ref. [6] to allow for the presence of a weak electric field. The action takes the form

$$S = -\frac{\pi v \hbar}{4} \int d\boldsymbol{x} \sum_{\alpha=-n_r+1}^{n_r} \Omega Q_{\alpha\alpha}(\boldsymbol{x}) \Lambda_\alpha - \frac{\pi v \hbar}{8} \int d\boldsymbol{x} \sum_{\alpha,\alpha'=-n_r+1}^{n_r} Q_{\alpha\alpha'}(\boldsymbol{x}) \left[\nabla \cdot D(\mu_x)\nabla\right] Q_{\alpha'\alpha}(\boldsymbol{x}). \tag{4}$$

Here α numbers $2n_r$ replicas, $\Lambda_\alpha = +1$ if $\alpha > 0$, and $\Lambda_\alpha = -1$ if $\alpha \leq 0$. The matrix field Q is defined on a $O(n_r, n_r)/O(n_r) \times O(n_r)$ Grassmann manifold and has the properties $Q^2 = 1$ and $\text{tr}\, Q = 0$. The integration over real space is understood to be restricted to the classically accessible region, defined by the requirement $\mu_x \equiv E - \boldsymbol{F} \cdot \boldsymbol{x} > 0$. The generalized diffusion coefficient is given by

$$D(\mu_x) = \tau \mu_x / m. \tag{5}$$

In deriving Eq. (4) we have restricted the consideration to small fields, which satisfy the relationship $Fl/\mu_x \ll 1$, where l is the mean free path. It can be shown that the two-point propagator of this field theory determines P.

To investigate the action (4) we parameterize the $2n_r \times 2n_r$ matrix Q in terms of real $n_r \times n_r$ matrices q, according to

$$Q = \begin{pmatrix} \sqrt{1 + qq^T} & q \\ -q^T & -\sqrt{1 + q^Tq} \end{pmatrix}, \tag{6}$$

and expand the action (4) in powers of q. For a one-loop calculation it suffices to keep terms up to $O(q^4)$.

3 The Gaussian fluctuations

We first consider the Gaussian approximation. The action (4) then yields a generalized diffusion equation for P,

$$(\Omega - [\nabla \cdot D(\mu_x)\nabla]) P(\boldsymbol{x}, \boldsymbol{y} \mid E, \Omega) = \delta(\boldsymbol{x} - \boldsymbol{y}). \tag{7}$$

This differential equation must be supplemented by boundary conditions. We require that the propagator vanishes at infinity in the classically accessible region, and that the current in the direction of the field vanishes at the classical turning point. The structure of the equation is a consequence of the symmetries of the action. It reflects the fact that the configuration averaged Green functions are symmetric with respect to exchange of $\boldsymbol{x}$ and $\boldsymbol{y}$, and invariant against generalized real-space translations $\boldsymbol{x} \to \boldsymbol{x} + \boldsymbol{a}$, $E \to E + \boldsymbol{F} \cdot \boldsymbol{a}$.

The general solution of Eq. (7) can be expressed in terms of hypergeometric functions. For our purposes, it is more illuminating to consider a special initial condition, namely, a homogeneous density n of charge carriers, all with energy μ. An electric field is then suddenly switched on at time $t = 0$, so

$$n_0(\boldsymbol{x}, E) = n\delta(E - (\mu + \boldsymbol{F} \cdot \boldsymbol{x})). \tag{8}$$

For such an initial charge carrier density the solution of Eq. (7) gives the probability for finding the charge carriers at time t with energy μ' if they had the energy μ at time $t = 0$,

$$\mathcal{P}(\mu', \mu \mid t) = \frac{\theta(\mu)\,\theta(\mu')}{D'F^2 t} \exp\left(-\frac{\mu + \mu'}{D'F^2 t}\right) I_0\left(\frac{2}{D'F^2 t}\sqrt{\mu\mu'}\right), \tag{9}$$

where I_0 is the modified Bessel function, and $D' = \tau/m$ is the Drude mobility. The first moment of this distribution, that is, the mean energy, increases with time according to

$$\epsilon_\mu(t) \equiv \int_0^\infty d\mu' \, \mu' \, \mathcal{P}(\mu', \mu \,|\, t) = \mu + D'F^2 t \,. \tag{10}$$

This heating is accompanied by an ohmic current,

$$\mathbf{j} = -D'n\mathbf{F} \,. \tag{11}$$

The generalized diffusion Eq. (7) thus describes heating of the charge carriers due to the work done on the system by the electric field. Equation (11) also shows that D' is indeed the mobility.

Fluctuations about the mean energy, calculated from the equation

$$\sigma_\mu^2(t) = \int_0^\infty d\mu' \, (\mu' - \epsilon_\mu(t))^2 \, \mathcal{P}(\mu', \mu \,|\, t) \,, \tag{12}$$

increase with time according to

$$\sigma_\mu^2(t) = (D'F^2 t)^2 + 2D(\mu) F^2 t \,. \tag{13}$$

Therefore, deviations from the mean energy are only negligible for small times, $t \ll t^*$, where

$$t^* = 2\mu^2/D(\mu) F^2 \,. \tag{14}$$

For $t \gg t^*$ the fluctuations about the mean energy are as large as the mean energy itself, so that the mean energy does no longer describe the state of the system adequately.

A more detailed analysis of the solution shows that t^* also sets the time scale for a change of the structure of a particle packet. At $t = 0$ the packet is a delta pulse in energy space, and for $t \ll t^*$ its spread is Gaussian. In this limit, the width of the particle packet increases with time in the same way as in the absence of the field. Therefore, the diffusion volume is not affected by the field for $t \ll t^*$. However, at $t \approx t^*$ the particle packet undergoes a restructuring. For $t \gg t^*$, the mean square deviation in the direction of the electric field increases with time according to

$$\langle (x - \langle x \rangle)^2 \rangle \approx (D'Ft)^2 \,. \tag{15}$$

Here $\langle \ldots \rangle$ denotes an average with respect to the distribution $\mathcal{P}$. For $t \gg t^*$, the diffusion volume thus increases much faster with time than in the absence of the field.

4 Scaling, and weak-localization corrections   We now perform a scaling analysis of the action (4), using the technique of Ref. [7]. To this end we extend the dimensionality of the system from 2 to $d = 2 + \epsilon$, and first consider the Gaussian fixed point that describes a diffusive phase [7]. At this fixed point, the Gaussian action is invariant against changes of scale of the form $q \to q' = b^d q$, $\Omega \to \Omega' = b^2 \Omega$ and $x' = x/b$, if the electric field is scaled according to $F \to F' = Fb$. Accordingly, F is a relevant operator with respect to the diffusive fixed point, with a scale dimension of 1. If we assign a coupling constant u to the terms quartic in q, we find that u decreases according to $u \to u' = b^{-(d-2)}u$, so u is an irrelevant operator with respect to the diffusive fix point for $d > 2$, and so are all terms of higher order in q. Since the scale dimension of the diffusion coefficient D at the diffusive fix point is zero, we have the scaling equation

$$D(\Omega, F, u) = D(\Omega b^2, Fb, ub^{-(d-2)}) \,. \tag{16}$$

Putting $\Omega = 0$, and $b = 1/F$, we find

$$D(0, F, u) \propto \text{const} + F^{d-2} \,. \tag{17}$$

Equation (17) suggests that in two-dimensions the weak-localization correction to the diffusion coefficient are logarithmic with respect to F. We have verified this by an explicit calculation of the one-loop corrections. The corrections obtained in this way are the same as those obtained in Ref. [3].

For $\Omega \neq 0$ there are two scaling regimes. For small Ω the scaling of the diffusion coefficient is governed by the field, and for large Ω, by the frequency. The scaling analysis shows that the cross-over between these scaling regimes occurs at $\Omega \approx \Omega^*$, where $\Omega^* \sim F^2$. As one would expect from these arguments, the explicit calculation of the one-loop corrections yields $\Omega^* = \kappa/t^*$, where $\kappa = O(1)$.

The critical fixed point, which describes the Anderson localization transition in the absence of an electric field, is more complicated. It cannot be found by power counting alone, but rather requires an explicit calculation within the framework of a loop expansion [1,7]. In particular, the scale dimension of F is determined by the loop expansion. However, in $d = 2 + \epsilon$ the scale dimensions at the diffusive and the critical fixed points, respectively, differ only by terms of $O(\epsilon)$. The leading contribution to the scale dimension $[F]$ of F is therefore still given by power counting, and we have

$$[F] = 1 + O(\epsilon). \tag{18}$$

The electric field is thus a relevant operator with respect to the Anderson localization fixed point. Strictly speaking, this discussion shows only that the localization fixed point is unstable in the presence of an electric field, and it does not tell what happens instead. However, explicit perturbative calculations to one-loop order suggest that there is a metallic phase in $d = 2$ for $F \neq 0$.

5 Conclusions   The arguments presented above show that in a weak electric field the scaling of the dc-conductivity at asymptotically low temperatures is governed by the electric field. However, any experiment effectively measures the conductivity at a frequency or temperature given by the inverse phase relaxation time, τ_ϕ^{-1}. Therefore, the impact of the field on the weak-localization corrections can only be observed if $\Omega^* \tau_\phi \gg 1$. Quantitative estimates show $\Omega^* < 10^4$ Hz, while a typical phase relaxation rate at dilution refrigerator temperatures is on the order of 10^{11} Hz. The field scaling is therefore not observable in current experiments.

These considerations solve the following paradox. While it is true that an arbitrarily small electric field destroys localization, as was found by Kirkpatrick [4], this effect manifests itself only at unobservably low temperatures. This explains why the experimental results are consistent with the zero-field theory, even though electric fields are present in the transport experiments.

We finally note that in our model inelastic collisions are not taken into account. Therefore, our results apply only to samples that are shorter than the energy relaxation length. The consideration of inelastic collisions is important in order to establish a nonequilibrium steady state characterized by an effective electron temperature. In our present treatment such a steady state is absent; the charge carriers are continuously heated up. We therefore plan further investigations that will focus on energy relaxation processes.

Acknowledgements   We would like to thank V. V. Bryksin and T. R. Kirkpatrick for helpful discussions. This work was supported by the DFG under grant No. Bl456/3-1, and by the NSF under grant No. DMR-01-32555.

References

[1] See., e.g., P. A. Lee and T. V. Ramakrishnan, Rev. Mod. Phys. **57**, 287 (1985).
[2] B. L. Altshuler, A. G. Aronov, and D. E. Khmelnitskii, Solid State Commun. **39**, 619 (1981).
[3] T. Tsuzuki, Physica **107B**, 679 (1981).
[4] T. R. Kirkpatrick, Phys. Rev. B **33**, 780 (1986).
[5] V. V. Bryksin, H. Schlegel, and P. Kleinert, Phys. Rev. B **52**, 16494 (1995).
[6] F. Wegner, Z. Phys B **35**, 207 (1979).
    A. J. McKane and M. Stone, Ann. Phys. (N.Y.) **131**, 36 (1981).
[7] D. Belitz and T. R. Kirkpatrick, Phys. Rev. B **56**, 6513 (1997).

phys. stat. sol. (c) **1**, No. 1, 63–66 (2004) / **DOI** 10.1002/pssc.200303610

Electron transport in the Anderson model

Andreas Alvermann[*1,2], **Franz X. Bronold**[1], and **Holger Fehske**[1]

[1] Institut für Physik, Ernst-Moritz-Arndt-Universität Greifswald, D-17487 Greifswald, Germany
[2] Physikalisches Institut, Universität Bayreuth, D-95440 Bayreuth, Germany

Received 1 September 2003, accepted 3 September 2003
Published online 28 November 2003

PACS 05.60.Gg, 71.30.+h, 72.15.Rn

Based on a selfconsistent theory of localization we study the electron transport properties of a disordered system in the framework of the Anderson model on a Bethe lattice. In the calculation of the dc conductivity we separately discuss the two contributions to the current-current correlation function dominating its behaviour for small and large disorder. The resulting conductivity abruptly vanishes at a critical disorder strength marking the localization transition.

Disorder strongly affects the motion of an electron, and even may fully suppress electron transport, as first revealed by P. W. Anderson [1]. The natural quantity to characterize the effect of disorder is the electric conductivity, whose calculation for a disordered interacting system should be understood as a primary goal in the field of localization physics. As a first preparative step, we consider the dc conductivity for a single (noninteracting) electron in a crystal with compositional disorder. Our work is directed towards an approach which is applicable for all values of disorder.

The generic model in the field of localization is the Anderson model

$$H = \sum_i \epsilon_i |i\rangle\langle i| - J \sum_{\langle ij \rangle} |i\rangle\langle j| \quad . \tag{1}$$

Here $-J \sum_{\langle ij \rangle} |i\rangle\langle j|$ is a tight-binding term with hopping matrix element J on a certain lattice and $\sum_i \epsilon_i |i\rangle\langle i|$ describes random local potentials. The ϵ_i are assumed to be identically independently distributed (i.i.d.) variables with distribution $p(\epsilon_i) = (1/\gamma)\Theta(\gamma/2 - |\epsilon_i|)$. γ specifies the strength of the disorder.

For noninteracting particles at $T = 0$, the Fermi function $f(\omega)$ in the Kubo formula

$$\sigma_{\mathrm{dc}} \propto \int d\omega \left(-\frac{df}{d\omega}\right) \chi_{JJ}(\omega) \tag{2}$$

for the dc conductivity becomes a δ-function, and σ_{dc} is solely determined by the disorder averaged current-current correlation function (here, $G_{ij}(\omega)$ denotes the retarted single-particle Green function)

$$\chi_{JJ}(\omega) = \sum_{ij} \sum_{\delta\delta'} (-\delta \cdot \delta') \langle \mathrm{Im}\, G_{i,j+\delta'}(\omega)\, \mathrm{Im}\, G_{j,i+\delta}(\omega) \rangle_{\mathrm{av}} \tag{3}$$

at the Fermi energy E_F (for interacting particles, this corresponds to a rigid band approximation). Below, we assume $E_F = 0$ (half filled band).

---

[*] Corresponding author: e-mail: Andreas.Alvermann@physik.uni-greifswald.de

To avoid the need to sum over all lattice sites in eq. (3), Girvin and Jonson [2] suggested to split this expression in two parts,

$$\chi_{JJ}(\omega) = P_1(\omega)\Lambda(\omega), \tag{4}$$

with the disorder averaged pair correlation function

$$P_1 = \langle \operatorname{Im} G_{ii}(\omega) \operatorname{Im} G_{jj}(\omega) - \operatorname{Im} G_{ji}(\omega) \operatorname{Im} G_{ij}(\omega) \rangle_{\mathrm{av}} \tag{5}$$

for adjacent lattice sites i, j, and a correction $\Lambda(\omega)$ accounting for the long range correlations. Note that in the regime of localized states $\Lambda(\omega)$ has no definite value since both $P_1(\omega)$ and χ_{JJ} vanish. Close to the localization transition χ_{JJ} is essentially determined by the pair correlation function, and $\Lambda(\omega)$ is nearly constant. For small disorder, on the other hand, $\Lambda(\omega)$ diverges. To cover the full range of disorder, we must therefore compute both $P_1(\omega)$ and $\Lambda(\omega)$ to a good approximation.

To calculate the pair correlation function $P_1(\omega)$, which captures the behaviour near the localization transition, we employ the self-consistent theory of localization developed by Abou-Chacra, Anderson, Thouless (AAT) [3]. This theory is based on a renormalized perturbation expansion [4] which sets up a closed set of recursion relations for local Green functions on the Bethe lattice. These recursion relations can be interpreted as a stochastic selfconsistency equation for the Green function $G_{jj}^{(i)}(\omega)$, which corresponds to the lattice with site i removed (and appears in the second step of the renormalized perturbation expansion). Solving this stochastic equation by a Monte-Carlo procedure a sample for $G_{jj}^{(i)}(\omega)$ is constructed from which the respective distribution can be calculated. On the Bethe lattice the full Green function as well as all nondiagonal Green functions may be expressed in terms of local Green functions $G_{jj}^{(i)}(\omega)$ as

$$G_{ii}(\omega) = \frac{1}{(G_{ii}^{(j)}(\omega))^{-1} - J^2 G_{jj}^{(i)}(\omega)} \quad , \qquad G_{ij}(\omega) = J G_{ii}(\omega) G_{jj}^{(i)}(\omega) \quad , \tag{6}$$

with j nearest neighbour i. Furthermore $P_1(\omega)$ can be calculated from the distribution of $G_{jj}^{(i)}(\omega)$

$$P_1(\omega) = \left\langle \frac{\operatorname{Im} G_{ii}^{(j)}(\omega) \operatorname{Im} G_{jj}^{(i)}(\omega)}{|1 - J^2 G_{ii}^{(j)}(\omega) G_{jj}^{(i)}(\omega)|^2} \right\rangle_{\mathrm{av}} . \tag{7}$$

Note that, since the Bethe lattice has no closed loops, $G_{jj}^{(i)}(\omega)$ and $G_{ii}^{(j)}(\omega)$ are i.i.d. random variables.

In figure 1 we depict, for the Bethe lattice with connectivity $K = 2$ and for bandwidth $W = 1$, the pair correlation function $P_1(\omega)$ at the band center ($\omega = 0$). For not too large disorder, $P_1(\omega)$ is finite, but abruptly vanishes for $\gamma \gtrsim 2.9$, indicating the transition from extended to localized states. As figure 1 moreover shows the behaviour of the pair correlation strongly depends on the imaginary part η in the energy argument $\omega + i\eta$ of the Green functions. To detect the localization transition it is necessary to perform the limit $\eta \to 0$ numerically (for details on this, see [5]).

Figure 2 shows a comparison between the pair correlation and the typical density of states (typDOS)

$$N^{\mathrm{typ}}(\omega) = \exp(\langle \ln N_{jj}^{(i)}(\omega) \rangle_{\mathrm{av}}), \qquad \text{with} \quad N_{jj}(\omega) = -\frac{1}{\pi} G_{jj}^{(i)}(\omega), \tag{8}$$

which has been suggested as an order parameter for localization [6]. Remarkably enough, the pair correlation follows the behaviour of the typDOS over the full range of energy, implying a close relation between these two quantities. In this sense, the typDOS might be itself understood as a kind of transport quantity.

In contrast to $P_1(\omega)$, the correction $\Lambda(\omega)$ contains contributions from all terms in eq. (3). Since directions on the Bethe lattice are ill-defined a direct calculation of these terms within AAT is not possible. A natural suggestion to overcome this obstacle is to use the coherent potential approximation (CPA). While

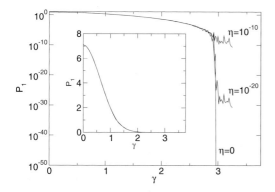

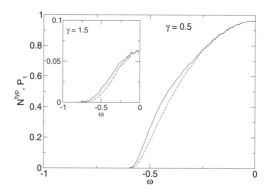

Fig. 1 Pair correlation $P_1(\omega)$ at the band center ($\omega = 0$). γ is measured in units of the bandwidth. η denotes the imaginary part of the energy argument in the Green functions.

Fig. 2 Comparison of typical DOS and pair correlations for two values of γ. The solid line shows the pair correlations $P_1(\omega)$, the dashed line the typDOS. Results are rescaled to their value at $\omega = 0$.

for small disorder the CPA surely produces a senseful result the situation for large disorder is less clear. To check whether a CPA calculation of $\Lambda(\omega)$ close to the localization transition is reasonable we have to estimate the qualitative behaviour of $\Lambda(\omega)$ within AAT. Running along a single non-retracing path in the Bethe lattice, all terms in eq. (3) along this path have the form

$$P_n(\omega) = \langle \operatorname{Im} G_{i,i+n-1}(\omega) \operatorname{Im} G_{i+1,i+n}(\omega) - \operatorname{Im} G_{i,i+n}(\omega) \operatorname{Im} G_{i+1,i+n-1}(\omega) \rangle_{\text{av}} \qquad (9)$$

similarly to the one-dimensional chain (of course, for $n = 1$ we get back the pair correlation function). On the Bethe lattice $(K + 1)K^{n-1}$ paths of length n originate from a single site, so that the magnitude of $\Lambda(\omega)$ can be estimated as [2]

$$\Lambda(\omega) \sim \sum_{n=1}^{\infty} (K + 1)K^{n-1}\frac{P_n(\omega)}{P_1(\omega)} \qquad . \qquad (10)$$

Without disorder ($\gamma = 0$) each term decays as $P_n(\omega) \sim K^{-n}$, and the sum diverges. With disorder, the $P_n(\omega)$ decay faster, resulting in a finite value for $\Lambda(\omega)$. Figure 3 demonstrates that $\Lambda(\omega)$ is non-zero and varies only slowly in the vicinity of the localization transition ($\gamma_c \approx 2.9$). Besides confirming the

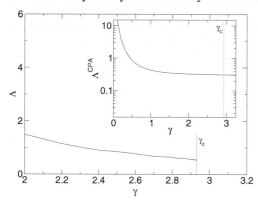

Fig. 3 $\Lambda(\omega)$-correction to the pair correlation functions at the band center ($\omega = 0$). The solid line is the AAT-estimate according to eq. (10), the dashed line shows the CPA-result. The vertical dotted lines indicate the position of the localization transition at $\gamma_c \approx 2.9$.

original expectation, this indicates the possibility to employ the CPA for the calculation of $\Lambda(z)$. From the CPA current-current correlation function $\chi_{JJ}^{\text{CPA}}(\omega)$ [7], the correction $\Lambda(\omega)$ can be approximated as $\Lambda^{\text{CPA}} = \chi_{JJ}^{\text{CPA}}/P_1^{\text{CPA}}$ (inset in figure 3). That $\Lambda^{\text{CPA}}(\omega)$ is non-zero even above the localization transition makes no problems in our approach: Since the pair correlation $P_1(\omega)$ is zero, $\chi_{JJ}(\omega)$ is as well, indicating localization of states.

Now we are in the position to put pieces together: If we calculate the pair correlation function $P_1(\omega)$ within AAT, and the correction $\Lambda(\omega)$ within CPA, the current-current correlation function $\chi_{JJ}(\omega)$ is expected to be obtained in good approximation. Although it is not necessary to split the current-current correlation function in the proposed way, it seems quite reasonable. First, we have identified the two contributions to the current-current correlation function which dominate its behaviour in the regimes of small respective large disorder. Second, since both contributions show different behaviour we can apply different methods for their calculation.

The pair correlation function $P_1(\omega)$ rules the behaviour of $\chi_{JJ}(\omega)$ for large disorder. While it is finite for all values of disorder, it is critical at the localization transition. Indeed the full effect of localization is contained in $P_1(\omega)$ but not in $\Lambda(z)$. Mean field treatments like the CPA, merely focusing on average values, are not sufficient for $P_1(\omega)$. The apparent relationship between the typical DOS and the pair correlation function makes this point especially clear: The pair correlation function does not agree with the averaged DOS ("arithmetic mean"), but with the typDOS ("geometric mean"). This failure of mean field treatments indicates the subtlety of the problem at hand. In constrast the selfconsistent theory of localization by Abou-Chacra et. al. [3] can be conveniently applied, and allows for a not too complicated calculation of the pair correlations. The correction $\Lambda(\omega)$ is in some sense opposite to the pair correlation function. While it is noncritical at the localization transition, it diverges for zero disorder. If one is only interested in the behaviour close to the localization transition one might completely forget about $\Lambda(\omega)$. But to correctly describe the regime of small disorder $\Lambda(\omega)$ has to be taken into account. Nevertheless, since $\Lambda(\omega)$ shows no critical behaviour, there is no need to go beyond a mean field treatment, e.g. provided by the CPA. As additional benefit, this avoids all possible problems arising from the lack of well defined directions on the Bethe lattice.

In conclusion, the proposed splitting of the current-current correlation function seems to be a successful first approximation. Since the pair correlation $P_1(\omega)$ is a transport quantity it is a natural localization criterion, whose calculation allows for a precise determination of mobility edges and the critical disorder for the localization transition. The relation between $P_1(\omega)$ and the typDOS shows that the latter one can indeed be used as a localization criterion as in [5, 8, 9]. The $\Lambda(\omega)$-correction, on the other hand, is of no importance to detect the localization transition, but accounts for the divergence of the current-current correlation function as $\gamma \to 0$. The final result for $\chi_{JJ}(\omega)$ correctly interpolates between the two limiting cases $\gamma \to 0$ and $\gamma \to \gamma_c$, where it matches the exact result.

Note that the AAT and the CPA can be combined to some extent with a treatment of interaction processes (statDMFA, DMFA/DCPA). Adopting the ideas presented here to these extensions might hint at some possible direction for future attempts to deal with electron transport in a disordered interacting system. Especially the selfconsistent theory of localization [3], which can be straightforwardly combined with a local treatment of interaction (statDMFA [5, 8, 9]), is a promising candidate for these studies. Work on this subject is in progress.

Acknowledgements   A.A acknowledges support from the European Graduate School, Bayreuth.

References

[1] P .W. Anderson, Phys. Rev. **109**, 1492 (1958).
[2] S. M. Girvin, and M. Jonson, Phys. Rev. B **22**, 3583 (1980).
[3] R. Abou-Chacra, P. W. Anderson, and D. J. Thouless, J. Phys. C **6**, 1734 (1973).
[4] E. N. Economou, Green's Functions in Quantum Physics, Springer series in solid-state sciences, Vol. 7 (Springer,Berlin, 1983).
[5] F. X. Bronold, A. Alvermann, and H. Fehske, to be published

     A. Alvermann, diploma thesis (in german), Universität Bayreuth (2003)
[6] V. Dobrosavljević, A. A. Pastor, and B. K. Nikolić, Europhys. Lett. **62**, 76 (2003).
[7] B. Velićky, Phys. Rev. **184**, 614 (1969).
[8] V. Dobrosavljević, and G. Kotliar, Phys. Rev. Lett. **78**, 3943 (1997).
[9] F. X. Bronold, and H. Fehske, Phys. Rev. B **66**, 073102 (2002).

phys. stat. sol (c) **1**, No. 1, 67–70 (2004) / **DOI** 10.1002/pssc.200303643

Conductivity of weakly and strongly localized electrons in a *n*-type Si/SiGe heterostructure

I. Shlimak[*, 1], **V. Ginodman**[1], **M. Levin**[1], **M. Potemski**[2], **D. K. Maude**[2], **A. Gerber**[3], **A. Milner**[3], and **D. J. Paul**[4]

[1] Minerva Center and Jack and Pearl Resnick Institute of Advanced Technology,
 Department of Physics, Bar-Ilan University, Ramat-Gan 52900, Israel
[2] High Magnetic Field Laboratory, Max-Planck-Institut für Festkörperforschung/CNRS,
 38042 Grenoble Cedex 9, France
[3] School of Physics & Astronomy, Raymond & Beverly Sackler Faculty of Exact Sciences,
 Tel Aviv University, 69978 Tel Aviv, Israel
[4] Cavendish Laboratory, University of Cambridge, Madingley Road, Cambridge CB3 0HE, UK

Received 1 September 2003, accepted 3 September 2003
Published online 28 November 2003

PACS 73.20 Fz, 73.40.Lq, 73.43.–f

We have investigated the temperature dependence of the longitudinal conductivity σ_{xx} and Hall resistance R_{xy} of *n*-type Si/SiGe heterostructures in the quantum Hall effect regime in magnetic fields up to 23 T. It is shown that for odd integer filling factors $i = 3,5,7,9$, when the Fermi level E_F is situated between the valley-split Landau levels, $\Delta\sigma_{xx}(T) \propto \ln T$, which is typical for weakly localized electrons. In the case of even i, when E_F lies between spin-split or cyclotron-split levels, $\sigma_{xx}(T)$ is characteristic of strong localization: activation of localized electrons from E_F to the nearest mobility edge: $\sigma_{xx} \propto \exp[-\Delta_i/T]$ for $i = 6, 10, 12$ or variable-range-hopping via localized states in the vicinity of E_F: $\sigma_{xx} \propto \exp[-(T_{0i}/T)]^{1/2}$ for $i = 4, 8$. For $i = 3, 6, 8, 10, 12$, the Hall resistance R_{xy} first overshoots the quantized plateau values h/ie^2 and then returns. The explanatory model involves the temporary parallel contribution of the delocalized and weakly localized electrons with different mobilities in the measurement of the Hall voltage V_{xy}.

1 Introduction

The measurement of the temperature dependence of the 2D conductivity $\sigma(T)$ in the quantum Hall effect regime is a very useful tool for the analysis of the density of states (DOS) of carriers at different filling factors ν. At integer filling factors, $\nu = i$, the Fermi level E_F lies in the middle of two Landau levels (LL) where the DOS is minimal and electron states are localized. In this case, the character of the longitudinal conductivity $\sigma_{xx}(T)$ is determined by the ratio of the energy distance between the two adjacent LLs E_i and the measured temperature interval. If $E_i \leq T$, one expects a weak non-exponential dependence for $\sigma_{xx}(T)$, while for $E_i \gg T$, the conductivity is strongly temperature-activated (see, for example, [1-3] and references therein):

$$\sigma_{xx}(T) = \sigma_0 \exp(-\Delta_i/T) \qquad (1)$$

Here, Δ_i is the energy of activation and $2\Delta_i$ reflects the mobility gap, which is less than E_i because of the non-zero width of the band of delocalized states in the center of each LL, the prefactor σ_0 is equal to $2e^2/h$, as predicted in [1]. For large $\Delta_i \gg T$, direct excitation of electrons to the mobility edge is unlikely and the preferable mechanism of conductivity is variable-range-hopping (VRH) via localized states in the vicinity of E_F [4–6]:

$$\sigma_{xx}(T) \propto \exp(-T_0/T)^m, \qquad (2)$$

[*] Corresponding author : e-mail: shlimai@mail.biu.ac.il, Phone: +972-3-531 81 76, Fax: +972-3-535 76 78

where $m = 1/2$ because of the existence of a Coulomb gap in the DOS at E_F [7, 8]. The parameter T_0 is connected with the localization radius $\xi(\nu)$ of the states for given ν: $T_0 = C_1 e^2/\kappa\xi(\nu)$. Here $C_1 \approx 6$ for two dimensions and κ is the dielectric constant of the host semiconductor.

Most previous measurements of $\sigma_{xx}(T)$ were performed on GaAs/AlGaAs heterostructures. The special feature of n- type Si/SiGe heterostructures lies in an additional splitting of the energy levels due to the lifting of two-fold valley degeneracy in a strong perpendicular magnetic field. As a result, in n-Si/SiGe heterostructures, odd ν corresponds to the location of E_F between valley-split LLs.

2 Experiment   The investigated sample was a Hall-bar patterned n-Si/Si$_{0.7}$Ge$_{0.3}$ double heterostructure, the electron concentration n and mobility μ at 1.5 K were $n = 9\cdot10^{15}$ m^{-2}, $\mu = 8$ m^2/V·s. The transverse R_{xy} and longitudinal R_{xx} resistances were measured using a standard lock-in technique, with the measuring current 20 nA at a frequency of 10.6 Hz. The general dependence in fields up to 23 T is presented in [9]. The plateau in R_{xy} were clearly observed at a portion $1/i$ of a quantized resistance $h/e^2 = 25.8$ kΩ. For ν approaching the plateau at $i = 3, 6, 8, 10, 12$, R_{xy} first overshoots the quantized plateau values h/ie^2 and then returns (Fig. 1a). When the filling factor ν reaches an integer value $\nu = i$, R_{xx} exhibits a deep minimum for even i (Fig. 1b).

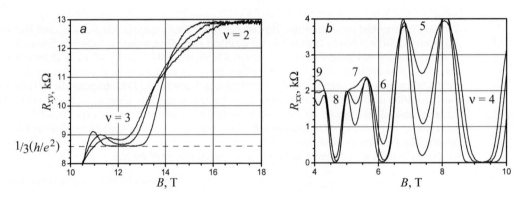

Fig. 1   Dependence $R_{xy}(B)$ (a) and $R_{xx}(B)$ (b) at $T = 1.2$, 0.6 and 0.2 K

3 Results and discussion   Let us first discuss the longitudinal conductivity $\sigma_{xx} = \rho_{xx}/(\rho^2_{xx} + \rho^2_{xy})$.

Odd integers ($i = 3,5,7,9$). In the case of a n-Si/SiGe heterostructure, odd filling factors correspond to the location of E_F between the valley-split LLs. It was shown in [10] that valley splitting could be observed only in the presence of a magnetic field normal to the interface and is given approximately by

$$\varepsilon_v \text{ [K]} \approx 0.174(N + 1/2) B \text{ [T]} . \tag{3}$$

Here, the valley-splitting energy ε_v is measured in Kelvin, the magnetic field B in Tesla, $N = 0,1,2...$ is the Landau index. Numerical estimation showed that the value of ε_v for magnetic fields 4T < B < 12T is about 1 K. Therefore, one cannot expect an activated character of $\sigma_{xx}(T)$ within the experimental temperature interval ($T = 0.2$-4.2 K). The analysis shows that $\Delta\sigma_{xx}$ depends weakly on T, with the best fit of experimental data being achieved by the logarithmic law: $\Delta\sigma_{xx}(T) \propto \ln T$ (Fig. 2a).

A logarithmic temperature dependence is usually interpreted as a manifestation of corrections to the conductance due to the quantum interference effects [11]. In strong perpendicular magnetic fields, weak-localization corrections to the conductivity are suppressed and $\Delta\sigma_{xx}$ is determined by quantum corrections due to the electron-electron interaction (see, for example, [12] and references therein). This leads to the following expression for $\Delta\sigma(T)$ [13]:

$$\Delta\sigma_{ee}(T) = [(\alpha p e^2)/(2\pi h)] \ln (T/T_{ee}) \tag{4}$$

where α is a constant of order unity and p is the exponent in the temperature dependence of the phase-breaking time $\tau_\varphi \propto T^p$. At low T, the phase is broken usually by the electron-electron interaction, leading to $p \approx 1$. Analysis shows that for $i = 5, 7, 9$, α is indeed of order unity, except for $i = 3$, where $\alpha \approx 0.4$. Thus, for odd integers $\sigma_{xx}(T)$ can be successfully described in terms of quantum corrections to the conductivity in strong magnetic fields caused by the electron-electron interaction.

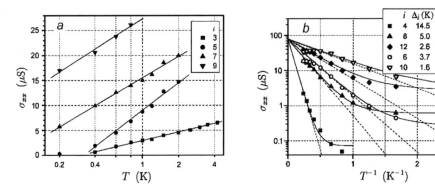

Fig. 2 (a) σ_{xx} for odd integers as a function of $\ln T$; (b) $\ln\sigma_{xx}$ vs. $1/T$ for even integers. The insert shows the experimental values Δ_i and calculated values E_i.

Even integers ($i = 4, 6, 8, 10, 12$). For even integers, there are two possibilities for the location of E_F: between cyclotron LLs ($i = 4,8,12$) and between spin-split levels ($i = 6,10$). Taking into account that for the strained Si valley, $m^* = 0.195m_0$, [13], the cyclotron energy is given by

$$\hbar\omega_C\,[K] = 6.86\,B\,[T]\,. \tag{5}$$

The energy of spin splitting $g^*\mu_B B$ depends on the effective g-factor g^*. The value of g^* increases for lower ν, oscillating between 2.6 and 4.2 [14]. For numerical estimates, we assume $g^* \approx 3.8$, giving

$$g^*\mu_B B\,[K\,] \approx 2.55\,B\,[T]\,. \tag{6}$$

In the calculation of E_i, all relevant splitting energies are taken into account. Because all $E_i \gg T$ (see insert in Fig. 2b), it is expected that $\sigma_{xx}(T)$ will be determined by the temperature-activated excitation of localized electrons to the mobility edge and characterized, therefore, by the constant energy of activation $\Delta_i < 1/2E_i$, Eq. (1). One can see from Fig. 2b that at high temperatures, the experimental points are in agreement with Eq. (1), which allows one to determine the values of Δ_i. The prefactor σ_0 for all curves is close to 78 μS $\approx 2e^2/h$, in accordance with the theoretical prediction [1].

However, it is seen, that with decrease of temperature, all dependences tend towards the residual conductivity $\sigma_i(0)$, which is almost temperature-independent. The possible origin of $\sigma_i(0)$ is discussed in [9].

It follows from Fig. 2b that the low-temperature experimental points for i = 4 and 8 do not fit well to the calculated curves. This can be explained by the fact that the values of Δi for i = 4 and 8 are substantially larger than for i = 6, 10, 12. As a result, direct thermal excitation of localized electrons to the mobility edge is unlikely and it is more probable that electron transport is due to the variable-range-hopping (VRH) conductivity via localized states in the vicinity of EF, Eq. (2). To check this assumption, we plot $\sigma xx(T)$ for i = 4 and 8 in the VRH scale: $\ln\sigma xx$ vs. $T^{-1/2}$. On this scale, all experimental points coincide with the calculated curves $\sigma xx(T) = \sigma 0 \exp(-Ti0/T)^{1/2} + \sigma i(0)$, where $\sigma 0$ and $Ti0$ are determined from experiment and $\sigma i(0)$ is the only adjustable parameter.

Overshoot of quantum Hall plateaus. We also investigate anomalous "hills" in the transverse Hall resistance R_{xy} in the quantum Hall effect regime, when R_{xy} overshoots the normal plateau value $(1/i)h/e^2$. However, if the magnetic field increases further, R_{xy} decreases to its normal value (Fig. 1a). This effect has been observed in different heterostructures: GaAs/Al$_{1-x}$Ga$_x$As [15, 16] and Si/Si$_{1-x}$Ge$_x$ of n- and p-type conductivity [17–19]. This means that the physical origin of this effect must have a generic character. We show that in our sample: (i) overshoot is observed for plateaus at all ν, except for $\nu = 2$ and $\nu = 4$, (ii) this effect is stationary: the amplitude of overshoot is strongly reproducible as magnetic field is turned up and down (Fig. 3); (iii) the relative amplitude of the overshoot $\Delta R_{xy} / [R_{xy}(\nu+1) - R_{xy}(\nu)]$ is about 10% for all ν. We assume that the overshoot can be explained by the model of parallel contributions to the Hall voltage of two electron bands with different mobilities: electrons from delocalized states in the center of LL and electrons from weakly localized states near the mobility edge. Two types of electrons may appear temporary when the Fermi level crosses the boundary between delocalized and localized states within LL. Existence of the weakly localized electrons depends on the shape of the LL which is crossed by the Fermi level at given magnetic field B. This can explain why the overshoot is not observed at all filling factors ν.

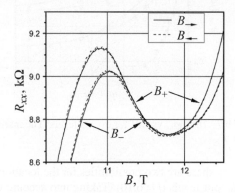

Fig. 3   The overshoot near the $\nu = 3$ measured at $T = 0.3$ K with two different directions of the magnetic field B and two different signs of dB/dt.

Acknowledgements   We thank B. I. Shklovskii for discussions, A. Belostotsky for help with the data analysis, and the Eric and Sheila Samson Chair of Semiconductor Technology for support. V. G. and M. L. thank the "KAMEA" Programme for financial support.

References

[1] D. G. Polyakov and B. I. Shklovskii, Phys. Rev. Lett. **74**, 150 (1995).
[2] M. M. Fogler, D. G. Polyakov, and B. I. Shklovskii, Surf. Sci. **361/362**, 255 (1996).
[3] M. Furlan, Physica B **249–251**, 123 (1998).
[4] D. G. Polyakov and B. I. Shklovskii, Phys. Rev. Lett. **70**, 3796 (1993)..
[5] F. Hohls, U. Zeitler, and R. J. Haug, Phys. Rev. Lett. **88**, 036802 (2002).
[6] D.-H. Shin et al., Semicond. Sci. Technol. **14**, 762 (1999).
[7] A. L. Efros and B. I. Shklovskii, J. Phys. C **8**, L49 (1975).
[8] B. I. Shklovskii and A. L. Efros, Electronic Properties of Doped Semiconductors (Springer, Berlin, 1984).
[9] I. Shlimak et al., preprint : cond-mat/0307695, to appear in Phys. Rev. B (2003).
[10] F. J. Ohkawa and Y. Uemura, J. Phys. Soc. Jpn. **43**, 917 (1977).
[11] B. L. Altshuler and A. G. Aronov, in Electron-Electron Interactions in Disordered Systems, edited by A.L. Efros and M. Pollak (North-Holland, Amsterdam, 1987).
[12] S. S. Murzin, JETP Lett. **67**, 216 (1998).
[13] S. Hikami, Phys. Rev. B **24**, 2671 (1981).
[14] S. J. Koester, K. Ismail, and J. O. Chu, Semicond. Sci. Technol. **12**, 384 (1997).
[15] C.A. Richter, R.G. Wheeler, R.N. Sacks, Surf. Sci. **263**, 270 (1992).
[16] H. Nii et al., Surf. Sci. **263**, 275 (1992).
[17] P. Weitz, R.J. Haug, K. von. Klitzing, F. Schäffler, Surf. Sci. **361/362**, 542 (1996).
[18] P.T. Coleridge, A.S. Sachrajda, H. Lafontaine, and Y. Feng, Phys. Rev. B **54**, 14518 (1996).
[19] R.B. Dunford et al., J. Phys.: Condens. Matter **9**, 1565 (1997).

phys. stat. sol. (c) **1**, No. 1, 71–74 (2004) / **DOI** 10.1002/pssc.200303647

Antiferromagnetic spin glass ordering in the compensated n-Ge near the insulator–metal transition

A. I. Veinger[*], **A. G. Zabrodskii**, **T. V. Tisnek**, and **S. I. Goloshchapov**

Ioffe Physico-Technical Institute, Russian Academy of Sciences, St. Petersburg 194021, Russia

Received 1 September 2003, revised 11 September 2003, accepted 18 September 2003
Published online 1 December 2003

PACS 71.30.+h, 76.30.Pk

The electron spin resonance investigation in the insulator phase of compensated n-Ge : As shows that near the insulator–metal transition the weakly localized electrons are connected in the pairs with the antiparallel coupled spins. As a result, the antiferromagnetic spin glass structure is produced. The main manifestations of the phenomenon are the fall of the (free) spin density and a g-factor shift to the strong magnetic field with the electron concentration growth in the pretransition range.

A widespread opinion that existed up to recent times was that the insulator to metal (IM) transition in doped semiconductors did not result necessarily in the spin coupling and thus in the essential change of the electron spin resonance (ESR) spectra. This opinion was based mainly on the Si : P system investigation (see for instance [1]). On the contrary, our recent studies of the ESR in compensated Ge : As [2] and in the 4H-SiC : N [3] showed the essential change of the ESR spectra near the IM transition caused by magnetic ordering. Here we discuss more in depth investigations into ESR spectra transformation resulting from the localized spin interaction in compensated n-Ge : As mainly in the pretransition range.

A suitable samples series was produced by the introduction of the compensating Ga impurity by neutron transmutation doping of the initially noncompensated in the n-Ge : As crystals. The thermal annealing of the attendant defects was executed at the temperature $T = 450$ °C in 60 h. As a result, the samples produced covered the compensation range $K = 0$–0.8 and the electron concentration range $n = 3.58 \times 10^{17}$–9.8×10^{16} cm^{-3} corresponding to $(0.98$–$0.26)$ n_C where the critical density $n_C = 3.7 \times 10^{17}$ cm^{-3} for $K = 0$. The ESR spectra of the noncompensated sample series were investigated for the clearing out of the compensation role.

ESR was observed in the insulator phase at the temperatures of the donor freezing out of the electron and in the hopping conductivity range. Figure 1 shows the qualitative changes of the ESR spectra near the IM transition in some noncompensated samples. One can see that the hyperfine structure disappears when the concentration grows from 2×10^{16} cm^{-3} to 1×10^{17} cm^{-3}. The new distortion of the ESR spectrum is observed near the IM transition at the concentration 3.58×10^{17} cm^{-3}: the spectrum symmetry here has the valley symmetry in Ge. The ESR spectrum in the metallic sample ($n = 5.75 \times 10^{17}$ cm^{-3}) has only one line. With the concentration growth, the g-factor of the line is decreased from 1.58 to 1.56. The same displacement of the g-factor is observed in the compensated samples (Fig. 2). The anisotropy disappears with the compensation growth in these samples and almost all of them have isotropic spectra. Only the line width depends on the magnetic field direction. The comparison of the g-factor shift for noncompensated and compensated samples shows that the compensation has little influence on the g-factor shift and it is determined mainly by the electron concentration.

---

[*] Corresponding author: e-mail: anatoly.veinger@pop.ioffe.rssi.ru, Phone: +7 812 247 91 52, Fax: +7 812 247 10 17

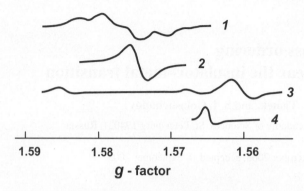

Fig. 1 ESR spectra of n-Ge:As in the insulator state (1–3) and in the metallic state (4); concentration n, 10^{17} cm^{-3}: 1–0.2, 2–1, 3–3.58, 4–5.75; for 3 the spectrum is shown when $H \parallel [110]$; other spectra are isotropic; $T = 3.2$ K.

The main parameters of the ESR spectra, which manifest antiferromagnetic order near the IM transition, are represented below in detail.

The first parameter is the "antiferromagnetic" g-factor shift. It is known that in the ferromagnet the resonance line displaces in the weak field because the internal magnetic field H_{in} adds to the external one H_{out} and the last decreases (g-factor grows) so conserving the sum:

$$H_{in} + H_{out} = H_{res} . \tag{1}$$

The corresponding H_{res} increases in our case. This shift can be explained by the following. The alternating magnetic field exists in the antiferromagnet due to the magnetic bond between the antiparallel and separated in the space spins. If the free spin moves in that space, it crosses the field and any adding frequency Δf influences this spin. When the velocity of the spin is v, the frequency $\Delta f = v/r_0$, where r_0 is the mean separated spin distance. Then the resonance condition can be written:

$$h(f + \Delta f) = g\beta(H + \Delta H) , \tag{2}$$

where ΔH is the adding magnetic field that corresponds the adding frequency.

Thus the presence of the antiferromagnetic pairs in the semiconductor can be the origin of the resonance field shift in the strong magnetic fields.

One could suggest that the explanation of the observed "antiferromagnetic" g-factor shift of the resonance line is the small change of the electron effective mass near the IM transition. It is known that the g-factor and the electron effective mass are connected by the relations [4]:

$$g_{\parallel}-2 = -(m_0/m_\perp-1)\delta/E_{13} ; \qquad g_\perp-2 = -(m_0/m_\parallel)\delta/E_{13} , \tag{3}$$

where δ is the parameter of spin-orbit splitting, E_{13} is the forbidden gap at $k = 0$.

It is seen from Eq. (4) that if $m_\parallel$ or $m_\perp < m_0$ and falls with the concentration growth then the g-factor falls too and the resonance magnetic field increases. The early experiments on Si:P [5] confirm the last origin, but the g-factor shift in Ge can be an "antiferromagnetic" one.

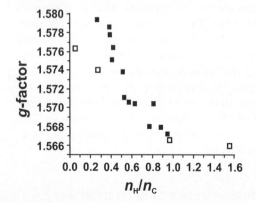

Fig. 2 g-factor for the n-Ge:As versus the concentration $n_H = N_D - N_A$ (the concentration of occupied impurity centers in the n_C units); open squares are the noncompensated samples; $T = 3.2$ K.

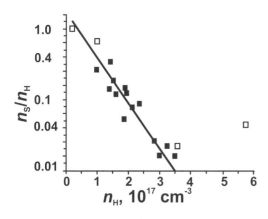

Fig. 3 Spin density for Ge:As samples in the pretransition range: open symbols correspond to the noncompensated samples and filled symbols to the compensated ones; $T = 3.2$ K.

The characteristic dependence of the spin density on the impurity concentration, shown in Fig. 3, indicates unambiguously that, in the insulating state in the vicinity of the IM phase transition, coupling of spins with the formation of an antiferromagnetic system of the spin-glass type occurs at low temperatures.

This coupling is characterized by the magnitude of the exchange interaction, which depends exponentially on the distance between spins in a pair. Evidently, with increasing concentration of a randomly distributed impurity, the spin coupling begins at pairs in which the spins are located closest together. If we assume that pairs are formed by centers the distance between which is smaller than some value r_0 and are occupied with electrons, we can estimate the density of such pairs by recognizing that the probability W of the nearest neighbor being located at a distance from r to $r + \delta r$ is determined by the expression

$$W = \exp\left(-4\pi r^3 n_H/3\right) 4\pi r^2 n_H \, dr, \tag{4}$$

where $n_H = N_D - N_A$ is the concentration of occupied impurity centers.

Then, the relative concentration of impurity centers forming coupled pairs, the distance between the spins of which is smaller than r_0, is determined by the integral

$$n/n_H = \int_0^{r_0} \exp\left(-(4/3)\pi r^3 n_H\right) \times 4\pi r^2 r_H \times dr = 1 - \exp\left(-4\pi r_0 n_H/3\right). \tag{5}$$

The relative density of free spins involved in resonance absorption exponentially decreases with increasing impurity concentration:

$$n_s/n_H = \exp\left(-4\pi r_0^3 n_H/3\right). \tag{6}$$

It can be seen from Eq. (6) that the dependence $\log\left(n_s/n_H\right) = f(n_H)$ should be linear. Figure 3 shows the corresponding portion of the concentration dependence of n_s. This confirms the statistical nature of the coupling mechanism considered above with $r_0^3 = 0.64 n_C^{-1}$.

In the vicinity of the IM transition, the characteristic distance r_0, which governs the spin interaction, would be modified due on the localization radius $a > r_0$.

$$a = a^*(1 - n_H/n_C)^{-\nu}, \tag{7}$$

where the typical values of the index ν are equal to 0.5–1 [6].

Then Eq. (6) can be rewritten in the form:

$$n_s/n_H = \exp\left[-(4/3)\pi a^{*3} n_H/(1 - n_H/n_C)^{3\nu}\right]. \tag{8}$$

The estimate of the parameters a^* and ν from the fit to the experiment gives: $a^{*3} = 0.37 n_C^{-1}$, $3\nu = 1.5$.

The new phenomenon that is revealed in the insulator state near the IM transition is the distortion of the ESR spectrum. It is well known that Ge what is the cubic crystal has the isotropic one. The lower

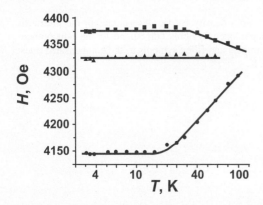

Fig. 4 Temperature dependency of the resonance fields for three lines of the noncompensated sample with concentration $n = 3.58 \times 10^{17}$ cm^{-3}; $H \parallel [110]$.

symmetry is observed under the pressure. Then it repeats the valley symmetry corresponding to the relations in Eq. (3) with $g_{\parallel} \approx 0.9$ and $g_{\perp} \approx 2$. In our case, the distortion is smaller with $g_{\parallel} \approx 1.68$ and $g_{\perp} \approx 1.50$ and the isotropic spectra are observed in the compensated samples with the $n < 3 \times 10^{17}$ cm^{-3}. Earlier we saw the large distortion in the compensated sample with $N_{\mathrm{D}} – N_{\mathrm{A}} = 3.85 \times 10^{17}$ cm^{-3}, that is almost exactly at the transition point [2]. However, that spectrum was observed at the temperature lower than 4 K.

In this sample series, anisotropic ESR spectra are observed in the wider temperature range. Figure 4 shows the temperature dependency of the resonance fields for the three lines that are indicated when the magnetic field has the [110] direction. It is seen that at low temperatures (<20 K) these fields are not dependent on the temperature In the higher-temperature range, the difference between the resonance fields decreases and they are mixed at the $T \approx 140$ K. This point is obtained by extrapolation of the temperature dependencies because the resonance lines are not indicated at this temperature.

The anisotropic ESR spectra appearance in the insulator state near the IM transition is the strangest effect from the observed ones in Ge : As. The origin of this remains unknown. This phenomenon needs more in-depth research.

Hence, dramatic peculiarities are found in the insulator state near the IM transition. Some are directly connected with the antiferromagnetic ordering in the spin subsystem and the others are the more complex result of this effect. The direct evidence of the antiferromagnetic spin glass ordering is the reduction of the free spins density in these conditions. As the connection of the g-factor shift as the distortion ESR spectra with the antiferromagnetic ordering are not so obvious and require more investigations.

Acknowledgement The work was supported by the Russian Foundation for Basic Research, grant 01-02-17813 and the President grant No. 2223.2003.02.

References

[1] N. F. Mott, Metal–Insulator Transitions (Tailor & Francis Ltd., London, 1974).
[2] A. I. Veinger, A. G. Zabrodskii, and T. V. Tisnek, Semicond. **34**, 45 (2000).
[3] A. I. Veinger, A. G. Zabrodskii, T. V. Tisnek, and E. N. Mokhov, Semicond. **37**, 846 (2003).
[4] L. M. Roth, Phys. Rev. **118**, 1534 (1960).
[5] K. Morigaki and S. Maekawa, J. Phys. Soc. Jpn. **32**, 462 (1972).
[6] A. G. Zabrodskii and K. N. Zinov'eva, Zh. Eksper. Teor. Fiz. **86**, 727 (1984).

phys. stat. sol. (c) **1**, No. 1, 75–78 (2004) / **DOI** 10.1002/pssc.200303602

Influence of the magnetic field and uniaxial pressure on the current instability in the hopping conductivity region

D. I. Aladashvili[*] and **Z. A. Adamia**

Department of Physics, Tbilisi State University, av. Chavchavadze 1, Tbilisi 0128, Georgia

Received 24 September 2003, accepted 2 October 2003
Published online 10 December 2003

PACS 72.20.Ee

In this specific work we present results of the investigation of the current oscillations in the hopping conductivity region at magnetic field and aniaxial stress, where the dependences of threshold field beginning of oscillations on temperature and on magnetic field and uniaxial pressure are determined . The model to explain the experimental results is presented.

1 Introduction   One of the interesting phenomena observed in the hopping conductivity region at strong electric fields is negative differential resistance [1, 2], which was predicted earlier by Nguen and Shklovskii [3]. The low frequency current oscillations observed in the hopping conductivity range in weakly compensated semiconductors are associated with "hopping" domain periodic formation and traveling in the bulk of the sample [4]. It was shown that the hopping domain is of triangular shape parallel to the external field, with equal leading and back edges and estimated velocity of the domain motion $c = 2.5 \times 10^{-3}$ m s^{-1} [5, 6].

In the present work the influence of uniaxial stress and magnetic field on the current instability in the hopping conductivity range has been studied.

2 Experimental details   Measurements were made on samples of lightly doped and very weakly compensated silicon with boron acceptors with the impurity concentration between 5×10^{22} and 8×10^{22} m^{-3} and the compensation ratio $K = (0.5$–$5) \times 10^{-4}$.

Samples were oriented along the [001] crystallographic axes . Some characteristic of the samples used in the experiment are listed in Table 1.

The samples under study were placed into a special holder which made it possible to change and measure the applied deformation. In the course of investigation the varying potential distribution along

Table 1

sample	basic impurity	type	concentration of basic impurity (m^{-3})	approximate compensation degree
A	B	p	5.9×10^{22}	4×10^{-5}
B	B	p	7.3×10^{22}	2×10^{-4}
D	B	p	7.0×10^{22}	2×10^{-5}
F	B	p	$6,0 \times 10^{22}$	8×10^{-4}
G	B	p	5.6×10^{22}	4×10^{-4}

[*] Corresponding author: e-mail: daladashvili@osgf.ge, Fax: 995 32 912849

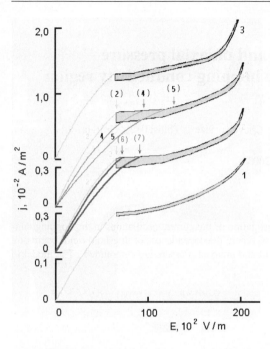

Fig. 1 Behaviour of the current density as E increases slowly with time for sample B:
(1) $T = 8.1$ K, $X = 0$, $B = 0$; (2) $T = 10.1$ K, $X = 0$, $B = 0$;
(3) $T = 11.8$ K, $X = 0$, $B = 0$;
(4) $T = 10.1$ K, $X = 10$ MPa, $B = 0$; (5) $T = 10,1$ K, $X = 16$ MPa, $B = 0$; (6) $T = 10,1$ K, $X = 0$, $B = 1$ T; (7) $T = 10.1$ K, $X = 0$, $B = 3.5$ T.

the sample was measured. This enabled one to control the regions of formation and travel of electric hopping domains. To restrict the travel range the samples had a form of a flat dumb-bell with smooth indium contacts at the edges. The pressure was applied to the crystal edges oriented along the [001] crystallographic axes. Voltage square pulses with duration somewhat higher than the domain travel time and with the E quantity sufficient for current oscillations to appear were applied to the samples. The shape of a periodically appearing current pulse was observed on the oscillograph and recorded on photographic paper.

In the course of the experiment the compression and the magnetic field were directed both along and across the [001] crystallographic axes.

3 Results and discussion   Figure 1 shows the dependence of the current density J on the electric field applied to sample B at different temperatures (curves 1, 2, 3). On the curves first the narrow Ohmic conductivity portion was observed at weak electric fields followed by the sublinear portion up to the onset of current oscillations. Then with increasing voltage (with scan velocity 1 V/s) the instability region is observed which passes into the impurity breakdown at higher electric fields.

As we can see from the figure the threshold field E_{th} for the onset of oscillations does not depend on temperature.

The theoretical threshold field in the hopping conductivity full saturation range is defined by the following formula [1]:

$$E_{th} = 2kT/eL, \tag{1}$$

where L is the characteristic length of the infinite cluster network [7].

As we see from equation (1) E_{th} must depend from temperature, as long as the mobility of electrons changes with temperatures. Here the number of mobile electrons was assumed constant at the percolation level. (Mobile electrons are electrons contributing to the hopping conductivity [2].)

The independence of E_{th} of temperature can be explained as follows: full saturation range in the temperature dependence of electric resistivity is not observed at the experiment, since at high temperatures the carrier transfer from impurity levels to the band becomes pronounced, while at low temperatures the "saturation portion" is limited by the Poole–Frenkel effect [8]. Therefore the number of mobile electrons on the percolation level changes both with temperature and electric field.

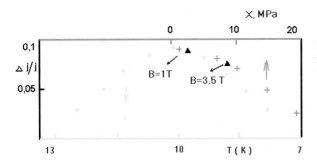

Fig. 2 Relative amplitude of the current oscillations near the threshold as functions of the temperature (•), stress (+) and magnetic field (▲) for sample 2.

For further investigation of this question we studied the influence of the elastic deformation and magnetic field on the threshold field for the onset of the oscillations.

In Fig. 1 the J vs. E dependences at different pressures and magnetic fields till the onset of oscillations are shown.

We can see the shift of threshold field of the beginning of the oscillations to the strong electric field side with application both of uniaxial stress X (curves 2, 4, 5) and magnetic field B (curves 2, 6, 7). Quantity of J_{max} of oscillations does not change with stress and magnetic field.

In Fig. 2 the dependences of the relative amplitude of current oscillations on temperature and on stress and magnetic field at T = const are given. The curve $\Delta J/J$ (T) has maximum near 10 K. The relative amplitude decreases with both an increase and a decrease in temperature. On the high temperature side it is caused by band conduction and on the low temperature side – by Poole–Frenkel effect, that is by excitation carriers at the percolation level. The relative amplitude decreases both with stress and magnetic field.

It is well known that when both the anisotropic deformation and magnetic field are applied in the pressure and magnetic fields range under investigation, the activation energy of hopping conductivity increases. (This happens due to acceptor state splitting which results to an increase in the energy required for part of the carriers to leap to the neighbouring impurity [9–11] , thus causing an increase in the activation energy of hopping conductivity. We neglect the change in the Bohr orbit in the pressure and magnetic field range under investigation.) Thus the range of Poole–Frenkel effect increases. We consider that the carrier mobility is constant. An increase in the hopping conductivity activation energy with uniaxial stress and magnetic field causes a decrease in the mobile electrons concentration at percolation level. Thus both with temperature decrease in hopping saturation region and application of pressure or magnetic field (at T = const) the concentration of mobile electrons decreases. However, when we change temperature, along with the concentration the carrier mobility also changes. Therefore the threshold field is constant with temperature.

But when we apply pressure or magnetic field (at T = const) only the concentration changes thus causing an increase in the threshold field at the onset of oscillations.

The dependences of the threshold field for the beginning of oscillations on pressure and magnetic field are given in Fig. 3.

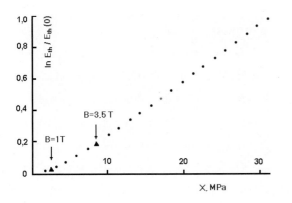

Fig. 3 Dependence of threshold field on stress (•) and magnetic field (▲) for sample 2. T = 10.1 K.

References

[1] D. I. Aladashvili, Z. A. Adamia, K. G. Lavdovskii, E. I. Levin, and B. I. Shklovskii, JETP Lett. **47**, 390 (1988).
[2] D. I. Aladashvili, Z. A. Adamia, K. G. Lavdovskii, E. I. Levin, and B. I. Shklovskii, Soviet Phys.–Semicond. **24**, 143 (1990).
[3] V. I. Nguyen and B. I. Shklovskii, Solid State Commun. **38**, 99 (1981).
[4] D. I. Aladashvili, Z. A. Adamia, and E. L. Tzakadze, Solid State Commun. **101**, 183 (1997).
[5] D. I. Aladashvili, Z. A. Adamia, N. V Gogoladze, and E. L. Tzakadze, phys. stat. sol. (b) **218**, 39 (2000).
[6] D. I. Aladashvili and Z. A. Adamia, Philos. Mag. B **81**, 1031 (2001).
[7] B. I. Shklovskii and A. L. Efros, Electronic Properties of Doped Semicond (Berlin, Springer, 1984).
[8] D. I. Aladashvili, Z. A. Adamia, K. G. Lavdovskii, E. I. Levin, and B. I. Shklovskii, Soviet Phys.–Semicond. **23**, 213 (1989).
[9] J. A. Chrobochek, F. H. Pollak, and H. F. Staunton. Philos. Mag. B **50**, 113 (1984).
[10] R. Buchko, P. Janiszewski, M. Stohr, J. A. Chroboczek, Phil. Mag. B **81**, 965 (2001).
[11] D. I. Aladashvili, Z. A. Adamia, K. G. Lavdovskii, and H. Fritzsche, Advances in Disordered Semicond., Vol. 3, Transport, Correlation and Structural Defects, edited by H. Fritzsche (World Scientific Publ. Co., Singapore 1990), p. 253.

phys. stat. sol. (c) **1**, No. 1, 79–83 (2004) / **DOI** 10.1002/pssc.200303608

Field dependence conductivity in disordered systems

D. Bourbie[*,1], **N. Ikrelef**[1], and **P. Nedellec**[2]

[1] Laboratoire des Sciences de la Matière Condensée, Université d'Oran, 31100 Es-sénia, Algeria
[2] Université d'Evry Val d'Essonne, Bd F. Mitterant, 91025 Evry Cedex, France

Received 1 September 2003, revised 20 October 2003, accepted 30 October 2003
Published online 10 December 2003

PACS 72.10 Bg, 72.20 Ee, 72.80 Le, 72.80 Ng

We present a model to describe the field dependencies of the hopping conductivity in disordered systems. In literature, the universal electric field dependence observed in the organic disordered systems has been interpreted by three models: Gaussian disorder model (GDM) which neglects the charge–dipole interaction, the correlated disorder model (CDM) valid in disordered materials with permanent dipole moments, and Geometry fluctuation model (GFM) recently proposed for conjugated polymers without permanent dipole moments. In our model, we show that the Poole Frenkel behaviour is due to the effect of the intensity of the electric field on the dimension of the transport path, that becomes one-dimensional in the high-fields region, or to the effect of the field on the tunnelling probability between the localized states. For the field dependence of the conductivity we find, $\ln \sigma \propto E^{\gamma}$, with $\gamma = 2$ in the low field region, $\gamma = 1$ in the moderate field region and $\gamma = 1/2$ in the high field region.

1 Introduction Transport phenomena in disordered systems under the influence of high electric field has recently become the object of intensive experimental and theoretical study. This was implied by observations of strong non-linearities in the field dependences of the conductivity. Time-of-flight (TOF) measurements in disordered organic solids, such as molecularly doped polymers, conjugated polymers and organic glasses, show that the electric field (E) dependence of electronic conductivity has approximately the Poole–Frenkel form, $\ln \sigma \propto E^{1/2}$ [1–3]. Two important models have been discussed in the literature with respect to the space-energy correlations of localized states responsible for transport, the Gaussian disorder model (GDM) and the correlated disorder model (CDM). In (GDM) such correlations are neglected [4] and in (CDM), it is assumed that correlation takes account of energy distributions for spatially close sites, the long-range interaction between the charged carriers and the dipole moments of the molecular dopants [5–7]. These models are valid in the materials with permanent dipole moments. This not the case of most conjugated polymers. Yu *et al.* recently proposed a model for the mobility of the conjugated polymers based on the fluctuations in molecular geometry [8].

In this paper we show that the behaviour is due to the effect of the intensity of the electric field on the dimension of the transport path, which becomes one-dimensional in the high-field region, and to the effect of the field on the tunnelling probability between the localized states.

We consider a finite array of N sites which are randomly distributed in position and energy. Each site is assumed to be occupied by no more than one electron, i.e. there is only one energy level (localized state) available to it at this site.

The time evolution of the occupational probability, called the rate equation, is given by

$$\frac{d\rho_i(t)}{dt} = -\sum_j \left\{ w_{ij}\rho_i(t)\left(1 - \rho_j(t)\right) - w_{ji}\rho_j(t)\left(1 - \rho_i(t)\right) \right\}, \tag{1}$$

where w_{ij} is the transition rate between the sites i and j.

---

[*] Corresponding author: e-mail: dbourbie@hotmail.com, Phone: +00 213 41514048, Fax: +00 213 41416021

In this form the rate equation is non-linear and very difficult to handle. For this reason, a linearization method has been proposed by Butcher [9]. Following this work, the frequency-dependent conductivity can then be written as

$$\sigma(\omega) = -\frac{\omega^2 e^2}{6\Omega k_B T} \left\langle \sum_{ij} F_i \, R_{ij}^2 G_{ij}(\omega) \right\rangle. \tag{2}$$

The bracket denotes a configurational average. Here R_{ij}^2 is the squared distance separating the sites i and j, k_B the Boltzmann constant and T the temperature of the system, $F_i = f_i (1 - f_i)$, with f_i is the thermal equilibrium distribution (Fermi–Dirac distribution). $G_{ij}(\omega)$ is the Green function corresponding to the linearized rate equation (LRE)

$$G_{ij}(\omega) = \frac{1}{i\omega - \Gamma_{ij}} \tag{3a}$$

with

$$\Gamma_{ij} = \tilde{\Gamma}_{ij}(1 - \delta_{ij}) - \delta_{ij} \sum_l \tilde{\Gamma}_{il} \,, \tag{3b}$$

where $\tilde{\Gamma}_{ij} = W_{ij} / F_i$, and $W_{ij} = W_{ji} = f_i (1 - f_j) w_{ij}$.

The propagator $G_{ij}(\omega)$ has been determined by two methods, the renormalized perturbative expansion (RPE) developed by Movaghar and Schirmacher [10] and the diagrammatic method developed by Gouchanour et al. [11, 12]. The two methods give the same expressions for the conductivity [10, 11],

$$\sigma(\omega) = \frac{e^2 \langle F(\varepsilon) \rangle N}{6\Omega k_B T} \left\langle \int d\mathbf{R} \frac{\tilde{\Gamma}(\varepsilon, \varepsilon', \mathbf{R})}{1 + G_1^\varepsilon(\omega) \left(\tilde{\Gamma}(\varepsilon, \varepsilon', \mathbf{R}) + \tilde{\Gamma}(\varepsilon', \varepsilon, -\mathbf{R}) \right)} \right\rangle. \tag{4}$$

where

$$G_1^\varepsilon(\omega) = \left(i\omega + \int d\varepsilon d\mathbf{R} N(\varepsilon) g(\mathbf{R}) \frac{\tilde{\Gamma}(\varepsilon, \varepsilon', \mathbf{R})}{1 + G_1^\varepsilon(\omega) \left(\tilde{\Gamma}(\varepsilon, \varepsilon', \mathbf{R}) + \tilde{\Gamma}(\varepsilon', \varepsilon, -\mathbf{R}) \right)} \right)^{-1} \tag{5}$$

with $\tilde{\Gamma}_{ij} = \tilde{\Gamma}(\varepsilon, \varepsilon', \mathbf{R})$, and $G_1^\varepsilon(\omega)$ averaged local Green function.
The conductivity expression (4) can be written as

$$\sigma(\omega) = \frac{e^2 \langle F(\varepsilon) \rangle N \langle R_{ij}^2(\omega) \rangle}{6\Omega k_B T} \int d\varepsilon N(\varepsilon) \left[\frac{1}{G_1^\varepsilon(\omega)} - i\omega \right]. \tag{6}$$

Expression (6) is identical to the formula of Sher and Lax [13], although it has been derived from a rather different formalism.

For the dc-conductivity ($\omega = 0$), The conductivity becomes inversely proportional to a generalized dwelling time $G_1^\varepsilon(\omega = 0)$, which characterizes the mean time that a carrier stays at site i, including the possibility of leaving the site and returning.
The transition rate between the site i and the site j separated by the distance $\mathbf{R}$ is given by

$$\tilde{\Gamma}(\varepsilon, \varepsilon', \mathbf{R}) = v_0 \exp\left(-2\alpha R - \beta(\Delta - e\mathbf{E}.\mathbf{R})\right), \quad \text{for } \Delta > e\mathbf{E}.\mathbf{R}. \tag{7}$$

For $\Delta < e\mathbf{E}.\mathbf{R}$, the tunnelling probability was strongly affected by the electric field [14].

$$\tilde{\Gamma}(\varepsilon, \varepsilon', \mathbf{R}) = v_0 \exp\left(-\beta e\mathbf{E}.\mathbf{R} - \left(\beta + \frac{4\alpha}{3eE}\right)\Delta\right), \quad \text{for } \Delta < e\mathbf{E}.\mathbf{R} \tag{8}$$

with $\Delta = \varepsilon - \varepsilon'$, $1/\alpha$ is the localization length and v_0 the characteristic frequency of phonons.

When supposing that $G_1^\varepsilon(\omega = 0)$ is independent of ε, then

$$\left(G_1^\varepsilon(\omega = 0)\right)^{-1} = v_0 e^{-\lambda} \tag{9}$$

where λ is a parameter to be determined.

Taking into account equations (7) and (9), and using the fact that, $G_1^\varepsilon(\omega = 0) \gg 1$, in the strongly localized systems, the equation (5) for ($\omega = 0$) lead to

$$1 \approx \frac{1}{\beta}\eta_d \int_0^\lambda dy N\left(\frac{y}{\beta}\right)(\lambda - y)^d \tag{10}$$

with $y = \beta\Delta$ and d the dimension of the system.

The temperature, the field and the dimension effects are contained in the function η_d

$$\eta_d = \int_0^{\theta_0} d\theta \frac{(2\pi \sin\theta)^{d-2}}{(2\alpha - \beta eE \cos\theta)^d}, \quad \text{for } d = 2,3, \tag{11a}$$

$$\eta_d = \frac{1}{(2\alpha - \beta eE)} \quad \text{for } d = 1, \tag{11b}$$

where θ_0 is the maximum angle of hopping.

For a constant density of states the equations (6) and (10) give for the dc-conductivity

$$\sigma(0) = \sigma_d \exp\left\{-\left[\left(\frac{d(d+1)}{N_0 \eta_d k_B T}\right)^{1/(d+1)}\right]\right\}. \tag{12}$$

For the one-dimensional systems, equation (11) conducts to

$$\sigma(0) = \sigma_1 \exp\left\{-2\sqrt{\frac{\alpha\beta}{N_0}}\left(1 - \frac{\beta eE}{2\alpha}\right)^{1/2}\right\}. \tag{13}$$

At low fields $\left(\beta eE / 2\alpha \ll 1\right)$, the conductivity increase as $\ln\sigma \propto E$, while for the high fields region it varies like $\ln\sigma \propto E^{1/2}$ (see Fig. 1). Such dependence of the conductivity may be linked to that observed in molecularly doped polymers [1–3, 15, 16].

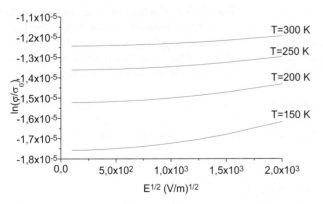

Fig. 1 The field dependence of one-dimensional conductivity, after equation (13), at different temperatures and for a constant density of states $N_0 = 10^{16}$ eV/cm and $\alpha^{-1} = 10$ Å.

At high-fields the carriers path becomes essentially one-dimensional along the applied field. Thus, one-dimensional transport models can be valid at high fields but not at low fields.

For three-dimensional systems $(d = 3)$ and isotropic hopping $(\theta_0 = \pi)$ and in the low field region $\beta eE / 2\alpha \ll 1$, the expressions (11) and (12) give

$$\sigma(0) = \sigma_3 \exp\left\{-\left[\left(\frac{T_0}{T}\right)^{1/4}\left(1 - \frac{\beta^2 e^2 E^2}{8\alpha^2}\right)\right]\right\}. \tag{14}$$

We rediscover the Apsley and Hughes expression [17],

For directed hopping $(\theta_0 = \pi / 2)$, and for the low field where $\beta eE / 2\alpha \ll 1$, the conductivity expression becomes

$$\sigma(0) = \sigma_3 \exp\left\{-\left[\left(\frac{T_0}{T}\right)^{1/4}\left(1 - \frac{3\beta eE}{16\alpha}\right)\right]\right\}. \tag{15}$$

This relation resembles to the Pollak and Riess expression [18].

For a parabolic density of states, which is the case of systems with inter-site electron–electron interaction, the conductivity expression gets the following form:

$$\sigma(0) = \sigma_d \exp\left\{-\left[\left(\frac{1}{k_B T}\left(\frac{d(d+1)(d+2)(d+3)}{2N_0\eta_d}\right)^{1/3}\right)^{3/(d+3)}\right]\right\}. \tag{16}$$

For three-dimensional systems $(d = 3)$ and isotropic hopping, and in the low region, $\beta eE / 2\alpha \ll 1$, we find

$$\sigma(0) = \sigma_3 \exp\left\{-\left[\left(\frac{T'_0}{T}\right)^{1/2}\left(1 - \frac{\beta^2 e^2 E^2}{12\alpha^2}\right)\right]\right\}. \tag{17}$$

For directed hopping, and for the low field $\beta eE / 2\alpha \ll 1$, it appears that

$$\sigma(0) = \sigma'_3 \exp\left\{-\left[\left(\frac{T'_0}{T}\right)^{1/2}\left(1 - \frac{\beta eE}{8\alpha}\right)\right]\right\}. \tag{18}$$

For a parabolic density of states, and at vanishing field, the $1/2$ law occurs [19].

We see that the expression (12) represents an unified relation of other results deduced by rather different formalisms (percolation theory, CTRW formalism, EMA technique). At low fields, the typical paths taken by carriers are three-dimensional (isotropic or directed hopping), whereas at high fields the paths become essentially one-dimensional along the applied field. Thus, one-dimensional transport models can be valid at high fields. This effect is due to the diminution of the number of percolation paths caused by the strong electric field.

In all these investigations, the tunnelling probability has been considered independent of the applied field. However this is not true in the high field region [14]. In this case the rate transition is given by equation (8). For a constant density of states, the conductivity expression reads [14]

$$\sigma(0) = \sigma'_d \exp\left\{-\left[\frac{d(d+1)}{N_0\eta_d}\left(1 + \frac{4\alpha}{3\beta eE}\right)\right]^{1/(d+1)}\beta(eE)^{d/(d+1)}\right\}. \tag{19}$$

In the one-dimensional systems, $\ln\sigma \propto -\beta\sqrt{E}$.

For a parabolic density of states, which is the case of three dimensional systems, we find

$$\sigma(0) = \sigma'_d \exp\left\{-\left[\frac{d(d+1)(d+2)(d+3)}{N_0\eta_d}\left(1+\frac{4\alpha}{3\beta eE}\right)\right]^{1/(d+3)}\beta(eE)^{d/(d+3)}\right\}. \tag{20}$$

In the three-dimensional systems, this relation reduces to, $\ln\sigma \propto -E^{1/2}$.

In the high field case, the conductivity decreases with increasing field as $\ln\sigma \propto -\beta E^{1/2}$, independently of the dimension of the system.

$$\sigma(0) \propto \exp\left\{-A_d\left(\frac{e^d}{N_0}\right)^{1/2d}\left(1+\frac{4\alpha}{3\beta eE}\right)^{1/2}\beta\sqrt{E}\right\}, \tag{21}$$

where A_d a constant depending on the dimension.

In summary, we have proposed a possible mechanism to describe electrical conductivity in disordered materials. The Main ingredients of our works consist in two mechanisms: the effect of the electrical field on the effective carrier dimension and the effect of the field on the tunnelling probability. The first mechanism give a field dependence identical to that observed in disordered polymers, $\ln\sigma \propto \beta E^{1/2}$. In the second mechanism, the conductivity decrease as $\ln\sigma \propto -\beta E^{1/2}$. Such phenomena have been observed in disordered polymers [1, 15, 16]. The expressions found can be generalized like, $\ln\sigma \propto E^\gamma$, with $\gamma = 2$ for isotropic hopping, $\gamma = 1$ for directed hopping and $\gamma = 1/2$ for one-dimensional hopping. This decrease is due to the effect of the electric field on the number of percolation paths. In increasing the field, certain paths become disallowed.

Acknowledgements This work has been presented at HRP10 Conference, thanks to the organisation committee and Abdus-Salam ICTP at Trieste for his financial support.

References

[1] M. A. Abkowitz, M. I. Rice, and M. Stolka, Philos. Mag. B **61**, 25 (1990).
[2] P. W. Blom, M. J. M. Dejong, and M. G. VanMunster, Phys. Rev. B **55**, R656 (1997).
[3] I. H. Campbell, D. L. Smith, C. J. Neef, and J. P. Ferraris, Apll. Phys. Lett. **74**, 2809 (1999).
[4] H. Bässler, phys. stat. sol. (b) **175**, 15, (1993).
[5] I. Bleyl, C. Erdelen, H. W. Schmidt, and D. Haarer, Philos. Mag. B **79**, 463 (1999).
[6] D. H. Dunlap, P. E. Parris, and V. M. Kenkre, Phys. Rev. Lett. **77**, 542 (1996).
[7] S. V. Novikov, D. H. Dunlap, V. M. Kenkre, P. E. Parris, and A. V. Vannikov, Phys. Rev. Lett. **81**, 4472 (1998).
[8] Z. G. Yu, D. L. Smith, A. Saxena, R. L. Martin, and A. R. Bishop, Phys. Rev. Lett. **84**, 721 (2000).
[9] P. N. Butcher, "*Linear and Nonlinear Electronic Transport in Solids*", edited by J. T. Devreese and V. E. Doren (Plenum, New York, 1976), p. 341.
[10] D. Bourbie, Philos. Mag. B **73**, 201 (1996).
[11] B. Movaghar and R. J. Shirmacher, J. Phys. C **14**, 859 (1981).
[12] C. R. Gouchanour, H. C. Andersen, and M. D. Fayer, J. Chem. Phys. **70**, 4254 (1979).
[13] L. Sher and M. Lax, Phys. Rev. B. **7**, 4491 (1973).
[14] D. Bourbie, N. Ikrelef, and P. Nedellec, Philos. Mag. Lett. **82**, 641 (2002).
[15] M. A. Abkowitz, Philos. Mag. B **65**, 817 (1992).
[16] M. Novo, M. Van der Auweraer, F. C. De Schryver, P. M. Borsenberger, and H. Bässler, phys. stat. sol. (b) **177**, 223 (1993).
[17] N. Apsley and H. P. Hughes, Philos. Mag. **30**, 963 (1974).
[18] M. Pollak and I. J. Riess, J. Phys. C **9**, 2339 (1976).
[19] A. L. Efros and B. I. Skhlovskii, J. Phys. C **8**, L49 (1975).

phys. stat. sol. (c) **1**, No. 1, 84–87 (2004) / **DOI** 10.1002/pssc.200303613

Spin transport in two-dimensional hopping systems

T. Damker[*,1], **V. V. Bryksin**[2], and **H. Böttger**[1]

[1] Institute for Theoretical Physics, Otto-von-Guericke-University, PF 4120, 39016 Magdeburg, Germany

[2] A. F. Ioffe Physio-Technical Institute, Politekhnicheskaya 26, 19526 St. Petersburg, Russia

Received 1 September 2003, revised 11 September 2003, accepted 16 September 2003
Published online 28 November 2003

PACS 71.38.Ht, 72.25.Dc, 72.25.Rb

An area of large current interest is the interplay of charge and spin transport of carriers in extended states. The study of this problem is of intrinsic scientific interest and is further motivated by the prospect of possible applications, e.g. a spin transistor. We perform a theoretical study of the interplay of charge and spin transport in the hopping regime. Specifically, we consider a two-dimensional (2D) system with Rashba spin-orbit interaction (SOI), where the localization of the electronic states is due to disorder and/or strong electron-phonon interaction. In this case, the state description of each site contains the spin orientation in addition to the occupation number. As a first step, we solve the rate equations of these quantities for an ordered (polaronic) system in two-site hopping approximation. Whereas the charge transport is not affected by SOI in this approximation, spin separation can occur if the crystal symmetry is sufficiently low. Furthermore, the total spin polarization normally decays exponentially. But, we find a critical threshold value for the in-plane electrical field, above which the spin polarization oscillates. Thus, this comparably simple approximation already yields a surprisingly rich phenomenology.

1 Introduction Whereas the interplay of spin and charge transport for itinerant electrons is a topic of many investigations, hopping spin transport has rarely been considered. The study of the influence of spin orbit interaction (SOI) on hopping transport was confined to the question of charge transport and neglected spin transport entirely [1–5]. Furthermore, randomly distributed SU(2) matrices were utilized for the description of spin rotation during a single hopping event [6–11]. This is possibly appropriate, if the SOI results from the motion of the electrons in electric fields of (randomly distributed) charged impurities. If the electric field leading to SOI is spatially constant over length scales relevant for hopping, the SU(2) matrices corresponding to hops between different pairs of sites are strongly correlated, and cannot be chosen randomly. This especially applies, if the perpendicular electric field is due to a structural anisotropy of the underlying crystal, or due to an external (confinement) field (Rashba SOI).

Our aim is to describe a system of electrons localized either by disorder or by polaron formation, which can move by hopping within a two-dimensional plane (e.g. a heterostructure quantum well) [12]. Furthermore, the spin dynamics due to Rashba SOI is taken into account.

2 Model To incorporate SOI in the hopping formalism we propose the following path. Developing the Dirac Hamiltonian up to order $1/c^2$ yields (among others) the SOI-term $-\dfrac{e\hbar}{4m^2c^2}\,\boldsymbol{\sigma}\cdot(\boldsymbol{E}\times\boldsymbol{p})$.

---

[*] Corresponding author: e-mail: thomas.damker@physik.uni-magdeburg.de

Here, e is the electron charge, $\hbar$ Planck's constant, m the (effective) electron mass, c the speed of light, σ the vector of Pauli matrices, E the electric field and p the momentum operator. This term can be incorporated into the kinetic term in the form of a SU(2) gauge potential $\frac{1}{2m}\left(p - \frac{e\hbar}{4mc^2}\sigma \times E\right)^2$, a transformation again correct to order $1/c^2$. For a homogeneous electric field, the effect of this gauge field on the physics of the problem can be treated in the following approximation. The eigen-values are unchanged in comparison to the problem without SOI, and the eigen-functions are decorated with the factor $e^{i\sigma \cdot (K \times r)}$, where we have introduced the quantity $K = \frac{e}{4mc^2}E$, which has the dimension of an inverse length. This approximation is valid to the lowest order in $|K \times r|$ and requires the condition $|K \times r| \ll 1$ to be fulfilled. This shows that it is inappropriate for extended states, but may be useful for localised states. For the SOI term it is immaterial whether the electric field E is due to an intrinsic crystal anisotropy or externally applied. Thus, introducing the usual Rashba SOI strength α and still allowing externally applied fields one can write

$$K = \frac{e}{4mc^2}E_{\text{ext}} + \frac{m}{\hbar^2}\alpha n, \tag{1}$$

where n is the vector normal to the plane (z direction).

This allows us to consider the hopping Hamiltonian

$$H = \sum_{m\sigma}\epsilon_m a^\dagger_{m\sigma}a_{m\sigma} + \sum_{mm'\sigma\sigma'}J^{m\sigma}_{m'\sigma'}a^\dagger_{m\sigma}a_{m'\sigma'} + \text{phonon terms} \tag{2}$$

in the presence of SOI. Here $a_{m\sigma}$ ($a^\dagger_{m\sigma}$) are annihilation (creation) operators of an electron on site m with spin σ. The designated phonon terms are taken to also include the electron-phonon interaction. Now, the spin structure of the hopping matrix elements J has to be determined. Proceeding analogously to Refs. [13, 14] while appropriately modifying the eigen-functions one obtains

$$J^{\text{SO}}_{m'm} = e^{-i\sigma \cdot (K \times R_{m'm})}J_{m'm}, \tag{3}$$

where $R_{m'm} = R_{m'} - R_m$ is the distance vector between the sites, and R_m is the coordinate vector of site m. $J_{m'm}$ is the hopping matrix element without SOI. The spin-dependent pre-factor for the transition matrix elements $e^{i\sigma \cdot (K \times r)}$ is identical to the one found in Ref. [4] to study the influence of Rashba-SOI on magneto-conductivity in the hopping regime and is valid to the lowest order in K. It is analogous to the Holstein transformation to treat the influence of a magnetic field on hopping.

Whereas the correlation between different sites can be neglected in a hopping context ($\langle a^\dagger_{m'\sigma'}a_{m\sigma}\rangle = 0$ for $m' \neq m$), the correlation between different spin states at the same site must be retained, since it contains the information about the spin orientation. The occupation number $\langle a^\dagger_m a_m\rangle$ as a central quantity of the (spinless) hopping theory is now replaced by the 2×2-matrix $\hat{\rho}_m|_{\sigma'\sigma} = \langle a^\dagger_{m\sigma'}a_{m\sigma}\rangle$. Using the one-particle approximation and calculating the transition probabilities in two-site approximation, a lengthy calculation yields the rate equation for this matrix quantity (in Laplace space s)

$$s\hat{\rho}_m(s) = \hat{\rho}_m|_0 + \sum_{m_1}\left\{e^{-i\sigma \cdot (K \times R_{m_1 m})}\hat{\rho}_{m_1}(s)e^{i\sigma \cdot (K \times R_{m_1 m})}W_{m_1 m} - \hat{\rho}_m(s)W_{mm_1}\right\}. \tag{4}$$

The term $\hat{\rho}_m|_0$ denotes the initial conditions at zero time. The hopping probability $W_{m_1 m}$ between sites m and m_1 is the same expression, which would be obtained without spin.

To assess the physical meaning of Eq. (4), it is advantageous to replace the 2×2-matrix $\hat{\rho}_m$ by a scalar and a vector quantity: the occupation probability $\rho_m = \text{Tr}(\hat{\rho}_m)$ and the spatial vector of the spin orientation $\boldsymbol{\rho_m} = \text{Tr}(\sigma\hat{\rho}_m)$. The corresponding rate equations read

$$s\rho_m = \rho_m|_0 + \sum_{m_1}\left\{\rho_{m_1}W_{m_1 m} - \rho_m W_{mm_1}\right\} \tag{5}$$

and

$$s\boldsymbol{\rho_m} = \boldsymbol{\rho_m}|_0 + \sum_{m_1}\left\{D_{m_1 m}\boldsymbol{\rho_{m_1}}W_{m_1 m} - \boldsymbol{\rho_m}W_{mm_1}\right\}. \tag{6}$$

Note, that the occupation number dynamics Eq. (5) is entirely unaffected by spin-orbit interaction in the two-site approximation considered here. The 3×3-matrices $D_{m_1 m}$ describe the rotation of the spin orientation during a hopping event and depend only on K and $R_{m_1 m}$.

3 Ordered two-dimensional system   As a first step we apply our theory to an ordered (polaronic) hopping system, where the electrons are confined to a two dimensional plane. We assume that the contribution to K of the in-plane electrical field $E^{\parallel}$ is negligible. Then $K \cdot R_{m_1 m} = 0$. By going to wave vector space and furthermore assuming $|K \times R_{m_1 m}| \ll 1$ Eq. (6) can be solved. The total spin polarization is then easily obtained in the limit of zero wave vector. The calculation of the spatial dependence requires an inverse Fourier transform.

For long wave lengths the Fourier transformed transition probability has the structure

$$W(q) = W(0) - q^2 D - i\mu q \cdot E, \tag{7}$$

where we have introduced the phenomenological diffusion constant D and the mobility μ.

In order to show explicit results, we introduce the dimensionless quantities

$$x = Kr, \qquad \tau = DK^2 t, \qquad \epsilon = \frac{\mu}{DK} E^{\parallel}. \tag{8}$$

Here, $K = |K|$. The spin polarization will furthermore be divided into the Cartesian coordinates perpendicular to the plane ($\rho^{\perp}$), along the in-plane electric field ($\rho^{\vdash}$) and perpendicular to both ($\rho^{\dashv}$).

$$\boldsymbol{\rho} = \rho^{\perp} \frac{K}{K} + \rho^{\vdash} \frac{E^{\parallel}}{|E^{\parallel}|} + \rho^{\dashv} \frac{K \times E^{\parallel}}{|K \times E^{\parallel}|} = \rho^{\perp} \frac{K}{K} + \rho^{\parallel}. \tag{9}$$

Using these quantities the total spin polarization for arbitrary initial conditions ρ_0 is obtained as

$$\rho^{\perp}(\tau) = e^{-6\tau} \left[\left(\cos\left(2\tau\sqrt{\epsilon^2 - 1}\right) - \frac{\sin\left(2\tau\sqrt{\epsilon^2 - 1}\right)}{\sqrt{\epsilon^2 - 1}} \right) \rho_0^{\perp} - \epsilon \frac{\sin\left(2\tau\sqrt{\epsilon^2 - 1}\right)}{\sqrt{\epsilon^2 - 1}} \rho_0^{\vdash} \right] \tag{10}$$

$$\rho^{\vdash}(\tau) = e^{-6\tau} \left[\left(\cos\left(2\tau\sqrt{\epsilon^2 - 1}\right) + \frac{\sin\left(2\tau\sqrt{\epsilon^2 - 1}\right)}{\sqrt{\epsilon^2 - 1}} \right) \rho_0^{\vdash} + \epsilon \frac{\sin\left(2\tau\sqrt{\epsilon^2 - 1}\right)}{\sqrt{\epsilon^2 - 1}} \rho_0^{\perp} \right] \tag{11}$$

$$\rho^{\dashv}(\tau) = e^{-4\tau} \rho_0^{\dashv} \tag{12}$$

Note that, when $\epsilon < 1$, the trigonometric functions become hyperbolic functions. Thus, there is a critical electrical field $\epsilon_c = 1$, which divides oscillatory and exponential behaviour. Replacing the SOI-quantity K by the electrical field $E^{\perp}$ perpendicular to the plane the critical field reads

$$E_c^{\parallel} = \frac{D}{4mc^2 \mu} E^{\perp} = \frac{k_B T}{4mc^2} E^{\perp}. \tag{13}$$

In the last equality, the Einstein relation between D and μ has been used. For vanishing electrical field $\epsilon = 0$, the total spin polarization develops according to

$$\rho^{\perp}(\tau) = e^{-8\tau} \rho_0^{\perp}, \qquad \boldsymbol{\rho}^{\parallel}(\tau) = e^{-4\tau} \boldsymbol{\rho}_0^{\parallel}. \tag{14}$$

One can see, that the z-component of the spin-polarization relaxes significantly faster than the parallel component.

Otherwise, allowing an electrical field ϵ, but starting with only a z-polarization (taken to be of value 1), one obtains

$$\rho^{\perp}(\tau) = e^{-6\tau} \left[\cos\left(2\tau\sqrt{\epsilon^2 - 1}\right) - \frac{\sin\left(2\tau\sqrt{\epsilon^2 - 1}\right)}{\sqrt{\epsilon^2 - 1}} \right] \tag{15}$$

$$\boldsymbol{\rho}^{\parallel}(\tau) = e^{-6\tau} \frac{\sin\left(2\tau\sqrt{\epsilon^2 - 1}\right)}{\sqrt{\epsilon^2 - 1}} \epsilon \tag{16}$$

This means, that the polarisation moves within the plane spanned by the z-axis and the in-plane electric field, and describes an inclined ellipse — apart from the relaxation factor. Below the critical field ($\epsilon < 1$) the oscillatory behaviour is changed into exponential behaviour (more precisely, the sum of two exponential terms), and the polarization monotonously relaxes with time.

For an inhomogeneous initial condition consisting of a single z-polarized spin at the origin, the total spin polarization is still described by Eqs. (15) and (16), but the local spin polarization is inhomogeneous. Its asymptotic behaviour for large times $\tau \to \infty$ and small distances $x \ll 2\sqrt{\tau}$ (Large distances will not be relevant here, since they are strongly suppressed by the "diffusion factor" $e^{-x^2/4\tau}$.) is

$$\rho^{\perp}(\tau,\boldsymbol{x}) = e^{-\frac{cx}{2}-\frac{c^2\tau}{4}-\frac{7}{4}\tau-\frac{x^2}{4\tau}} \frac{3}{8\sqrt{\pi\tau}} J_0\left(\frac{\sqrt{15}}{2}x\right) \tag{17}$$

Note, that in this approximation (long times and short distances), the sign of the polarization — determined by the argument of the Bessel function J_0 — only depends on the distance to the origin, and is furthermore not time dependent.

4 Outlook Experimental relevance: Published experimental values of the Rashba SOI strength are of the order of $\alpha \approx 10^{-8} \ldots 10^{-9}$ eV cm [4], which corresponds to $K \approx m/m_e \left(\frac{1}{76\,\text{nm}} \cdots \frac{1}{7.6\,\text{nm}}\right)$, where the factor m/m_e allows for a modified effective electron mass. This means that the relevant hopping lengths have to be smaller than a value in the range 760 Å to 76 Å (due to the assumption $KR \ll 1$). The corresponding perpendicular electric field in this case is $E^{\perp} \approx (m/m_e)^2 \, 10^{14}$V/m, which is of the order of crystal fields, or up to a few orders of magnitude lower, if the effective electron mass is small. The critical field at a temperature of 1 K is (see Eq. (13)) $E_c^{\parallel} \approx (m/m_e)^2 \, 10^3$V/m. With the values as given, this means that the spin coherence length is of the order of 100 Å and the critical field is 10^4 V/m. For an effective electron mass of 1/10, the corresponding quantities would be 1000 Å and 10^2 V/m. If the SOI effects are due to an externally applied field, the critical field might well be quite small. Also, the relevant hopping lengths and the spin coherence length can then be very much larger than the value given above.

For a crystal with low symmetry, the transition probability does not have the isotropic form Eq. (7). Even though this $W(\boldsymbol{q})$ is not spin dependent, it can lead to anisotropic spin transport, because the quantity D in Eq. (6) depends on $\boldsymbol{R}_{m_1 m}$, and thus couples anisotropies of W to spin orientations.

There are efforts under way to incorporate three-site terms into to the formalism. Preliminary results show the occurrence of the spin Hall effect in this approximation, which is consistent with expectations. Another line of work is the introduction of site disorder. The effects considered above are expected to survive in principle, but the disorder will probably affect the relaxation.

References

[1] B. Movaghar and L. Schweitzer, phys. stat. sol. (b) **80**, 491 (1977).
[2] B. Movaghar and L. Schweitzer, J. Phys. C: Solid State Phys. **11**, 125 (1978).
[3] Y. Osaka, J. Phys. Soc. Japan **47**, 729 (1979).
[4] T. V. Shahbazyan and M. E. Raikh, Phys. Rev. Lett. **73**, 1408 (1994).
[5] G.-H. Chen, M. E. Raikh, and Y.-S. Wu, Phys. Rev. B **61**, R10539 (2000).
[6] Y. Asada, K. Slevin, and T. Ohtsuki, Phys. Rev. Lett. **89**, 256601 (2002).
[7] Y.-L. Lin, Europhys. Lett. **43**, 427 (1998a).
[8] Y.-L. Lin, Phys. Rev. B **58**, 13544 (1998b).
[9] E. Medina and M. Kardar, Phys. Rev. B **46**, 9984 (1992).
[10] E. Medina and M. Kardar, Phys. Rev. Lett. **66**, 3187 (1991).
[11] Y. Meir, N. S. Wingreen, O. Entin-Wohlman, and B. L. Altshuler, Phys. Rev. Lett. **66**, 1517 (1991).
[12] Y. Takahashi, K. Shizume, and N. Masuhara, Phys. Rev. B **60**, 4856 (1999).
[13] T. Holstein, Phys. Rev. **124**, 1329 (1961).
[14] H. Böttger and V. V. Bryksin, Hopping Conduction in Solids (Akademie-Verlag, Berlin, 1985).

phys. stat. sol. (c) **1**, No. 1, 88–91 (2004) / **DOI** 10.1002/pssc.200303628

Hopping transport through nanoconstriction controlled by a single hop

V. I. Kozub* and **A. A. Zuzin**

A. F. Ioffe Physico-Technical Institute, 194021 St.-Petersburg, Russia

Received 1 September 2003, accepted 3 September 2003
Published online 28 November 2003

PACS 72.20.Ee;73.40.Lq; 73.63.Rt

The hopping transport through the point contact between two semiconductors is considered for the case when the conductance is controlled by a single hop through the constriction. The choice of the "critical pair" of the hopping sites depends on temperature, and the temperature dependence of the conductance exhibits exponentially large mesoscopic fluctuations. For large biases the conductance is strongly nonlinear and also exhibits giant fluctuations as a result of "switching" between different "critical pairs". The contribution of each of such pairs to the I-V curve has a step-like form with a region of negative differential resistance related to resonant tunneling which appears to be effective at some "resonant" values of the bias.

The point contacts are widely used to obtain important information concerning electron transport. To the best of our knowledge, until now these methods were mainly used for the studies of the metallic state. As for the insulating side of the metal-insulator transition we can mention only several publications. The paper [1] reported the studies of nanofabricated Si point contact with the size larger than the hopping length. The hopping conductivity of the Mott type was observed; at finite biases it was strongly nonlinear and was interpreted in terms of the theory [2] . Recently the paper [3] reported an observation of hopping transport through the break junction with a size (according to the authors statement) comparable with the hopping length. Strongly nonlinear behavior was ascribed to the features of the Coulomb gap in the density of states.(Some doubts concerning this interpretation were formulated in [4], [5]). Much greater attention was paid to the gated low-dimensional semiconducting structures of small size (for the review see [6]). The main result of these studies were predictions and observations of giant mesoscopic fluctuations of the conductance as a function of the gate voltage. Nevertheless the size of the structure was considered to be larger than the hopping length. At the same time the present experimental technique can be readily used to fabricate the devices with a size less than the typical distance between the hopping sites. Thus a profound theoretical investigation of hopping transport in nanoscale devices is of definite interest.

Let us model the point contact by an orifice within a thin energy barrier with an infinite height. Let us consider the electron hopping between two sites at the positions $\mathbf{R_1}$ and $\mathbf{R_2}$ on the opposite sides from the barrier where the distances R_1 and R_2 are assumed to be larger than the orifice radius D. The overlapping integral for the states centered at $\mathbf{R_1}$ and $\mathbf{R_2}$ can be estimated as

$$I \sim \frac{e^2}{a}\frac{R_1+R_2}{a}\left(\frac{D}{a}\right)^{1/2} e^{DA/a} \exp -\frac{R_1+R_2}{a} \equiv I_0 \exp -\frac{R_1+R_2}{a} \tag{1}$$

So one obtains for the conductance of the resistor connecting the sites 1 and 2:

$$G_{1,2} = G_{0;1,2} \exp -\frac{2}{a}(R_1+R_2) \exp -\frac{|\varepsilon_1 - \varepsilon_2|}{T} \tag{2}$$

---

* Corresponding author: e-mail: ven.kozub@mail.ioffe.ru

The probability to find the two sites 1 and 2 within the energy band $\Delta\varepsilon$ is

$$W(R_1, R_2, \Delta\varepsilon) = F(x(1))F(x(2)), \qquad F(x) = xe^{-x}; \qquad x = \frac{2}{3}\pi R^3 g\Delta\varepsilon \tag{3}$$

Here we assumed the density of states g to be constant. The most probable pair corresponds to

$$R_1 = R_2 = \left(\frac{3x_1}{2\pi g\Delta\varepsilon}\right)^{1/3} \tag{4}$$

where $x_1 = 1$ corresponds to the maximum of $F(x)$. Substituting these values into Eq.2 we obtain

$$G_{12} = G_0 \exp - \left(\frac{T_0}{T}\right)^{1/4}; \qquad T_0 = \frac{2^{13}}{9\pi ga^3} \tag{5}$$

Thus we have the Mott law, although T_0 is 16 times larger than the value $2^9/9\pi ga^3$ obtained with a help of similar procedure for 3D hopping. The probability of the "hop through the orifice" is strongly suppressed with respect to hop in the bulk. So we indeed deal with a "critical" pair. One has in mind that Eq.5 holds only for "most probable" realizations within an ensemble of the point contacts which differ by a spatial distribution of the localized states. Actually the change of a "critical pair" needs a finite temperature variation. One has a set of "critical resistors", the temperature variation leads to "switching" between them. One deals with a succession of activation laws giving rise to giant mesoscopic fluctuations of $G(T)$. At at the "switching" temperature the actual phase volume ($x = (2/3)g\Delta\varepsilon\pi R^3$) at least for the one of the half spaces should contain two relevant sites. The probability for such an event is

$$W = \frac{1}{2}x^2 e^{-x} \tag{6}$$

which is maximal at $x = x_2 = 2$ (in comparison to the value $x_1 = 1$ giving the maximal probability for the only one site to be present). As is easily seen, the replacement of x_1 by x_2 in Eq.4 gives the value of T_0 2 times larger than estimated before. Correspondingly, one concludes that the characteristic temperature scale for the fluctuation is given as $\delta T \sim T$ while the magnitude of the fluctuation is

$$|\delta(\log G(T))| \sim 2^{1/4}|\log G(T)| \tag{7}$$

Let us turn to the nonlinear conductance in the case when $eV >> \Delta\varepsilon$. In this case one deals with a "directed hopping": the electrons hop from the half-space with higher chemical potential to the half-space with lower chemical potential. The energy eV plays a role of $\Delta\varepsilon$ in Eq.3, that is the actual "phase volume" is given as $x = \frac{2}{3}\pi R^3 geV$ The "most probable" tunneling current is

$$\log I(eV) = \frac{2}{a}\left(\frac{x_1}{(2/3)\pi geV}\right)^{1/3} \tag{8}$$

Actually one deals with giant "mesoscopic" fluctuations of $I(V)$ since an increase of the bias leads to a switch on of additional current-carrying channel. (Note that qualitatively similar considerations were given by Shklovskii [8] for "point" tunnel junctions).

If we restrict ourselves by exponential factor, the current carried by a given "critical pair" does not depend on V. Thus the $I(V)$ curve would correspond to a staircase of "steps" related to different "critical resistors". The ratio of the currents for the successive steps and the characteristic scale of V are

$$\frac{\log I(e(V+\delta V))}{\log I(eV)} \sim 2^{1/3} \qquad \frac{\delta V}{V} \sim \frac{x_2}{x_1} = 2 \tag{9}$$

Note that qualitatively similar results were obtained by Shklovskii for the case of small tunnel junctions [8].

Actually a contribution of a given "resistor" has more complex form. 1) The "switching" happens when the resistor with smaller ΔR still has activation contribution. So the left step edge is given as

$$-\frac{2\Delta R}{a} - \frac{\Delta\varepsilon}{T}sign(\Delta\varepsilon); \qquad \delta\varepsilon = \varepsilon_2 - \varepsilon_1 - eV \tag{10}$$

(saturated at $\delta\varepsilon(eV) < 0$. 2) The "plateau" region of the step is modified due to the bias dependence of the preexponential factor. 3)If $\delta\varepsilon \to 0$, the resonant tunneling is possible. While at $V = 0$ such a "resonant" configuration seems to be improbable, the bias "tunes" the "critical" resistor to the resonant configuration. We assume that the main mechanism of dephasing of the resonant process is related to escapes of an electron from the sites 1 and 2 to the sites in the surrounding - to say from the site 2 to the closest site (3), the escape rate will be denoted as τ_s^{-1}. For very small $|\Delta\varepsilon|$ the rate of the "resonant hopping" τ_r^{-1} is

$$\tau_r \sim (\tau_s/I^2\hbar^2)^{1/3} \tag{11}$$

(where I is the overlapping integral within the "critical pair" and $E = \left((\delta\varepsilon)^2 + I^2\right)^{1/2}$) while for larger $|\delta\varepsilon|$

$$\tau_r \sim \tau_s(E/I)^2 \tag{12}$$

By the order of magnitude this estimate coincides with the estimate for probability of hopping between sites 1 and 3 mediated by scattering center 2 [7]. The crossover between two limiting cases corresponds to $\hbar I^2/E^3 \sim \tau_s$. The "resonant" channel dominates over the "sequential hopping" $1 \to 2$, $2 \to 3$ if the hopping time

$$\tau_{1,2} = \gamma\frac{I^2}{\hbar E} \tag{13}$$

is larger than τ_r. ($\gamma \propto E^2$ is the parameter accounting for the electron-phonon coupling). As a result one obtains the following behavior of $G(V)$ (specifying $\varepsilon_2 > \varepsilon_1$). The exponential increase of conductance at the edge of the step related to some "effective resistor" where $\delta\varepsilon = \varepsilon_2 - \varepsilon_1 - eV < 0$ crosses over to the "resonant tunneling" with an increase of $|eV|$. Initially the hopping rate is controlled by Eq.12, and

$$G \propto (\varepsilon_2 - \varepsilon_1 - eV)^{-2} \tag{14}$$

increases with an increase of $|eV|$. Then G saturates in accordance with Eq.11. However the further increase of $|eV|$ lead to an increase of $|\Delta\varepsilon|$ and according to Eqs.12, 14 G decreases. So one concludes that at this region the differential resistance is *negative*. When the regime of sequential hopping is restored the hopping rate is described by Eq.13. In this case the rate increases with an increase of $|eV|$ due to an increase of the phase volume of the phonons involved and the differential resistance is again positive.

Until now we assumed that the bias is concentrated on the "critical" pair. However an increase of V leads to a decrease of "critical resistance" due to a decrease of $R_1 + R_2$. At large V this separation starts to be smaller than the typical hopping length l_h within the bulk. It can be shown that an increase of V by 8 times corresponds to a decrease of $R_1 + R_2$ by 2 times which is comparable to l_h. So the picture described above holds only for $\Delta\varepsilon(T) < eV < 8\Delta\varepsilon(T)$, that is the number of "mesoscopic steps" does not exceed 3. At higher biases the effect of redistribution of the bias between the "critical resistor" and the surrounding becomes important. For the biases as large as to ensure the nonlinear conductivity at the distances from the orifice larger than the correlation length of the percolation cluster in the bulk the resistance is formed in the remote region. Thus one deals with the local conductivity σ and electric field E. The value $\sigma(E)$ is expected to increase exponentially for the electric fields larger than some critical field E_c. As a result, with a logarithmic accuracy E is nearly constant ($\sim E_c$) around the constriction up to the distances R_c ($E_c R_c \sim V/2$). So one has $I \sim 2\pi R_c \sigma_c E_c$ where $\sigma_c = \sigma(E_c)$ and $I \propto V^2$ (see e.g. [1]).

Let us compare our results with the results for small gated semiconducting structures (see e.g the review paper [6].) Raikh and Ruzin considered the systems where G was controlled by some punctures. For the small systems G was shown to be strongly size-dependent. However Raikh and Ruzin concentrated on the systems with many punctures or hopping sites (like 1D hopping channels) In the papers [9], [10] mesoscopic transport via localized states in dielectric layer situated between metallic banks was studied. It was shown that an important role is played by the pairs of sites allowing resonant tunneling. The main attention was paid to short contacts with relatively large area. The number of the possible tunneling channels was large while contributions of different channels were of the same order of magnitude. Indeed, in the systems considered the total tunneling length for different channels was the same and equal to the distance between the banks. The average conductance exhibited power-law temperature and bias behavior while the conductance fluctuations were small enough. In contrast to the systems mentioned above, the point contact between 3D semiconductors does not imply any fixed spatial scale for the tunneling. As a result, an increase of T or V in our case leads to hierarchical structure of the hopping channels since the "more close" pair which becomes available at larger T or V allows exponentially larger G. Correspondingly, the $G(T)$ and $G(V)$ laws are described by exponentials while the fluctuations are also exponentially large.

Let us discuss the results of [3] where an observation of the Coulomb gap with a help of break junction technique was reported. The interpretation was based on standard tunnel spectroscopy equations implying that the tunneling probability W_t does not depend on V. We emphasize that for the point contact geometry W_t is exponentially dependent on V. This effect is much stronger than the possible effect of the density of states and can hardly be ignored if transport is dominated by a single "effective resistor". Another important feature is "mesoscopic" character of hopping which was emphasized in the discussion given above. At the same time the device reported in [3] exhibits no pronounced mesoscopic fluctuations. These facts evidence to our opinion that this device does not exhibit "single hop" transport. Note that the interpretation can not be valid even if the constriction contains an insulating layer. As it was demonstrated by Larkin and Shklovskii [5], the devices with large area of the tunnel barrier can nor reveal the features of the Coulomb gap while the devices with small area should exhibit large mesoscopic fluctuations and nonlinear conductivity of the same type as discussed above [5]. So the interpretation of the experimental results obtained in [3] needs some further insight to the problem.

To summarize, we studied hopping transport through semiconductor nanoconstriction when the conductance is controlled by a single hop. It was shown that the temperature and bias dependence of the conductance exhibits exponentially large mesoscopic fluctuations. $I(V)$ curve contains regions of the negative differential conductance because at some V the "critical" pair of sites becomes resonant.

We are grateful to Yu.M.Galperin and to B.I.Shklovskii for valuable discussions. A.A.Z. is grateful to International Center for fundamental physics in Moscow and to the Fund of non-commercial programs "Dynasty". V.I.K acknowledges a financial support by RFFI, grant N 03-02-17516.

References

[1] J.W.H.Maes, J.Caro, V.I.Kozub, K.Wemer and S.Radelaar) J.Vac.Sci.Technol.B **12**, 3614-3618 (1994)
[2] B.I.Shklovskii, Sov. Phys. - Semicond., **6**, 1964 (1973)
[3] B.Sandow, K.Gloos, R.Rentzch, A.N.Ionov, W.Schirmacher. Phys. Rev. Lett., **86**, 1845 (2001)
[4] V.I.Kozub Phys. Rev. Lett., **89** , 229701-1 (2002)
[5] A.I.Larkin, B.I.Shklovskii Phys. Stat. Sol. (b), 189 (2002)
[6] Raikh M.E., Ruzin I.M. In: Hopping transport in solids, ed. by M.Pollak and B.Shklovskii, Elsevier, 1991, p. 271
[7] Shklovskii B.I., Spivak B.Z. In: Hopping transport in solids, ed. by M.Pollak and B.Shklovskii, Elsevier, 1991, p. 271
[8] Shklovskii B.I., unpublished
[9] A.I.Larkin, K.A.Matveev. Sov. Phys.: JETP, **66**, 580 (1987)
[10] L.I.Glazman, K.A.Matveev. Sov. Phys.: JETP, **67**, 1276 (1988)

phys. stat. sol. (c) **1**, No. 1, 92–95 (2004) / **DOI** 10.1002/pssc.200303639

Tunneling spectroscopy in the hopping regime

B. Sandow[*, 1], **O. Bleibaum**[2], and **W. Schirmacher**[3]

[1] Institut für Experimentalphysik, Freie Universität Berlin, 14195 Berlin, Germany
[2] Institut für Theoretische Physik, Otto-von-Guericke Universität Magdeburg, 39016 Magdeburg, PF 4120, Germany
[3] Physik-Department E13, Technische Universität München, James-Franck-Straße 1, 85747 Garching, Germany

Received 1 September 2003, revised 8 October 2003, accepted 9 October 2003
Published online 28 November 2003

PACS 72.20.Ee, 73.40.Gk, 73.40.Lq

Charge transport across tunneling junctions of n-doped Ge has been investigated experimentally and theoretically. Using tunneling spectroscopy we were able to observe the density of states and the effect of the electron-electron interaction on the excitation spectrum of samples, in which hopping is the transport mechanism close to equilibrium. To analyze the data of the measurements we derive an expression for the tunneling current in the hopping regime. We use our expression for the tunneling current in order to analyze the character of the transitions. Doing so, we show that in the experiment only transitions with at most small energy transfer were relevant.

Tunneling spectroscopy is one of the methods for obtaining information on the electronic structure of solids [1]. Usually planar tunnel junctions are used to probe the density of states (DOS) of a metal or of a semiconductor. A superconducting counter-electrode with a known quasi-particle DOS serves then as reference to *verify* tunnelling. If the tunneling takes place between electrons of the same energy (elastic or resonant tunneling) the tunneling current is a convolution of the densities of states of the contact materials [2]. Thus, the density of states of one material can be extracted if the density of states of the other material is known. In the case of inelastic tunneling the electronic transitions occur with emission or absorption of phonons. In this case the tunneling current becomes also sensitive to the phonon density of states [3].

However, this method requires knowledge and skill to prepare the interfaces, especially the thin insulating layer. For about two decades the mechanical-controllable break junction technique has been used to study clean interfaces of metals over a wide range of lateral contact sizes from bulk transport to vacuum tunneling. For a break junction device it is essential to bring a considerable force to the low temperature region of the refrigerator, to break a sample into parts, and to adjust the vacuum gap between both electrodes.

Mechanical-controllable break (MCB) junctions [4–6] offer an improved alternative spectroscopic method because those junctions can be prepared in situ at ultra-high vacuum conditions to yield clean interfaces. For clean electrodes the scattering from defects in the barrier region should become less important than the reflection and transmission characteristics of the proper vacuum potential barrier, allowing one to systematically study the characteristics of the junction.

In the hopping regime tunneling spectroscopy has proven useful in investigating the impact of the electron-electron interaction on the density of states of impurity bands. In this transport regime tunnel-

---

[*] Corresponding author: e-mail: sandow@physik.fu-berlin.de, Phone: +49 030 838 53048, Fax: +49 030 838 55487

ing experiments have been performed with conventional metal-barrier-semiconductor contacts [7] and mechanically controllable break junctions [8].

Here we report tunneling spectroscopy measurements with MCB-junction on disordered semiconductors and present a theoretical investigation of the nature of break-junction tunneling current for doped semiconductors in the hopping regime [9].

In tunneling spectroscopy experiments on Ge:As using break junctions [8], the traditional formula $I(U) \propto \int N(V) N(V - eU) [f(V) (1 - f(V - eU))] dV$ for the tunneling current has been applied, where $N(V)$ is the density of states and $f(V)$ the Fermi function. According to this formula the tunneling current is a convolution of the density of states of the contact materials. However, since the nature of the electronic states in such system is very different from that in metals or superconductors it is not obvious that the traditional expression for the calculation of the tunneling current can also be applied to tunneling spectra in the hopping regime. The theoretical investigations published in the literature lead to controversial results [10, 11]. Therefore it is not completely clear which concrete current-voltage relationship can be expected in the hopping regime and to what extent disorder induced fluctuations manifest themselves in the tunneling current.

In the standard theory of hopping transport [12, 13] the impurities are assumed to provide localized electronic states at sites $\boldsymbol{R}_m$ with localization length α^{-1} and characteristic energies V_m. The dynamics of the charge carriers can be described by the simple rate equation with transition probabilities

$$W_{nm} = \nu(|V_{nm}|) \exp\left\{-2\alpha R_{nm} + \frac{\beta}{2} (V_{nm} - |V_{nm}|)\right\}. \tag{1}$$

For a hop from the site n to the site m, $\beta = 1/k_BT$, $R_{nm} = |\boldsymbol{R}_n - \boldsymbol{R}_m|$ is the distance between the sites, and $V_{nm} = V_n - V_m$ is the difference between the site energies. In a close-to-equilibrium transport problem the pre-exponential factor $\nu(|V_{nm}|)$ in Eq. (1), characterizing the ability of the phonon to induce a transition, is usually replaced by a constant, since it affects the transport coefficients only a little. In the description of a tunneling experiment, however, the structure of this quantity is essential, as also pointed out in Ref. [11], and thus has to be taken into account.

For a break junction we consider two samples separated by a distance l. We denote the labels of the left sites with lower case letters m, n and those of the right sites with upper case ones M, N. Both samples have the same densities of states in the absence of the electric field, that is $N_L(V)|_{E=0} = N_R(V)|_{E=0}$. If both samples are separated very far from each other, there are no transitions between left sites and right sites and both samples are in equilibrium. If we now decrease the sample separation, tunneling transitions between the left and the right sample become possible with transition probabilities

$$W_{mM} = \nu(|V_{mM}|) \exp\left\{-2\kappa l_{mM} + \frac{\beta}{2} (V_{mM} - |V_{mM}|)\right\} \tag{2}$$

The tunnel transitions between the left and the right sample do *not* contain the localization length α and the site separation R_{mM}, but the *vacuum tunneling parameter* κ between the two bulk samples and an effective tunneling length $l_{mM} = l + \delta l_{mM}$. The quantity δl_{mM} is a small correction to the tunneling distance between the two samples taking into account the different wave function amplitudes for a given pair (m, M) of localized states. κ^{-1} is of the order of a few Ångstroms, whereas α^{-1} typically takes values around 10 nm. The current j cross the junction can be calculated by averaging the quantity

$$j = -\frac{e}{\Omega}\left[\sum_m \boldsymbol{R}_m \frac{d\rho_m}{dt} \sum_M \boldsymbol{R}_M \frac{d\rho_M}{dt}\right]. \tag{3}$$

Here ρ is the occupation probability of the site m and M. Ω is the total volume which contains the sites in the summation in Eq. (3). We now use the fact that the tunneling rates (2) are orders of magnitudes smaller than the hopping rates (1) so that separate equilibriums are established in the samples. ρ_m and ρ_M are therefore Fermi functions $f_{L/R}(V)$ with chemical potential $\mu_{L/R}$. Applying the principle of detailed balance and performing the configuration average using of the densities of states, we obtain the expression

$$\langle j_x \rangle = e\tilde{W} \int\limits_0^{eU} dV' \int\limits_0^{V'} dV\, N_L(V)\, N_L(V' - eU)\, \nu(V' - V). \tag{4}$$

Here $\tilde{W} \propto \exp(-2\kappa l)$ is given by an integral over the transition probabilities. Furthermore, $N_L(V)$ is the density of states in the left sample and μ_L has been chosen as the zero point of the energy axis.

In investigating the expression (4) we focus on a density of states of the type

$$N_L(V) = N_0 + N_\gamma |V|^\gamma, \tag{5}$$

where $\gamma \approx 2$ for three-dimensional systems at zero temperature and N_0 vanishes at zero temperature [13, 14]. The results for tunneling experiments on systems with such a density of states are depicted in Fig. 1. The experimental data in Fig. 1 show a dip in the density of states for small V. These data indicate that the derivative of the tunneling current in the vicinity of $V \approx 0$ is not zero. However, our Eq. (4) shows that in the hopping regime the derivative of the tunneling current with respect to the voltage at $V = 0$ is always zero, if the transitions are inelastic. If, e.g, we use the deformation potential approximation of Ref. [13], which yields

$$\nu(V) = \nu_0 \frac{|V|}{\left(1 + \left(\dfrac{V}{2\hbar s \alpha}\right)^2\right)^4} \tag{6}$$

(s is the velocity of sound and ν_0 a constant), we find that

$$\langle j_x \rangle \propto V^{3+\zeta}, \tag{7}$$

for small V. Here $\zeta = 2\gamma$ if $N_0 = 0$ and $\zeta = 0$ otherwise. This implies $d\langle j_x \rangle/dV|_{V=0} = 0$, which is contradiction to the experiments (see Fig. 1).

On the other hand, if we assume that the transitions are elastic, so that $\nu(|V - V'|) \propto \delta(V - V')$ then the expression (4) reduces to the result by Bardeen, according to which the derivative of the tunneling current with respect to the voltage is non-zero if $N_0 \neq 0$. A similar result can, however, also be obtained from quasi-elastic approximation, in which

$$\nu(V) = \nu_0 \theta(\omega - |V|). \tag{8}$$

Here ν_0 is a constant frequency and ω is the maximum amount of energy transferred in one hop. In this approximation the formula for the tunneling current takes also the simple form

$$\langle j_x \rangle = e\nu_0 \tilde{W} \omega \int\limits_{\mu_L}^{\mu_L + eU} dV\, N_L(V)\, N_L(V - eU), \tag{9}$$

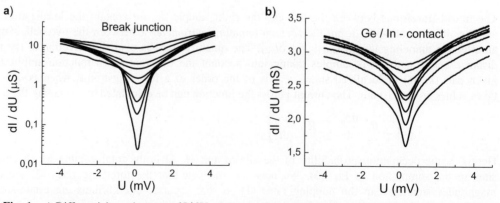

Fig. 1 a) Differential conductance dI/dV of Ge:As for several temperatures, measured by the break-junction device. b) The same, measured by conventional tunnel spectroscopy using an indium contact.

if $\omega \ll eU$. Evaluating the current explicitly to the density of states (5) we obtain

$$\langle j_x \rangle = e v_0 \tilde{W} \omega e U \left[N_0^2 + \frac{2}{\gamma + 1} N_0 N_\gamma |eU|^\gamma + \frac{(\Gamma(1 + \gamma))^2}{\Gamma(2 + 2\gamma)} N_\gamma^2 |eU|^{2\gamma} \right] .$$

Thus the results for the elastic approximation agree with those of the quasi-elastic approximation if the experimental resolution is smaller than the quasi-elastic energy, as it is expected. Accordingly, our results show that the transitions in the experiment of Ref. [8] are either elastic, or quasi-elastic transition with very small energy transfer.

In order to be sure that the break-junction measurements can be interpreted in a similar way as that of a conventional metal-barrier-semiconductor tunnel junction we have performed tunnel spectroscopy of Ge:As in the hopping regime for various temperatures using both techniques. The results are shown in Fig. 1. It is clear that in both cases the Coulomb gap is filled by thermal excitations as the temperature is increased [8]. Although the overall trends in both data sets are the same, there are some differences to be noticed. These difference can be attributed to the fact that in the case of the break-junction the differential conductance is proportional to the convolution of two shifted impurity band spectra whereas in the case of the conventional junction it is proportional to that of the DOS of the semiconductor and the metal.

In conclusion we have shown that break-junction spectroscopy can be used for samples with hopping transport in a similar fashion as it is done for metals.

Acknowledgements O. B. and W. S. are grateful for hospitality at the University of Oregon. Illuminating discussions with H. Böttger, V. L. Bryksin, M. Pollak, A. L. Efros, B. I. Shklovskii, Z. Ovadyahu, K. Gloos, R. Rentzsch, A. N. Ionov are gratefully ackowledged. O. B. acknowledges financial aid by the DFG (Deutsche Forschungsgemeinschaft) under the grant Bl 456/3-1.

References

[1] D. A. Bonnel, Scanning Probe Microscopy and Spectroscopy: Theory, Techniques and Applications, (Wiley, NY, 1993).
[2] J. Bardeen, Phys. Rev. Lett. **6**, 57 (1961).
[3] J. M. Rowell, W. L. McMillan, and W. L. Feldmann, Phys. Rev. **189**, 658 (1969).
[4] J. Moreland and P. K. Hansma, Rev. Sci. Instr. **55**, 399 (1984).
[5] J. Moreland and J. W. Ekin, J. Appl. Phys., Rev. **58**, 3888 (1985).
[6] C. J. Muller, J. M. van Ruitenbeek, and L. J. de Jongh, Physica C **191**, 485 (1992).
[7] J. G. Masey and Mark Lee, Phys. Rev. Lett. **75**, 4266 (1995); **77**, 3399 (1996).
[8] B. Sandow, K. Gloos, R. Rentzsch, A. N. Ionov, and W. Schirmacher, Phys. Rev. Lett. **86**, 1845 (2001); **89**, 229702 (2002).
[9] O. Bleibaum, B. Sandow, and W. Schirmacher, to be published in Phys. Rev. B.
[10] V. I. Kozub, Phys. Rev. Lett. **89**, 229701 (2002).
[11] A. I. Larkin and B. I. Shklovskii, phys. stat. sol. (b) **230**, 189 (2002).
[12] H. Böttger and V. V. Bryksin, Hopping Conduction in Solids (Akademie-Verlag, Berlin, 1985).
[13] A. L. Efros and B. I. Shklovskii, Electronic Properties of Doped Semiconductors (Springer Verlag, Berlin, 1984).
[14] M. Pollak and M. Ortuno, in: Electron–Electron Interaction in Disordered Systems, edited by. A. L. Efros and M. Pollak, North-Holland, Amsterdam, p. 287 (1985).
[15] see Ref. [13], p. 84, formula (4.2.18).

phys. stat. sol. (c) **1**, No. 1, 96–100 (2004) / **DOI** 10.1002/pssc.200303646

A study of the bolometric capabilities of p-type GaSb and InSb:Ge for phonon pulse experiments

B. M. Taele[*,1] and **R. Mukaro**[2]

[1] School of Physical Sciences and Engineering, National University of Lesotho, P.O. Roma 180, Lesotho
[2] Department of Physics, Bindura University of Science, P. Bag 1020, Bindura, Zimbabwe

Received 1 September 2003, revised 18 September 2003, accepted 30 October 2003
Published online 10 December 2003

PACS 29.40.Vj, 72.20.Ee, 73.61.Ey

The potential of p-type GaSb and InSb:Ge for use as low-temperature hopping bolometric detectors of particles and photons has been investigated. In the range $1.4\ \text{K} \le T \le 4.2\ \text{K}$, GaSb conductivity followed $\ln \rho(T) \propto \exp (T_0/T)^{1/4}$ law with sensitivity, $\alpha(T) = (1/R)dR/dT$, varying from $0.6\ \text{K}^{-1}$ to $0.2\ \text{K}^{-1}$. As a detector at 1.4 K, phonon pulses generated in a silicon absorber bonded to GaSb were easily distinguished with a response time of $1\ \mu\text{s}$. Conductivity in InSb detectors exhibited thermally activated-hopping conduction, $\rho(T) \propto \exp(\varepsilon_3/k_\text{B}T)$, with an activation energy of $\varepsilon_3 \sim 0.12\ \text{meV}$. In the range $1.4\text{K} \le T < 4.2\text{K}$, $\alpha(T)$ of InSb varied from $0.7\ \text{K}^{-1}$ to $0.2\ \text{K}^{-1}$ with an abrupt increase above 4.2K rising to a maximum value of 8.1K^{-1} at 4.5K. For the range $5\ \text{K} \le T \le 10\ \text{K}$, $\alpha(T)$ had an exponential decay.

1 Introduction The rapid temperature dependence of the hopping conduction of doped semiconductors is widely used as a bolometric mechanism for low-temperature detectors of X-rays and exotic dark-matter particles [1]. In this conduction regime, charge carriers *hop* from an occupied dopant site to an unoccupied one. The uniformity of the dopant distribution, degree of compensation and dopant concentration are critical parameters. The temperature dependence of the device resistance in this regime is of the form $R(T) = R_0(T)\exp (T_0/T)^n$ (in the low-field limit) where the parameters n, T_0 and $R_0(T)$ relate to the particular type of hopping that is active [2]. $R_0(T)$, T_0 and n depend on the details of the density of states (DOS) around the Fermi energy, $N(E_\text{F})$. The performance of a hopping bolometric detector is related to the sensitivity $\alpha(T)$ defined by $(1/R)\partial R/\partial T$. The higher the sensitivity, the higher is the response to an external excitation and the signal-to-noise ratio. Since $\alpha(T) = (n/T)(T_0/T)^n$ and T_0 is a decreasing function of the doping level it is possible, in principle, to make $\alpha(T)$ high by making the doping level sufficiently small.

Two of the most commonly used hopping systems are neutron transmutation-doped germanium and ion-implantation-doped silicon. These bolometers have many desirable characteristics, such as; high purity, sensitivity and reproducibility. However, the speed of response and hence detector dead time, are limited by the timescale of thermalisation. This in turn, depends on the magnitude of the anharmonic interaction determining phonon frequency down-conversion, and the electron–phonon coupling strength through which the electrons and phonons come to thermal equilibrium. Since Ge and Si bolometers get strong electron–phonon decoupling at low temperatures, so no thermal equilibrium is attained [3]. For a given phonon frequency both these quantities are larger in III–V semiconductors, because of the lower Debye frequency and the additional coupling mechanism of piezoelectric and deformation potentials. Because of the additional channels for the electron–phonon interactions, GaSb, InSb and other III–V bolometers may be expected to have faster response times and hence to resist electron–phonon decoupling down to lower temperatures. Used as absorbers, the narrow-band gaps of GaSb (0.72 eV) and InSb

---

[*] Corresponding author: e-mail: bm.taele@nul.ls

(0.17 eV) make them more suitable for X-ray absorption compared to Si and GaAs due less charge trapping effect. Indium is also a candidate material for detecting solar neutrinos through the inverse β-decay reaction in [115]In [1]. Since GaSb and InSb are thoroughly investigated semiconductors, and the technology of preparation of single crystals is highly advanced. There are few previous studies of these materials as low temperature bolometers. For this reason we have made preliminary investigations exploring general principles. Previous work carried out on the low temperature dependence of resistance of these semiconductors suggests that they would have responsivities and sensitivities, which compare favourably with ion-implanted Si and NTD Ge bolometers [4].

2 Experiment Bolometers have been constructed using Czochralski-grown undoped GaSb ⟨100⟩ single crystals and Ge-doped InSb $(4.5 \times 10^{14} \, \text{cm}^{-3})$. Undoped bulk grown GaSb is always p-type with a hole concentration of the order $(1-2) \times 10^{17} \, \text{cm}^{-3}$ at 300 K. The native defect responsible is Ga on an Sb site. The centre is doubly ionised with the first and second levels lying 80 and 33 meV above the valence band edge [5]. Amongst the narrow-gap III–V semiconductors, InSb has the highest mobility with the smallest electron effective mass and the largest negative g-factor. Ge in InSb is an acceptor with ionisation energy 10 meV. Samples were then cut and lapped to size (approximately $0.5 \times 2 \times 4 \, \text{mm}^3$). The substrates were thoroughly cleaned with organic solvents, rinsed in de-ionised H_2O and blown dry. Oxidised layers were eliminated by soaking samples in 10% HCl, and Ohmic contacts were made by vacuum-deposition of Au.

The detectors were characterised by heat pulse technique [6], usually associated with studies of non-equilibrium phonons than of particles. However, the basic process, a change in device resistance resulting from phonon excitation is the same in both applications. Heat pulses were generated by electrical excitation (a few milliwatts) of a thin Al film evaporated on the backside of a polished Si wafer, to which a bolometer is acoustically bonded with a thin epoxy resin. The Al film was designed to have a resistance of 50Ω at liquid helium temperatures. Phonon pulses reaching the bolometer were typically microwatts in power and nanoseconds in duration, having a Planckian distribution with an approximate temperature of around 17 K which corresponds to a dominant frequency of 1 THz. The phonon spectrum emitted by a thin film generator is a Planck distribution at an effective temperature, T_{eff}, related to the power dissipated per unit area, $P(T)$, through the expression, $P(T) \approx B_0 T_{\text{eff}}^4$ where B_0 has the value of 203 $W K^{-4} m^{-2}$ [7]. Typical power levels required to give acceptable sensitivity are a few $mWmm^{-2}$. In operation, the bolometer is provided with a bias current from a dc source through a load resistor R_L.

3 Results, analysis and discussion Low-temperature dependence of the dc resistance of GaSb bolometer is shown on a logarithmic scale as a function of $T^{-1/4}$ in Fig. 1. Using a least-squares method to fit the data to the VRH formula, we found that a $T^{-1/4}$ law gave a considerably better fit than the $T^{-1/2}$ law. The

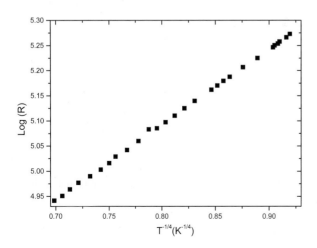

Fig. 1 Temperature dependence of GaSb bolometric detector in the range 1.4–4.2 K at a bias current of 1 μA.

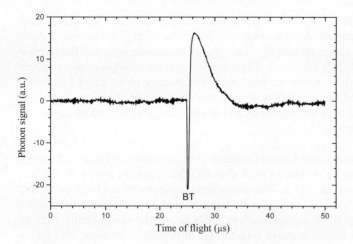

Fig. 2 GaSb bolometric phonon pulse signal detected at 1.40 K.

uncertainty in the resistance measurements was typically less than 0.1%. The $T^{-1/4}$ law in p-type GaSb corresponds to Mott VRH conduction, and has been reported previously [4]. The slope of the $\ln R$ versus $T^{-1/4}$ plot determined in a measurement at thermal equilibrium gives T_0 directly, and in this case was found to be 140 K. $\alpha(T)$ deduced from the derivative of the plot of $\ln R$ versus T is given by $-(-0.42 - 3.46 \ln T)^{-1} \mathrm{K}^{-1}$, and is 0.63 K^{-1} at 1.40 K. The magnitude of $\alpha(T)$ falls by a factor of more than 3 as the bath temperature is increased from 1.4 K to 4.2 K. Using the standard expression for Mott VRH, $k_B T_0 \approx 18/N(E_F)\xi^3$ [where ξ is the localisation length] and combining it with the Mott's expression for the hopping length, $R_{\mathrm{hop}} = \{(8\pi/9)k_B T \times N(E_F)/\xi\}^{-1/4}$ [8] one obtains $R_{\mathrm{hop}}/\xi \approx 0.4(T_0/T)^{1/4}$, which shows that the ratio R_{hop}/ξ is of order of unity in VRH regime studied.

Figure 2 shows the detected phonon-pulse versus time after traversing a 500 μm thick Si absorber following heater pulse excitation power of approximately 10 mW, pulse duration 50 ns, using a bias of current 1.0 μA at 1.4 K. A sharp initial negative spike due to an inductive coupling between the generator and the bolometer marked the real time of flight zero point and this was followed by a broader feature due to the arrival of phonons after multiple reflections in the Si substrate. If we consider that arriving phonon pulses of energy ΔW raise the temperature of the detector by ΔT ($T \gg \Delta T$) where $\Delta W = C\Delta T$ and the specific heat $C \sim T^3$, the voltage excursion ΔV_B, developed across R_L is given by $\Delta V_B \approx I(R_L/R_B) \times (dR_B/dT)\Delta T$. Hence the signal voltage induced by the phonon pulse is proportional to $\Delta W/T^3$ and a change $\Delta V_B/\Delta T$ in the voltage with temperature is proportional to the change $\Delta R_B/\Delta T$ in bolometer resistance with temperature. Thus from Fig. 1, we can relate directly the output signal to the change in temperature.

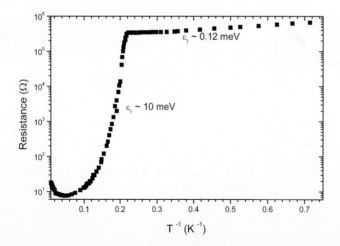

Fig. 3 Temperature dependence of the InSb:Ge bolometer resistance plotted as ln R versus 1/T.

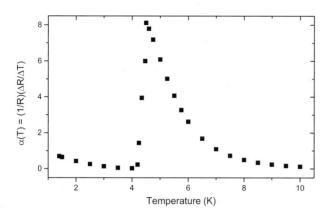

Fig. 4 Sensitivity (α) of the InSb:Ge bolometric detector versus temperature in the range 1.4 K to 10 K.

The temperature dependence of $R(T)$ determined in the range $1.4\text{K} \leq T \leq 10\text{K}$ for InSb(Ge) detector is shown in Fig. 3. Two activated conduction mechanisms are clearly seen. At these temperatures the extrinsic activation regime is attained and thermal freeze-out is achieved at around 5 K. The $R(T)$ curve shows a very rapid change in resistance in the range 5 K to 6 K, where the resistance changed from $25\,\text{k}\Omega$ to $350\,\text{k}\Omega$. In the range $1.4\,\text{K} \leq T \leq 4.2\,\text{K}$, the variation in the resistance is weaker, only changing by a factor of 2 (from about $350\,\text{k}\Omega$ at 4.20 K to just under $700\,\text{k}\Omega$ at 1.4 K).

All the data fall on the curve $R^{-1}(T) = [R_1 \exp(\varepsilon_1/k_B T)]^{-1} + [R_3 \exp(\varepsilon_3/k_B T)]^{-1}$ following the Fritzsche's formula [9]. The activation energy ε_1 associated with carriers excited in the valence band from acceptor impurity centres was found to be 10meV. In the range $1.40\,\text{K} \leq T \leq 4.2\,\text{K}$, ε_3 was found to be 0.12 meV. The conductivity fits the T^{-1} dependence, suggesting hopping between neighbouring acceptor centres, with the pre-factors R_1 and R_3 given by $2.3 \times 10^{-6}\,\Omega$ and $1.6 \times 10^5\,\Omega$, respectively. The overall behaviour and orders of magnitudes of the pre-factors and energies are in agreement with the work of Gershenzon [10]. Figure 4 shows $\alpha(T)$ as a function of temperature in the range $1.40 \leq T \leq 10\,\text{K}$. As seen from the graph, there are two distinct regions of the curve. Following a smooth decrease in $\alpha(T)$ from 1.4 K to 4.2 K, as would be expected, an abrupt increase was observed reaching a maximum value of about $8.1\,\text{K}^{-1}$ at 4.5 K, after which an exponential decay down to 10 K followed. This value is to be compared with $0.7\,\text{K}^{-1}$ obtained at 1.4 K, the temperature of the phonon-pulse experiment. $\alpha(T) = 0.7\,\text{K}^{-1}$ is comparable with $\alpha(T) = 0.63\,\text{K}^{-1}$ of GaSb which suggest a comparable performance as a VRH phonon detector. For InSb detectors, the regime where the resistance is a very strong function of temperature, and the magnitudes $\alpha(T)$ is highest is not accessible for our phonon-pulse measurements. Temperatures just above ^{4}He are usually very difficult to control with accuracy, and there is no efficient cooling of the bolometer system, hence unsuitable for particle and photon detection experiments.

In summary, the feasibility of p-type GaSb and InSb for low temperature phonon detection has been demonstrated. Ideally, it should be possible to improve the performance for both significantly by making smaller detectors and operating them at temperatures lower than currently used. With further optimisation, promising applications of these bolometers are in nuclear spectroscopy, solar neutrinos and for investigating phonon interactions in low-dimensional semiconductor structures/devices grown on GaSb and InSb substrates.

Acknowledgements We especially want to thank Professor J. K. Wigmore for his continued support and for providing me the appropriate environment for performing this work.

References

[1] D. Twerenbold, Rep. Prog. Phys. **59**, 349 (1996).
[2] B. I. Shklovskii and A. L. Efros, Electronic Properties of Doped Semiconductors (Springer-Verlag, Berlin, 1984).
[3] D. McCammon, M. Galeazzi, and D. Liu, et al., phys. stat. sol. (b) **230**, 194 (2002).

[4] S. V. Demishev, Int. J. Mod. Phys. B **8**, 865 (1994).
[5] K. Nakashima, Jpn. J. Appl. Phys. **20**, 1085 (1981)
[6] B. M. Taele, N. S. Lawson, and J. K. Wigmore, Nucl. Instrum. Methods A **444**, 46 (2000).
[7] B. M. Taele, PhD Thesis, Lancaster University, 2000.
[8] N. F. Mott, Conduction in Non-Crystalline Materials (Oxford University Press, Oxford, 1986).
[9] H. Fritzsche, J. Phys. Chem. Solids **6**, 69 (1958).
[10] E. M. Gershenzon, I. N. Kurilenko, and L. B. Litrak-Gorskaya, Sov. Phys. – Semicond. **8**, 689 (1974).

phys. stat. sol. (c) **1**, No. 1, 101–104 (2004) / **DOI** 10.1002/pssc.200303652

Variable range hopping in hydrogenated amorphous silicon

I. P. Zvyagin[*], **I. A. Kurova**, and **N. N. Ormont**

Faculty of Physics, Moscow State University, 119992 Moscow, Russia

Received 1 September 2003, revised 16 September 2003, accepted 16 September 2003
Published online 28 November 2003

We study the effect of high-temperature annealing on the conductivity temperature dependence of a-Si:H films. Two types of films were studied: uniform and nanolayered prepared by cyclic repetition of CVD and hydrogen plasma treatment. It is shown that for annealing temperatures above 450 °C, the low-temperature conductivity obeys the law $\ln \sigma = \text{const} - (T_0/T)^n$, where the exponent is $n \cong 1/4$ for uniform and $n \cong 1/3$ for layered films indicating two-dimensional conduction for layered films. It is argued that the measurements of the conductivity temperature dependence can provide information about spacial distribution of defects produced by HTA and about the defect creation mechanism.

1 Introduction   Variable range hopping (VRH) conduction was currently observed in sputtered amorphous silicon (a-Si) films (both as-deposited and hydrogenated), where the density of states in the gap is sufficiently large (about $10^{19} \text{eV}^{-1} \text{cm}^{-3}$) [1–3]. This conduction was usually attributed to hopping between deep localized states at the dangling bonds (DBs) near the Fermi level. On the other hand, for undoped glow-discharge hydrogenated amorphous silicon (a-Si:H) films hopping conduction is not observed, due to a low density of DB localized states (about $10^{16} \text{ eV}^{-1} \text{cm}^{-3}$).

In what follows we report on the observation of VRH in glow-discharge a-Si:H films after high-temperature annealing (HTA) producing hydrogen effusion (dehydrogeniztion). We show that for nominally uniform films the temperature dependence of the conductivity is well described by the three-dimensional Mott's law, whereas for films obtained by cyclic CVD with periodic hydrogen plasma (HP) treatment, the observed conductivity temperature dependence indicates the two-dimensional hopping conduction. We argue that the measurements of hopping conduction can provide information about the defects introduced by HTA.

2 Experimental   We studied two types of a-Si:H films that we call uniform and layered. The uniform films were obtained as usual by continuous plasma chemical vapor deposition (CVD) form silane – argon mixture in a diode h.f. system; the resulting DB concentration was below 10^{16}cm^{-3}. The layered films were obtained by alternatively repeating thermal CVD from silane – argon mixture and *in situ* hydrogen plasma (HP) treatment at the deposition temperature (250 °C) (the deposition process is described in more detail in [4]). A similar procedure of alternatively using CVD from silane – argon mixture and HP treatment was earlier used (under different preparation conditions: gas mixture composition, deposition temperature) with the aim of improving photoelectric properties of the material [5, 6]. It was shown that the *in situ* HP treatment of CVD a-Si:H surface can result in chemical annealing producing a decrease in dangling bond concentration and an increase in photoconductivity. We study multilayer films consisting of narrow alternating nanolayers of thicknesses 15 nm and 3 nm, respectively (the total number of periods in our films was 28). The thinner layers created by hydrogen diffusion from the surface during HP treatment had a structure different from that of thick layers and a much higher hydrogen content (above 20 at.%) [4]. We studied the effect of vacuum HTA on the film properties.

---

[*] Corresponding author: e-mail: zvyagin@sc.phys.msu.su, Phone: 007 095 939 37 31, Fax: 007 095 939 37 31

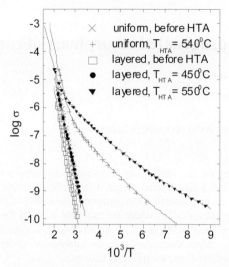

Fig. 1 Temperature dependence of the conductivity of a-Si:H films: the uniform film before and after annealing at 540 °C and the layered film before and after annealing at 450 °C and 550 °C for 30 min. Solid lines correspond to Eq. (2), where the parameters of the hopping and band contributions are determined as described in the text.

The temperature dependence of the conductivity of both uniform and layered films was measured before and after HTA. In Fig. 1 we plot the temperature dependence of the conductivity of a uniform film before and after annealing at 540 °C and the temperature dependence of the conductivity of layered films before and after annealing for 30 min. at 450 °C and 550 °C (for all films the annealing time was 30 min.). We see that in contrast to activated behaviour of unannealed films and films annealed at low temperatures, the film conductivity after HTA becomes nonactivated and large in the region of low temperatures. This behavior is interpreted as the onset of hopping. It is known that variable range hopping conductivity is well described by the relation

$$\ln \sigma_h = \ln \sigma_0 - (T_0/T)^n, \tag{1}$$

where σ_0 is the hopping conductivity prefactor, T_0 is the Mott parameter and the exponent is $n = 1/4$ in the three-dimensional case (Mott's law) and $n = 1/3$ in the case of two dimensions in the absence of Coulomb gap effects [7, 8].

It can be verified that our data in Fig. 1 at low temperatures are well described by Eq. (1). To find the exponent n, we use the procedure suggested by Zabrodskii (e.g., see [9]) and plot $\ln w$, where $w = d\ln /d\ln T$, as a function of $\ln T$. A straight line $\ln w = a - b \ln T$ indicates a dependence of the form (1), where the slope is $b = n$ and a is expressed in terms of T_0. The dependence $\ln w$ vs. $\ln T$ is shown in Fig. 2 for the uniform and layered films after HTA. We obtain that for the uniform film for temperatures below 340 K, the conductivity σ obeys Eq. (1) with $n = 0.23\pm0.03$ whereas for the layered film $n = 0.35\pm0.03$. This result indicates that for the layered film after HTA conduction is two-dimensional. Moreover, for uniform and layered films setting n equal to 1/4 and 1/3, from the data of Fig. 1 we obtain $T_0 = T_{03} = 3.4\times10^8$ K and $T_0 = T_{02} = 2.0\times10^6$ K, respectively.

We can use a standard procedure to decompose the conductivity into a sum of the band and hopping contributions,

$$\sigma = \sigma_b + \sigma_h, \tag{2}$$

where the hopping term is given by Eq. (2) and σ_b is the band contribution.

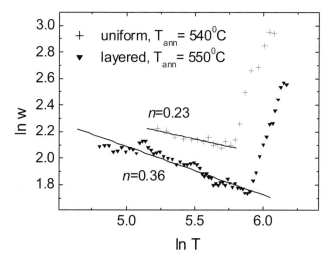

Fig. 2 ln w vs. ln T-plots for the uniform and layered films after HTA. Straight lines show the best fits of the low-temperature data and the corresponding exponents n.

The procedure is illustrated in Figs. 3a,b; in Fig. 3a we plot log σ as a function of $(T_0/T)^{1/4}$ for the uniform film annealed at 540^0C. We fit the low-temperature nonactivated parts of the curves by Eq. (1) and then after subtracting the hopping contribution determined by Eq. (1) from the measured total conductivity, we deduce the band contribution σ_b. The results of this procedure are shown in Fig. 3b; the activation energy of the band conductivity thus obtained is 0.79 eV. Similarly, we can find the temperature dependence of the band contribution of the layered films after HTA; the activation energy of the band conductivity in these films is 0.80 eV.

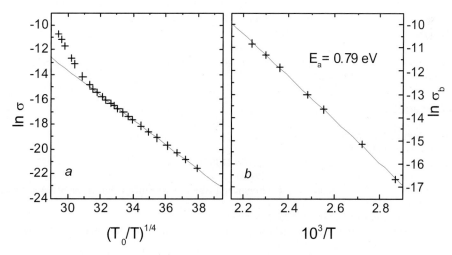

Fig. 3 a) conductivity as a function of $(T_0/T)^{1/4}$ for a uniform film after HTA at 540^0C; solid line corresponds to the hopping contribution extrapolated to higher temperatures according to Eq. 1 with $n = 1/4$; b) the band contribution calculated from the high-temperature data.

3 Discussion In Fig. 3b we see that for uniform films, the activation energy for band conduction after HTA is essentially the same as before HTA. This shows that HTA introducing DBs and giving rise to hopping in the low-temperature range does not substantially change the position of the localized states in the gap that pin the Fermi level with respect to the conduction band edge. By measuring hopping conduction, one can evaluate the localized state parameters after HTA by using the above values of T_{03}. For annealed uniform films, setting the localization radius equal to 5×10^{-8}cm, we derive that the density of states at the Fermi level is $4.8\times10^{18}\,\text{eV}^{-1}\text{cm}^{-3}$. Thus we see that HTA and hydrogen effusion can increase the density of localized states by more than two orders of magnitude compared to the material before HTA.

For layered films before HTA the activation enegies are higher (about $E_\sigma = 0.87$, see Fig. 1); indeed, layered films are characterized by the presence of layers with high hydrogen content; it is known that the material with hydrogen concentration exceeding about 20 at.% has a larger bandgap attributed to the increase in the concentration of SiH_2-bonds [6] and greater dark conductivity activation energies E_σ. The activation enegy decreases (to about 0.8 eV) after HTA and becomes almost the same as for uniform films after hydrogen effusion.

Another characteristic feature of layered films is the observation of two-dimensional hopping after HTA; this implies a different mechanism of defect generation in these films compared to the generation of DBs in the bulk of uniform films. For layered films, using the same value of the localization radius, we find that the effective two-dimensional density of states at the Fermi level is $1.3\times10^{13}\text{eV}^{-1}\text{cm}^{-2}$; the observation of two-dimensional hopping shows that the DBs created by HTA in the thin (hydrogen-rich) layers control the low-temperature conductivity. As noted above, the characteristic feature of hydrogen-rich a-Si:H is a large concentration of SiH_2 bonds, which control many of the material properties. Therefore, it can be assumed that DB creation during HTA accompanied by hydrogen effusion is related to SiH_2-bond breaking. In films with variable hydrogen content, this gives rise to DB creation predominantly in hydrogen-rich regions. For the layered films studied, it implies highly nonuniform distribution of DBs created by HTA: even for moderate annealing temperatures, the concentration of DBs in these layers after HTA becomes so high that at low temperatures the two-dimensional hopping conduction along the layers dominates over the bulk hopping.

Thus, the study of the temperature dependence of the conductivity in a-Si:H films after HTA indicates the onset of hopping at low temperatures. The conductivity measurements can be used not only to determine the variation of material parameters (such as the Fermi level position and the density of states at the Fermi level) but also to obtain information about the mechanisms of structural modification of the material by HTA.

Acknowledgements The work was supported by the Russian Foundation for Basic Research, the programs "Integration" and "Universities of Russia".

References

[1] M. I. Knotek, Solid State Coomun. **17,** 1431 (1975).
[2] A. Lewis, Phys. Rev. B **13,** 2505(1976).
[3] A. Stesmans and Y. Wo, Solid State Commun. **62,** 435 (1987).
[4] V. P. Afanas'yev et al., Semiconductors **36,** 38 (2002).
[5] I. Sakata et al., in: Solid State Phenomena, Hydrogenated Amorphous Silicon, Ed. H. Neber-Aeschbacher, Vol. 44-46, 127 (1995).
[6] K. Fukutani et al., J. Non-Crystal. Solids **227–230,** 63 (1998).
[7] B. I. Shklovskii and A. L. Efros, Electronic Properties of Doped Semiconductors, Springer-Verlag, Heidelberg, 1984.
[8] I. P. Zvyagin, Transport Phenomena in Disordered Semiconductors, Moscow, MSU Publ., 1984.
[9] A. G. Zabrodskii, Philos. Mag. B **81,** 1153 (2001).

phys. stat. sol. (c) **1**, No. 1, 105–108 (2004) / **DOI** 10.1002/pssc.200303653

Superlattices with intentional disorder: impurity-assisted vertical hopping

I. P. Zvyagin[*] and **K. E. Borisov**

Faculty of Physics, Moscow State University, 119992 Moscow, Russia

Received 1 September, revised 17 September 2003, accepted 17 September 2003
Published online 28 November 2003

PACS 72.20.Ee, 73.63.–b

We discuss a new mechanism of vertical hopping transport in superlattices with intentional disorder. It is shown that at low temperatures impurity-assisted hopping can dominate over the phonon-assisted processes considered in conventional hopping theories.

1 Introduction Superlattices with intentional disorder (SLID) are multiple quantum-well structures in which a given random size quantization level distribution is realized by controlling the well width variation during the structure growth [1]. Such structures are convenient for studying the problem of localization, in particular, the interplay between disorder and Coulomb effects. Most of the experiments were made on silicon-doped GaAs/GaAlAs SLIDs both by optical methods and by direct measurements of vertical (in the direction of the growth axis) conductivity [2]. For sufficiently large vertical disorder, electron transport is known to be hopping, i.e., it is due to inelastic tunelling between electronic states localized in the vertical direction at different wells of the SL [3].

For the standard situation of hopping between simple localization centers with a single (or a few) levels per site, only inelastic electron hops are possible (usually, involving phonons). The situation is different for SLIDs. Indeed, in this case the problem is reduced to that of macrosites with continuous energy spectrum that are localized at quantum wells for strong disorder. Since intrasite transition rates are usually much larger than the rates of intersite (vertical) transitions, the electron distribution in the macrosites is quasiequilibrium and vertical transport is determined by the integral transition rates $\Gamma_{\lambda\lambda'}$ obtained by summing the individual transition rates over the initial and final states in the macrosites λ [4],

$$\Gamma_{\lambda\lambda'} = \sum_{k_\| k_\|'} W_{\lambda k_\| \lambda' k_\|'} f_{\lambda k_\|}^{(0)} (1 - f_{\lambda' k_\|'}^{(0)}). \tag{1}$$

Here the electron states are $\{\lambda \ k_\|\}$, λ is the index for the solutions of the one-dimensional problem with the potential corresponding to the band edge variation along the growth axis, $k_\|$ is the two-dimensional wave vector for the free electron propagation along the well planes, $W_{\lambda k_\| \lambda' k_\|'}$ is the transition probability and $f_{\lambda k_\|}^{(0)}$ is the equilibrium (Fermi) distribution. The electron path is a succession of hops that includes both intrasite inelastic (phonon-assisted) transitions with no spatial displacement and intersite rate-limiting transitions, which determine the electron transfer in space. Intersite transitions can be both inelastic (phonon-assisted) and elastic (phonon or impurity-assisted). In what follows, we

---

[*] Corresponding author: e-mail: zvyagin@sc.phys.msu.su, Phone: 007 095 939 37 31, Fax: 007 095 939 37 31

evaluate the integral impurity-assisted transition rates between states localized at different macrosites of a SLID and show that under reasonable conditions at low temperatures hopping transport can be determined by impurity-assisted transitions rather than by the standard mechanism of phonon-assisted hopping.

2 Impurity-assisted transition rates As usual, in the envelope function approximation the wave functions of electronic states in SLIDs can be written in the form,

$$\psi_{\lambda k_{\parallel}}(z) = S^{-1/2} \varphi_{\lambda}(z) \exp(i\, k_{\parallel}\rho), \tag{2}$$

where S is the cross-sectional area, z is the coordinate in the SLID growth direction, ρ is the in-plane position vector and $\varphi_{\lambda}(z)$ is the wave function for an eigenstate λ with energy ε_{λ} corresponding to the solution of the one-dimensional problem with the potential describing the modulation of the conduction band edge. The energies of the states $\{\lambda\, k_{\parallel}\}$ are $E_{\lambda k_{\parallel}} = \varepsilon_{\lambda} + \hbar^2 k_{\parallel}^2 / 2m$. For the structures considered, typical well widths are such that for a single-well problem, upper size quantization levels in the wells lie much higher than the lowest levels. Therefore, in what follows we neglect contributions from all dimensional subbands except the lowest one. Moreover, we consider the case of sufficiently strong disorder in which the states $\{\lambda_{\parallel}k_{\parallel}\}$ are strongly localized in the z-direction (along the growth axis) and the $\varphi_{\lambda}(z)$ are close to the "atomic-like" wave functions $\chi_n(z)$ (n is the well number) with small admixture of wave functions of neighboring wells and the state energies ε_{λ} are close to the eigenvalues ε_n of the corresponding one-well problem.

For simplicity, we describe the impurity potential by using the model of zero-radius potential

$$U(\mathbf{r}) = U_0 \delta(\mathbf{r} - \mathbf{R}_i)\{1 - (\mathbf{r} - \mathbf{R}_i)\nabla\}, \tag{3}$$

where U_0 is the potential parameter and $\mathbf{R}_i = \{\rho_i, z_i\}$ is the position vector of impurity atom i.

Now the integral impurity-assisted transition rates between quantum wells can be explicitly calculated. Note that when calculating the matrix elements in Eq. (1), we choose the hybridized wave functions as a basis, since the hybridization time τ_h (or the time of coherent electron propagation between the wells) is much smaller than the phase-breaking time τ_{φ}. Indeed, the hybridization time can be estimated as the inverse of the period of density oscillations in a double-well structure with different energy levels, $\tau_h \approx \hbar / \Delta$, where Δ is the interlevel distance. For $\Delta = 10$ meV we obtain $\tau_h \approx 10^{-13}$ s. On the other hand, the phase-breaking time τ_{φ} is mainly due to intrasite inelastic scattering by phonons; this time was estimated to be about 10^{-10} s [5].

Now we easily obtain

$$\Gamma^{(\text{imp})}_{\lambda\lambda'} = \frac{mU_0^2}{16\pi\hbar^3} |\varphi_{\lambda}(z_i)|^2 |\varphi_{\lambda'}(z_i)|^2 kT \exp\left\{-\frac{\varepsilon - \mu}{kT}\right\}, \tag{4}$$

where μ is the Fermi level and ε is the greater of the two energies, ε_{λ} and $\varepsilon_{\lambda'}$. Equation (3) describes quasielastic interwell transitions (with $E_{\lambda k_{\parallel}} \approx E_{\lambda' k'_{\parallel}}$), which are accompanied by substantial variation of the kinetic energy, i.e., of $k_{\parallel}$. The activation factor here appears from the statistical factors describing the population of the initial or final state.

The dependence of the integral transition rates (4) on the impurity position along the Oz-axis is mainly related to the form of electron wave functions describing the initial and final states. The maximum of the transition rates is obtained for impurities lying inside the wells where the impurities are neutral at low temperatures; this indicates that Eq. (3) can be a reasonable approximation for impurity potential under these conditions.

Summing over the impurity atoms with bulk uniform concentration N, we obtain the total integral transition rate

$$\Gamma_{\lambda\lambda'}^{(imp)} = \frac{SNmU_0^2}{16\pi\hbar^3}kT \int |\varphi_\lambda(z)|^2 \cdot |\varphi_{\lambda'}(z)|^2 \, dz \cdot \exp\left(-\frac{\varepsilon-\mu}{kT}\right). \tag{5}$$

3 Impurity-assisted versus phonon-assisted hopping It is known that at high temperatures (above about 70 K) intersite transition rates are dominated by optical phonon-assisted processes [1, 3]; in what follows we are mostly interested in the lower temperature region, where transitions involving acoustic phonons become important.

Acoustic phonon-assisted transition rates were analyzed in [6]. The result is
(a) for $T < T_0$,

$$\Gamma_{\lambda\lambda'}^{(ph)} \approx \frac{E_{ac}^2 \pi^2 S \, I_1}{4\rho s a^2}kT \exp\left(-\frac{\varepsilon-\mu}{kT}\right) \tag{6}$$

and
(b) for $T > T_0$,

$$\Gamma_{\lambda\lambda'}^{(ph)} \approx \frac{E_{ac}^2 \pi \, S \, I_2}{4\rho\hbar s^2 a} (kT)^2 \exp\left(-\frac{\varepsilon-\mu}{kT}\right). \tag{7}$$

Here $T_0 = 2\pi\hbar s/a$, E_{ac} is the deformation potential constant, ρ is the material density, s is the velocity of sound, a is the average width of quantum wells, $I_\alpha = \int x^{2-\alpha} M(x) dx$ and $M(x) = \left|\langle \varphi_\lambda(z) | \exp(i\pi xz/a) | \varphi_{\lambda'}(z) \rangle\right|^2$.

As shown in [6], the acoustic phonons that take part in vertical electron transitions are almost two-dimensional; their energies $\hbar\omega$ are typically smaller than kT_0 and their effective density of states is $D_{ph} \sim \omega \cdot kT$. Acoustic phonon-assisted transition rates are proportional to the number of phonons, i.e., to $D_{ph}N_{ph}$, where N_{ph} is the Bose-Einstein function. For $T > T_0$ we have $\hbar\omega < kT$ and $D_{ph}N_{ph} \sim (kT)^2$. In the low-temperature range for typical phonons $\hbar\omega \geq kT$ and $D_{ph}N_{ph} \sim \omega \, kT \exp(-\hbar\omega/kT)$; note that the factor $\exp(-\hbar\omega/kT)$ is cancelled by the exponential factor in the occupation probability of electronic states and we arrive at the expression (6) for the prefactor.

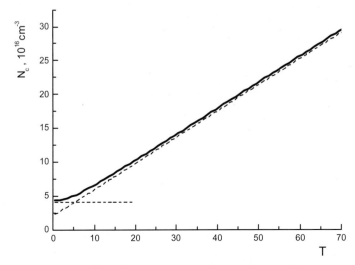

Fig. 1 Temperature dependence of the critical bulk concentration N_c for a uniformly doped SLID.

© 2004 WILEY-VCH Verlag GmbH & Co. KGaA, Weinheim

Though the main exponential temperature dependence of the integral impurity-assisted transition rate (5) is the same as for phonon-assisted transitions, the preexponential factors have a significantly different temperature dependence. By using Eqs (5), (6) and (7), we can estimate the critical impurity concentration for which the impurity-assisted and phonon-assisted transition rates become comparable.

For a uniformly doped SLID, the temperature dependence of the critical bulk concentration N_c is shown in Fig. 1 (the following values of the parameters were used in the calculation: $E_{ac} = 8.6$ eV, $\rho = 1.53$ g/cm^3, $s = 3.7 \times 10^5$ cm/s, the conduction band offset $V_0 = 0.3$ eV, the well widths $a_1 = 7$ nm and $a_2 = 5.9$ nm, the barrier width $b = 4$ nm and the parameter $U_0 = 6.25 \times 10^{-21}$ eV cm^3).

4 Discussion   Thus we see that, in contrast to conventional hopping, for $N > N_c$, for the low-temperature vertical conduction mechanism considered, the impurity-assisted quasielastic tunnelling transitions can be rate-limiting at reasonable impurity concentrations. Under these conditions, the actual hopping path is a succession of fast intrasite phonon-assisted inelastic transitions and slow intersite quasielastic impurity-assisted transitions. The prefactor in the expression for the interwell transition rate is just the well-known "attempt-to-jump" frequency, whose temperature dependence is usually neglected [7]. However, this dependence can actually play an important role: it is different for the two mechanisms, so that the relative contribution of impurity-assisted processes increases with lowering temperature (see Fig. 1).

Interface roughness can play a role similar to that of impurities [5]. If the interaction of an electron with interface roughness is taken into account, $k_\parallel$ is no more conserved for interwell transitions. Therefore, roughness scattering-assisted transitions are possible between states of equal energies with different $k_\parallel$, i.e., they are quasielastic and are similar to impurity-assisted transitions. The corresponding transition rates have the same activation factors as those for impurity-assisted transition rates. Therefore, the temperature dependence of the d.c. conductivity is practically the same for these situations.

Acknowledgements   The work was supported by the Russian Foundation for Basic Research, the programs "Integration" and "Universities of Russia".

References

[1] A. Chomette et al., Phys. Rev. Lett. **57**, 1464 (1986).
[2] G. Richter et al., Superlattices and Microstructures **22**, No. 4, 475 (1997).
[3] Lin-Wang Wang et al., Phys. Rev. B **53**, 2010 (1996).
[4] I. P. Zvyagin et al., Nanotechnology **11**, 375 (2000).
[5] R. Ferreira and G. Bastard, Phys. Rev. B **40**, 1074 (1989).
[6] K. E. Borisov and I. P. Zvyagin, Moscow University Phys. Bull., Phys. Astr. **44**, No. 4 (2003).
[7] N. F. Mott and E. A. Davis, Electron processes in non-crystalline materials (Clarendon Press, Oxford, 1979).

phys. stat. sol. (c) **1**, No. 1, 109–112 (2004) / **DOI** 10.1002/pssc.200303604

Hopping relaxation of excitons in GaInNAs/GaNAs quantum wells

H. Grüning[1], **K. Kohary**[2], **S. D. Baranovskii**[*1], **O. Rubel**[1], **P.J. Klar**[1], **A. Ramakrishnan**[3], **G. Ebbinghaus**[3], **P. Thomas**[1], **W. Heimbrodt**[1], **W. Stolz**[1], and **W. W. Rühle**[1]

[1] Department of Physics and Material Sciences Center, Philipps-University Marburg, D-35032 Marburg, Germany

[2] Department of Materials, University of Oxford, Parks Road, Oxford OX1 3PH, UK

[3] Infineon Technologies, Corporate Research CPR 7, D-81730 Munich, Germany

Received 1 September 2003, accepted 3 September 2002
Published online 28 November 2003

PACS 73.20.Mf; 73.21.Fg; 73.63.Hs; 78.55.Cr

Exciton photoluminescence (PL) in a GaInNAs/GaNAs quantum well was measured in the temperature range from 15 K to 300 K. Two striking features of the PL were observed: the nonmonotoneous temperature dependence of the Stokes shift and the abrupt increase of the PL linewidth in a rather narrow temperature range. These features are known to be strong indications of the hopping relaxation of excitons via localized states distributed in space and energy. Computer simulations of the hopping relaxation of excitons were carried out. Comparison between the simulation results and the experimental data provides an important and reliable information on the energy shape of the density of states and also on the energy range, in which localized states for excitons are distributed.

1 Introduction

Semiconductor quantum-well (QW) structures are intensively studied because of their potential for opto-electronic devices. QW's based on semiconductor alloys are of particular interest, since they allow one to fabricate structures with optimized optical properties. All semiconductor heterostructures possess a certain degree of disorder due to their alloy structure and/or imperfect interfaces. In particular, in narrow QW's the interface roughness creates a disorder potential giving rise to band tails composed from localized states. These tails affect the dynamics of the Coulomb-correlated electrons and holes and they influence essentially the optical properties of devices based on QW alloy heterostructures. In order to control the optical properties of devices, one should have a technique for characterization of the band tails caused by disorder. The crucial questions are: what is the energy scale of the band tails in QW's and what is the energy density of the tail states (DOS). We suggest such a characterization technique for quaternary GaInNAs-based QW's. The technique is based on the comparison between the experimental data and the results of the Monte Carlo computer simulation for the temperature dependences of the PL Stokes shift and that of the PL linewidth.

---

* Corresponding author: e-mail: baranovs@staff.uni-marburg.de, Phone: +49 6421 2825582, Fax: +49 6421 2827076

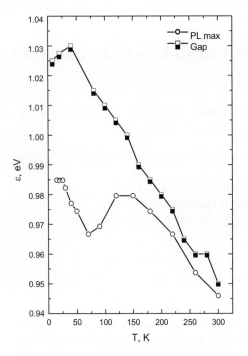

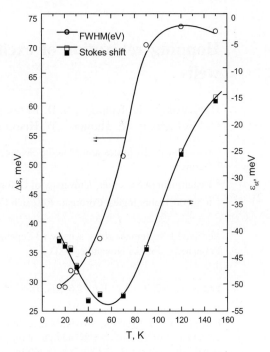

Fig. 1 Experimental results for both the PL peak and the gap as a function of temperature.

Fig. 2 Experimental results for the PL linewidth (FWHM) and Stokes shift as a function of temperature.

2   Results and discussion

The 6.2 nm wide $Ga_{0.63}In_{0.37}N_{0.01}As_{0.99}$ QW with $GaN_{0.017}As_{0.983}$ barriers was grown by metal-organic vapour-phase epitaxy on a (100) GaAs substrate. The temperature-dependent PL measurements were performed using a HeNe laser and low excitation densities on the sample. The corresponding band gap shift was determined by PL excitation spectroscopy at temperatures below 80 K and by photomodulated reflectance spectroscopy above 80 K using standard set-ups.

In Fig. 1 the results are given for the measured temperature dependence of the PL peak position and for the temperature dependence of the band gap. The anomalous increase of the gap with temperature below 40 K is within the experimental error. While the band gap decreases monotonously with raising temperature, the temperature dependence of the PL peak is essentially nonmonotonous. The temperature dependence of the Stokes shift is shown in Fig. 2 along with the temperature dependence of the PL linewidth [full width at half maximum (FWHM)]. One can see an abrupt increase of the linewidth in a rather narrow temperature range between 15 K and 90 K, while above 90 K the linewidth looks almost temperature-independent.

Both observed striking effects – the non-monotonous temperature dependence of the Stokes shift and the abrupt increase of the PL linewidth with temperature – are known in the literature for GaInNAs QW's [1, 2], as well as other alloy QW's, for example, for InGaAs/InP [3] and for InGaAs/AlInAs [4]. The qualitative explanation of both effects has already been given by Skolnick *et al.* [3]. They were attributed to the motion of excitons via localized states, induced in QW's by disorder. With rising temperature excitons become more mobile and they can be trapped by centers with lower energies and, hence, the lower-energy states become increasingly more populated with rising temperature. This enhances the Stokes shift. With the further increase of temperature the motion of excitons over localized states becomes even faster and their thermal equilibrium distribution can be achieved leading to higher PL energies and to the smaller

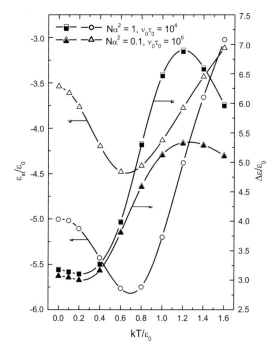

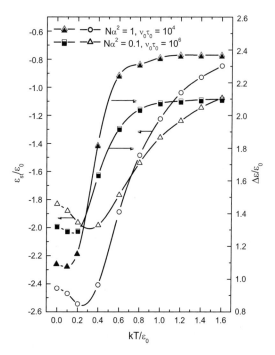

Fig. 3 Temperature dependence of the PL peak Stokes shift and PL linewidth (FWHM) for the exponential DOS (computer simulation).

Fig. 4 Temperature dependence of the PL peak Stokes shift and PL linewidth (FWHM) for the Gaussian DOS (computer simulation).

Stokes shift. Concerning the linewidth, its abrupt increase with T was also attributed to the increase in the mobility of excitons. More mobile excitons can recombine from a broader energy distribution of the localized states than in the situation of lower temperature.

In our work we put this argumentation on a quantitative basis by carrying out Monte Carlo computer simulations of the exciton hopping motion via localized states. The simulation algorithm was similar to that described in detail in [5]. In this algorithm a hopping motion of excitons via localized states characterized by some energy distribution is simulated with using Miller-Abrahams transition rates. Besides the energy shape of the DOS, there are three essential parameters in the model: (kT/ϵ_0, $N\alpha^2$, and $\nu_0\tau_0$). Here ϵ_0 is the DOS energy scale, T is the temperature, k is the Bolzmann constant, N is the concentration of localized states, α is the decay length of the exciton wave function in the localized states, ν_0 is the attempt-to-escape frequency, and τ_0 is the typical exciton lifetime.

First we carried out the simulations of the PL with the exponential DOS

$$g(\epsilon_0) = \frac{N}{\epsilon_0} \exp\left(\frac{\epsilon}{\epsilon_0}\right). \tag{1}$$

The simulated results for the PL Stokes shift ϵ_{st} and for the FWHM $\Delta\epsilon$ are given in Fig. 3. The best way to identify the DOS energy scale, ϵ_0, from a comparison with experimental data is to find such a PL characteristic feature which depends on the ϵ_0 and does not depend (or very weakly depends) on other parameters. The best candidate is the PL linewidth $\Delta\epsilon/\epsilon_0$ at low T. As evident from Fig. 3, this quantity at low T is practically independent of $N\alpha^2$ and $\nu_0\tau_0$. In the experiment at low T, the FWHM is $\Delta\epsilon = 30$ meV (see Fig. 2). Computer simulations on the other hand give at low T $\Delta\epsilon \approx 3\epsilon_0$. Straightforward evaluation gives the DOS energy scale $\epsilon_0 \approx 10$ meV.

There is another characteristic feature of the simulated PL data which is also stable against the variation of parameters $N\alpha^3$ and $\nu_0\tau_0$. This is the temperature at which the Stokes shift has its minimum value. In the experiment this temperature is $T \approx 60$ K (see Fig. 2). The computer simulation gives $kT \approx 0.6\epsilon_0$. Hence, the DOS scale can be estimated as $\epsilon_0 \approx 9$ meV, which is consistent with the value $\epsilon_0 \approx 10$ meV obtained above from the FWHM data. Furthermore, the temperature kT/ϵ_0 at which the FWHM $\Delta\epsilon/\epsilon_0$ reaches its maximum value is also consistent with such an estimate $\epsilon_0 \approx 9$ meV. The simulation shows that the maximum of the FWHM corresponds to the temperature $kT \approx 1.2\epsilon_0$. In the experiment the FWHM maximum is observed for the temperature 120 K. From this comparison one comes to the energy scale $\epsilon_0 \approx 8 - 9$ meV. The next step is to check the possibility to describe with the exponential DOS the abrupt increase of the FWHM from 30 meV to 70 meV observed in the experiment for the temperature range from 15 K to 90 K. Since the DOS energy scale was roughly estimated as $\epsilon_0 \approx 10$ meV the simulation should provide the increase of the quantity $\Delta\epsilon/\epsilon_0$ from 3 to 7. As follows from Fig. 3 such increase is provided with very reasonable values of parameters $N\alpha^2 = 1$ and $\nu_0\tau_0 = 10^4$.

Computer simulations were also carried out with a Gaussian DOS

$$ g(\epsilon) = \frac{N}{\epsilon_0\sqrt{2\pi}} \exp\left(-\frac{\epsilon^2}{2\epsilon_0^2}\right). \tag{2} $$

Corresponding results for the FWHM, $\Delta\epsilon$, and for the Stokes shift $\epsilon_{\max}$ are shown in Fig. 4. In contrast to the case of the exponential DOS, all PL characteristic features for the Gaussian DOS are sensitive to the choice of parameters $N\alpha^2$ and $\nu_0\tau_0$. Moreover, for the case of a Gaussian DOS, it appears not possible to describe consistently various observed PL features with a single set of model parameters. For example, analytical calculations for the Gaussian DOS show that the upper limit of the PL linewidth is $\Delta\epsilon \approx 2.4\epsilon_0$ [5]. In the experiment the upper limit of the FWHM is $\Delta\epsilon = 73$ meV. Hence, the DOS energy scale ϵ_0 can be estimated as $\epsilon_0 \approx 30$ meV. On the other hand, the simulation with a Gaussian DOS predicts the abrupt increase of $\Delta\epsilon$ at $kT \approx 0.6\epsilon_0$, i.e., $T \approx 210$ K for suggested $\epsilon_0 \approx 30$ meV. This temperature is approximately twice as large as the temperature $T \approx 90$ K observed in the experiment. Therefore we have to conclude that the Gaussian DOS is not suitable for description of the energy disorder in the studied QW's.

3 Conclusions

The PL linewidth and Stokes shift were measured in GaInNAs/GaAs quantum wells in the temperature range between 15 K and 300 K. The experimental results were compared with computer simulations of the exciton hopping relaxation via localized states. It is shown that the exponential energy distribution of localized states with a characteristic energy scale $\epsilon_0 \approx 9$ meV provides the best agreement with all pronounced experimentally observed features of the PL spectrum. In contrary, the Gaussian energy distribution fails to fit experimental data with any choice of model parameters. It is shown herewith that the study of the exciton luminescence in the hopping regime provides a strong tool to characterise the disorder-induced effects in III-V quantum wells.

Acknowledgements Financial support of the Deutsche Forschungsgemeinschaft and of the Founds der Chemischen Industrie is gratefully acknowledged. Autors also like to thank the Optodynamic Center of the Philipps-University Marburg and the European Graduate College "Electron-Electron Interactions in Solids".

References

[1] A. Hierro et al., J. Appl. Phys. **94**, 2319 (2003).
[2] I. A. Buyanova, W. M. Chen, and C. W. Tu, Semicond. Sci. Technol. **17**, 815 (2002).
[3] M. S. Skolnick et al., Semicond. Sci. Technol. **1**, 29 (1986).
[4] S. T. Davey, E. G. Scott, B. Wakefield, and G. J. Davies, Semicond. Sci. Technol. **3**, 365 (1988).
[5] S. D. Baranovskii, R. Eichmann, and P. Thomas, Phys. Rev. B **58**, 13081 (1998).

phys. stat. sol. (c) **1**, No. 1, 113–116 (2004) / **DOI** 10.1002/pssc.200303605

Potential fluctuations in disordered semiconductors measured by transport and optical methods

P. Bozsoki[*, 1], **S.D. Baranovskii**[1], **P. Thomas**[1], and **S.C. Agarwal**[2]

[1] Department of Physics and Material Sciences Center, Philipps-University Marburg, 35032 Marburg, Germany
[2] Department of Physics, Indian Institute of Technology, Kanpur 208016, India

Received 1 September 2003, accepted 3 September 2003
Published online 28 November 2003

PACS 72.80.Ng, 78.40.Pg

From the recent analysis of the potential fluctuations in disordered semiconductors on the basis of optical and transport measurements [1] it was concluded that these two different kinds of phenomena evidence extremely different energy scales of the random potential in the same sample. We resolve this puzzle using for the analysis of experimental data the well-known theories of transport and optical absorption in a disordered system with long-range potential fluctuations, caused by charged impurities [2, 3]. The key point in our consideration is the essential difference between the density of states caused by the long-range fluctuations and the shape of the absorption coefficient. The latter is known to depend essentially not only on the fluctuation probability but also on the tunnelling efficiency of the optically excited electrons in the potential relief provided by the fluctuations [2].

1 Introduction

Recently a comprehensive experimental study of transport phenomena and of the optical absorption in hydrogenated doped amorphous silicon (a-Si:H) was carried out [1]. The results were interpreted in the framework of the theories exploiting random potential fluctuations caused by the random spatial distribution of impurity atoms in the host semiconductor matrix. In order to interpret the data for the drift mobility of charge carriers, a theory of Overhof and Beyer [3] was used. In this approach it is assumed that random potential fluctuations have a Gaussian distribution with some characteristic scale V_0

$$g(V) = g_0 \exp\left\{-\frac{V^2}{V_0^2}\right\}. \tag{1}$$

According to the transport theory for charge carrier mobility limited by potential fluctuations [3], the slope E_Q of the sum $\ln\sigma + e/kS$ versus inverse temperature 1/T can be related to the scale V_0 of the potential distribution (1) via $E_Q = 1.25\ V_0$. Here σ and S are the measured conductivity and thermoelectric power, respectively, e is the elementary charge and k is the Boltzmann constant. Measured values of E_Q in a-Si:H doped with phosphorous lead to the magnitudes of V_0 between 0.04 and 0.08 eV, depending on the doping level.

---

[*] Corresponding author: e-mail: Peter.Bozsoki@Physik.Uni-Marburg.de, Phone: +49 64212824241,
Fax: +49 64212827076

Optical absorption was also measured in the same samples of doped a-Si:H [1]. The data for the absorption coefficient α can well be fitted by the Urbach formula

$$\alpha(\omega) = \alpha_0 \exp\left\{-\frac{E_0 - \hbar\omega}{E_\alpha}\right\}, \tag{2}$$

where $\hbar\omega$ is the photon energy, E_α is the tailing parameter and the absorption threshold E_0 is chosen as the photon energy at which the absorption coefficient attains a given value α_0. The quantity of primary interest for us is the tailing parameter E_α. Very often the theory of John et al. [4] is used to describe the absorption coefficient. This theory presumes that the absorption coefficient follows the energy dependence of the density of localised states (DOS) and it is based on the Gaussian potential fluctuations described by Eq. (1). It relates the tailing parameter E_α of the absorption coefficient to the characteristic scale of the potential distribution V_0 via equation $V_0^2 = 20 E_\alpha \varepsilon_L$, where $\varepsilon_L \approx 0.5$ eV in doped a-Si:H [1]. Using this equation and measured values of E_α one obtains for V_0 estimates of about 1 eV, which is more than an order of magnitude higher than the estimate for V_0 obtained from the measurements of the drift mobility in the same samples. This puzzle was recently raised as a challenging problem in the field of doped amorphous semiconductors [1]. The suggested solution was based on the assumption that the disorder potential limiting transport differs from that limiting the light absorption [1]. We try to resolve this puzzle on another basis, namely, by analysing the relation between E_α and V_0 using the theory of Shklovskii and Efros for light absorption [2,5], which drastically differs from that of John et al. [4].

2 Results and discussion

The theory of light absorption in doped semiconductors developed by Shklovskii and Efros [2, 5] can briefly be described as follows. Assume that charged donors and acceptors with total concentration N are randomly distributed in the amorphous matrix. Due to spatial fluctuations in the distribution of charge impurities, there are potential fluctuations with various length scales up to the screening length, which is assumed larger than the essential length for light absorption determined below. Absorption of a photon with frequency ω and energy deficit Δ with respect to the absorption threshold occurs via a process involving electron tunnelling into the nearest potential well. This tunnelling process causes the frequency dependence of the absorption coefficient to be essentially different from the energy dependence of the DOS in the tail region. It is this issue, which makes the essential difference between the theory for light absorption of Shklovskii and Efros [2, 5] and that of John et al. [4]. We would like to emphasise this difference between the two theories. While the DOS function in the deep tail is determined by the optimal fluctuation with respect to the creation of a localized state with a given energy, the absorption event with a given frequency deficit $\Delta/\hbar$ is provided by a fluctuation with optimal product of its probability and the probability for electron to tunnel under the corresponding potential barrier. Hence, the contribution of the above fluctuation to the absorption coefficient is proportional to the product [2]

$$\exp\left\{-\frac{3\kappa^2\Delta^2}{8\pi e^4 NR}\right\} \exp\left\{-\frac{R\sqrt{m\Delta}}{\hbar}\right\}, \tag{3}$$

where the first exponential factor is the probability of having the necessary number of extra impurities in the volume of the fluctuation for a potential amplitude Δ in a solid with dielectric constant κ. The second factor is the probability for an electron (assumed to be lighter than a hole) with mass m to tunnel over a distance R under the barrier Δ. This exponent has its maximum at

$$R = R_{\max} = a\left(\frac{\Delta}{E_0}\right)^{3/4}\left(\frac{3}{8\pi Na^3}\right)^{1/2} \tag{4}$$

with

$$a = \frac{\hbar^2 \kappa}{me^2} \quad , \quad E_0 = \frac{me^4}{\hbar^2 \kappa^2} \, . \tag{5}$$

Substituting Eqs. (4) and (5) into Eq. (3), one obtains

$$\ln \frac{\alpha_A(\varDelta)}{\alpha_0} = -\beta\gamma \left(\frac{\varDelta}{E_0} \right)^{5/4} \frac{1}{(Na^3)^{1/2}} \, , \tag{6}$$

where $\gamma = (3/(2\pi))^{1/2}$ and β is a numerical factor shown in [5] to be $\beta = 2/(5\sqrt{\pi})$, if electrons are much lighter than holes. The above formulae differ from the previously published results [2,5] by numerical coefficients, in particular, by the coefficient γ in Eq. (6).

The obtained energy dependence for the absorption coefficient

$$\ln \frac{\alpha_A(\varDelta)}{\alpha_0} = -\left(\frac{\varDelta}{E_a} \right)^{5/4} \tag{7}$$

cannot be distinguished experimentally from the dependence

$$\ln \frac{\alpha_A(\varDelta)}{\alpha_0} = -\left(\frac{\varDelta}{E_a} \right) \tag{8}$$

indicated in the experiments with doped a-Si:H [1, 6] and a-Ge:H [7]. Therefore, we think that equation (6) should be used for the description of the experimental data. Comparing equation (7) with equations (5) and (6), one arrives at the following dependence of the Urbach parameter E_a on the concentration of dopant atoms N :

$$E_a = (\beta\gamma)^{-4/5} E_0 (Na^3)^{2/5} \, . \tag{9}$$

As can easily be seen from equations (5) and (9), the dependences of E_a on parameters m and κ are rather weak. Therefore, we take the free electron mass for m and assume $\kappa = 10$ for our numerical estimates. The concentration of impurities N needed to fit the experimental value of $E_a \approx 100$ meV [1] is equal to $\approx 0.8 \cdot 10^{20}$ cm^{-3} [8], which seems very reasonable [6].

Now one can estimate the typical energy scale V_0 of the disorder potential in Eq. (1) that leads to the above frequency dependence of the absorption coefficient. The values of the energy deficit Δ in the experiments [1,6,7] are of the order of $\Delta \approx 0.2$ eV. Inserting this value $\Delta = 0.2$ eV along with numerical parameters specified above into Eq. (4), one obtains the typical length scale of potential fluctuations responsible for such a light absorption of the order of $R = R_{max} \approx 13$ Å. Representing the first exponential term in Eq. (3) in the form of Eq. (1) with replacing Δ by V one obtains from such an estimate the value of $V_0 \approx 0.06$ eV. This estimate of V_0 is in good agreement with the value between 0.04 and 0.08 eV obtained earlier from the transport measurements [1]. Of course, this estimate of V_0 via Eqs. (3) - (9) is not exact and it can be considered only as an order of magnitude estimate. Nevertheless, we believe that the agreement between the V_0 values obtained from transport measurements and from the light absorption measurements solve the problem raised in [1].

3 Conclusions

The puzzling problem of the huge discrepancy in the estimates for the scale of potential fluctuations obtained on one hand from transport measurements and on the other hand from the measurements of light absorption in the same samples of doped a-Si:H has been resolved by using an appropriate theory of light absorption [5]. The puzzle has been caused by identifying the frequency dependence of the light

© 2004 WILEY-VCH Verlag GmbH & Co. KGaA, Weinheim

absorption with the energy dependence of the density of localized tail states [4]. Following previous considerations [5] we claim that the energy dependence of the transition matrix element in the absorption event can drastically affect the frequency dependence of the absorption coefficient and hence it should be appropriately taken into account.

Acknowledgements   Financial support from the Fonds Der Chemischen Industrie, the Deutsche Forschungsgemeinschaf, the Optodynamic Center Marburg and that of the European Graduate College "Electron-electron interactions in solids" is gratefully acknowledged.

References

[1] A. K. Sinha and S.C. Agarwal, J. Vac. Sci. Technol. B **18**, 1805 (2000).
[2] B. I. Shklovskii and A.L. Efros, Electronic Properties of Doped Semiconductors (Springer, Heidelberg, 1984).
[3] H. Overhof and W. Beyer, Philos. Mag. B **43**, 433 (1981); ibid. B **47**, 377 (1983).
[4] S. John, C. Soukoulis, M.H. Cohen, and E.N. Economou, Phys. Rev. Lett. **57**, 1777 (1986).
[5] B. I. Shklovskii and A.L. Efros, Sov. Phys. JETP **32**, 733 (1971).
[6] J. A. Howard and R.A. Street, Phys. Rev. B **44**, 7935 (1991).
[7] I. Chambouleyron and D. Comedi, J. Non-Cryst. Solids **227-230**, 411 (1998).
[8] S.D. Baranovskii, K. Kohary, P. Thomas, and S. Yamasaki, J. Mater. Sci.: Materials in Electronics, in print.

phys. stat. sol. (c) **1**, No. 1, 117–120 (2004) / **DOI** 10.1002/pssc.200303629

Hopping-randomwalk of electrons at localized band tail states in random fractal fluctuation

Kazuro Murayama[*] and **Yoriko Ando**

College of Humanities & Sciences, Nihon University, Tokyo 156-8550, Japan

Received 1 September 2003, revised 9 October 2003, accepted 9 October 2003
Published online 28 November 2003

PACS 05.40.Fb, 05.45.Df, 71.23.An, 72.20.Ee, 78.55.Qr

Power law decays in dispersive transient current and photoluminescence in amorphous semiconductors are explained from hopping-randomwalk of carriers at localized band tail states in random self-affine fluctuation of the 3.9 fractal dimension. The hopping-randomwalk is reported. The mean square distance is given by $\langle r(t)^2 \rangle \sim t^\kappa$ with $\kappa < 1$, which shows anomalous diffusion. The hopping area where electrons pass through is complicated and has a fractal dimension. The fractal and fracton dimensions of the randomwalk are calculated at different temperatures. The fractal and fracton dimensions of the hopping-randomwalk in amorphous semiconductors is shown to be able to be obtained from the dispersive transient current and photoluminescence decay. It is shown that the temperature behavior of both dimensions in a-Si:H agrees with that in the simulation of the hopping-randomwalk.

© 2004 WILEY-VCH Verlag GmbH & Co. KGaA, Weinheim

1 Introduction Photoluminescence decay and dispersive transient current in amorphous semiconductors are described with the following two different power law decay functions [1-5],

$$f(t) \sim t^{-\gamma}, \qquad (t < t_0),$$
$$t^{-\delta}, \qquad (t > t_0), \qquad (1)$$

where the exponent γ is less than 1 and δ is larger than 1. In the dispersive transient current, dispersion parameters α_i and α_f are used instead of the exponents, which are related by the equations $\gamma = 1 - \alpha_i$ and $\delta = 1 + \alpha_f$. A localized state at the band tail in amorphous semiconductors has a different energy at a different site for the structural disorder and namely energies of the localized states fluctuates randomly. The power law behavior of the transient current is considered to be responsible for the static energy fluctuation of the localized band tail states. It is probable that the energy fluctuation of the localized band tail states is self-affine such as surface of earth [6]. A random fluctuation with the self-affinity can be described by a fractal dimension [6]. One of the authors shows with the Monte Carlo simulation that the dispersive transient current is explained from hopping-randomwalk of electrons at localized band tail states in the random self-affine fluctuation

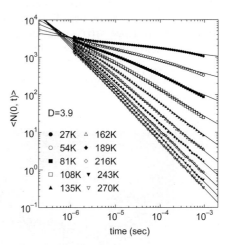

Fig. 1 The number of electrons at the starting site after the hopping-randomwalk for t sec at localized band tail states in the random fluctuation of D = 3.9. The results simulated at different temperatures from 27 to 270 K are shown. Straight lines show eq. (2).

---

[*] Corresponding author: e-mail: murayama@phys.chs.nihon-u.ac.jp

and concludes that the fluctuation with the fractal dimension D = 3.9 is the best to explain the transient current in amorphous semiconductors [7]. Furthermore, the power law decays in the photoluminescence are shown to be reproduced from the model that electrons which walk randomly by hopping at the localized band tail states recombine geminately with trapped holes [8]. Here, the details of the hopping randomwalk are reported.

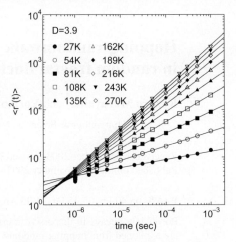

Fig. 2 Mean square distance of an electron from the starting site after the hopping-randomwalk for t sec at localized band tail states in the random fluctuation of D = 3.9. The results simulated at different temperatures from 27 to 270 K are shown. Straight lines show eq. (3).

2 Hopping-randomwalk at localized band tail states in self-affine fluctuation
In the simulation, sites of localized band tail states in amorphous semiconductors are assumed to form a simple cubic lattice with the lattice constant equal to the mean hopping distance a_H (a_H = 2 nm in a-Si:H) [9, 10] for simplifing the calculation. The 50 simple cubic lattices of 128x128x128 are prepared for the simulation. Energies of localized states at lattice sites are calculated with Fourier Transform filtering [6] to have a static self-affine fluctuation of the fractal dimension D = 3.9. The site energy distribution, namely the distribution of localized band tail states, of $\exp(-E^2/\Sigma^2)$, (Σ = 105 meV), has been used [7, 8]. The hopping frequency, ν_{ij}, from a lattice site i with the energy, E_i, to the nearest neighbor site j with E_j, has been calculated according to $\nu_{ij} = \nu_0 \exp \{-(\Delta E_{ij} + |\Delta E_{ij}|)/2kT\}$ where $\Delta E_{ij} = E_j - E_i$ [11]. ν_0 is the attempt-to-escape frequency and $\nu_0 = 10^7$ sec^{-1} is used. An electron starts the hopping-randomwalk at t=0 from a lattice site chosen at random in the simple cubic lattice. After the randomwalk has been repeated for 200 electrons for each of the 50 simple cubic lattices, the averaged number of electrons <N(0,t)> existing at the starting site of r = 0 after time t and the mean square distance <r(t)^2> calculated at different temperatures are shown in Figs. 1 and 2, respectively. The figures show that <N(0,t)> and <r(t)^2> are described with the following power law functions,

$$<N(0,t)> \sim t^{-\xi}, \tag{2}$$

$$<r(t)^2> \sim t^{\kappa}. \tag{3}$$

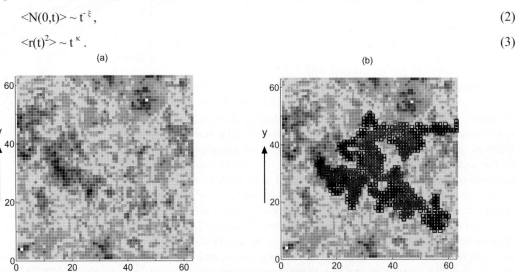

Fig. 3 (a) The random fluctuation of D=2.9 formed on a 64x64 square lattice. The deeper green shows the sites with the higher energies and the deeper brown the sites with the lower energies. (b) The areas where electrons pass through by the hopping-randomwalk. The red and black squares show the areas where electrons pass through for 1 and 10 sec, respectively.

© 2004 WILEY-VCH Verlag GmbH & Co. KGaA, Weinheim

The exponents ξ and κ are less than 1 at low temperatures, increase with increasing temperature and at high temperatures approach 1.5 and 1, respectively, which are the values in the 3 dimensional uniform space. Small values of ξ and κ at low temperatures show that a large part of electrons exists still around the starting site even after long time. The temperature behavior of the exponents ξ and κ approaching 3/2 and 1 at high temperatures shows that electrons have enough thermal energy to diffuse over the random fluctuation to anywhere in the lattice. The equation (3) with κ less than 1 shows that the diffusion due to the hopping-randomwalk is an anomalous diffusion.

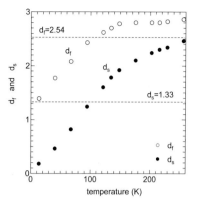

Fig. 4 Temperature dependence of the fractal dimension d_f and the fracton dimension d_s in the hopping area where electrons pass through by hopping at localized band tail states in random fluctuation of D = 3.9. The fractal and fracton dimensions of percolation cluster in the 3 dimension are shown with dotted lines.

3 Fractal and fracton dimensions We have calculated the electron distribution in space induced by the hopping-randomwalk at localized band tail states in the static random fluctuation with the self-affinity of D=2.9 on a 64x64 square lattice shown in Fig. 3(a). The deeper green shows sites with the higher energies and the deeper brown shows sites with the lower energies. An electron starts the hopping-randomwalk from the center of the lattice. It is repeated for 100 electrons. The red and black squares in Fig. 3(b) show the areas where the electrons pass through for 1 and 10 sec, respectively. The areas have complicated structures like fractal. This suggests that the hopping-randomwalk can be analyzed with the fractal dimension of the area and the fracton dimension of randomwalk.

In the case of randomwalk in fractal structures such as percolation clusters, $<N(0,t)>$ and $<r(t)^2>$ are given by $<N(0,t)>\sim t^{-d_s/2}$ and $<r(t)^2>\sim t^{d_s/d_f}$, where d_f is the fractal dimension of the hopping area and d_s the fracton dimension of the randomwalk [12,13]. Therefore, the following relations are derived from eqs.(2) and (3),

$$d_s = 2\xi, \qquad d_f = 2\xi/\kappa . \qquad (4)$$

The fractal and fracton dimensions calculated from the exponents ξ and κ in the simulation are shown in Fig. 4. Both dimensions are less than 2 at low temperatures and increases with increasing temperature and finally approach at high temperatures to 3 of the Euclidean dimension.

The photoluminescence decay at 108 K calculated by assuming that an electron can recombine radiatively with a hole when the electron which walks randomly by hopping at the localized band tail states in the random self-affine fluctuation arrives at the site of the hole is shown in Fig. 5 [8]. The photoluminescence decay calculated is described with

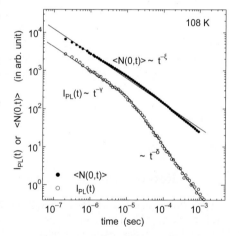

Fig. 5 Photoluminescence decay $I_{PL}(t)$ [8] and electrons $<N(0,t)>$ at the starting site after the hopping randomwalk for t sec at localized band tail states in random fluctuation of D=3.9 simulated at 108 K. The straight lines show eqs. (1) and (2).

eq. (1). Electrons at the starting site $<N(0,t)>$ after time t at 108 K are shown together in Fig. 5. It is seen that the exponent γ at short times in the power law decay of the photoluminescence agrees with the exponent ξ of $<N(0,t)>$. This shows that the fracton dimension d_s of the hopping-randomwalk in amorphous semiconductors can be obtained from the exponent γ of the photoluminescence decay with the following equation,

$$d_s = 2\gamma . \tag{5}$$

It is derived with the Einstein relation that the mean square distance $\langle r(t)^2 \rangle$ of carriers in amorphous semiconductors is given with the dispersion parameter α_i by $\langle r(t)^2 \rangle \sim t^{\alpha i}$ [14]. This shows that the fractal dimension d_f of the hopping area in amorphous semiconductors can be obtained from the exponent γ of the photoluminescence decay and the dispersion parameter α_i by using the following equation,

$$d_f = 2\gamma / \alpha_I . \tag{6}$$

The exponent γ of the power law decays of the photoluminescence observed at different temperatures in a-Si:H are shown in Fig. 6(a) together with the dispersion parameter α_i of the dispersive transient current observed in the electron time-of-flight experiment. The electron time-of-flight experiment has been executed in low electric fields because the Einstein relation requires that carriers are in thermal equilibrium. The fracton and fractal dimensions of electron hopping-randomwalk obtained from the exponent γ and the dispersion parameter α_i in a-Si:H by using eqs. (5) and (6) are shown in Fig. 6(b) together with those obtained from the simulation. The figure shows that the temperature behavior of the fracton and fractal dimensions in electron hopping-randomwalk in a-Si:H almost agrees with those of the simulation of the hopping-randomwalk at localized band tail states in the random self-affine fluctuation of D=3.9.

Acknowledgements This work was partially supported by a research fund of the College of Humanities and Sciences, Nihon University.

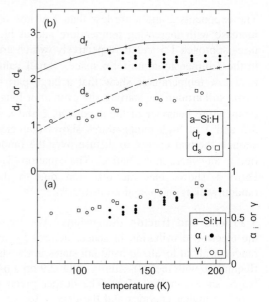

Fig. 6 (a) Temperature dependence of the exponent γ of the photoluminescence decay and the dispersion parameter α_i in a-Si:H and (b) temperature dependence of the fractal dimension d_f and the fracton dimension d_s in a-Si:H calculated from γ and α_i. Open circles and squares in (a) show the exponent γ obtained from the total light decay of the photoluminescence observed and the photoluminescence decay in ref. [4] in a-Si:H, respectively. Those in (b) are d_s calculated with eq. (5). Solid circles in (a) show the dispersion parameters α_i obtained from electron time-of-flight experiment in a-Si:H and solid circles in (b) show d_f calculated with eq. (6). d_f and d_s obtained from the simulation are shown with solid and dashed lines, respectively.

References

[1] G. Pfister and H. Scher, Adv. Phys. **27** (1978) 747.
[2] T. Tiedje, "The Physic of Hydrojenated Amorphous Silicon", edited by J.D. Joannopoulos and G. Lucovsky, (Springer-Verlag, Berlin Heidelberg New York Tokyo, 1984), p. 261.
[3] H. Oheda, S. Yamasaki, A. Matsuda, and K. Tanaka, J. Non-Cryst. Solids **59/60** (1983) 373.
[4] K. Murayama and T. Ninomiya, J. Non-Cryst. Solids **77/78** (1985) 699.
[5] K. Murayama and T. Ninomiya, Solid State Commun. **53** (1985) 125.
[6] R.F. Voss, "The Science of Fractal Images", edited by H. Peitgen and D. Soupe (Springer-Verlag, 1988), p. 21.
[7] K. Murayama and J. Kuwabara, Solid State Commun. **103** (1997) 591.
[8] Y. Ando, A. Sasaki, M. Shingai, and K. Murayama, phys. stat. sol.(b) **230** (2002) 15.
[9] K. Murayama, H. Oheda, S. Yamazaki, and A. Matsuda, Solid State Commun. **81** (1992) 887.
[10] K. Murayama, K. Ohno, Y. Ando, and A .Matsuda, phys. stat. sol. (b) **230** (2002) 221.
[11] N.F. Mott and E.A. Davis, "Electronic Processes in Non-crystalline Materials, 2nd edition" (Clarendon Press, Oxford, 1979), p.65.
[12] S. Alexander and R. Orbach, J.Physique **43** (1982) L-625.
[13] R. Rammal and G. Toulouse, J. Physique **44** (1983) L-13.
[14] K. Murayama and M. Mori, Philos. Mag. B **65** (1992) 501.

phys. stat. sol. (c) **1**, No. 1, 121–125 (2004) / **DOI** 10.1002/pssc.200303601

Manifestation of Coulomb gap in two-dimensional p-GaAs–AlGaAs structures with filled lower or upper Hubbard bands

N. V. Agrinskaya[*], **V. I. Kozub**, **A. V. Chernyaev**, and **D. V. Shamshur**

Ioffe Physico-Technical Institute, Russian Academy of Sciences, Politekhnicheskaya ul. 26, St. Petersburg 194021, Russia

Received 1 September 2003, revised 12 September 2003, accepted 16 September 2003
Published online 1 December 2003

PACS 72.20.Ee, 73.40.Kp

We have studied 2D hopping conductivity within the impurity bands formed by shallow acceptors (Be) in structures of GaAs/AlGaAs. Strong selective doping allowed contributions of both lower (A^0) and upper (A^+) Hubbard bands to be observed in different samples. The observed conductivity was of the Efros–Shklovskii type, however, the effective temperature (T_0) strongly depended on the impurity concentration (while only weak changes of the localisation radii were observed). Since the dielectric constant in 2D does not depend on concentration, the observed variation of T_0 can be explained as a result of variation of the coefficient C entering the equation for the above mentioned T_0. So the value of C is not a universal one and depends on impurity concentration. In particular the coefficient C differs by an order of magnitude for A^0 and A^+ bands for samples with the same acceptor concentration.

1 Introduction   In recent years, much interest was paid to studies of two-dimensional (2D) structures. This was to a large extent caused by the observation of the transition from the insulator type of conduction to the metallic type in such structures [1]. We have recently proposed an explanation of the experimental data assuming that the "tail" of localized states (originated due to disorder potential) and, in particular, the states of the upper Hubbard band play an important role in transport [2]. So the hopping transport over the localized states in 2D structures is of particular interest. It is known that an important feature of the hopping transport is the presence of the Coulomb gap near the Fermi level, which leads to the Efros–Shklovski (E–S) law for the temperature behavior of resistance:

$$\rho = \rho_0 \exp\left(\frac{T_0}{T}\right)^{1/2} . \tag{1}$$

Due to the fact that κ, the dielectric constant, does not depend on impurity concentration in 2D, the Coulomb gap becomes more pronounced at high impurity concentration near the metal–insulator transition. Until now most of the observations of the Coulomb gap in 2D were related to gated structures [3]. In the presence of the gate, the Coulomb interaction is screened at distances larger than the distance from 2D layer to the gate. Accordingly, as the temperature tends to zero, i.e., as the hopping distance increases, the Coulomb gap becomes irrelevant and Mott's law prevails. In this respect, the system without the gate is of considerable interest. Recently, we reported an observation of the Coulomb gap in low-temperature hole-hopping transport in GaAs/AlGaAs multiwell structuress (for the A^+ impurity band) [4]. The special interest was paid to the effective temperature T_0 in the law defined in Eq. (1). According to theoretical

[*] Corresponding author: e-mail: nina.agrins@mail.ioffe.ru

considerations:

$$T_0 = Ce^2/\kappa a, \tag{2}$$

where a is localization radius, κ is dielectric constant. If values of a have been estimated from positive quadratic magnetoresistance, coefficient C can be evaluated with an assumption that κ is constant for 2D. Note that the estimation of the pre-exponential factor ρ_0 in the Coulomb-gap regime is of fundamental importance. It was assumed in [5] that in the case of electron-assisted hopping, it should have a universal form $\rho_0 = h/e^2$ for samples close to metals where $T_0 \sim T$.

In the GaAs/AlGaAs system studied, the well width was $d \sim 15$ nm and the localization radius of the acceptor Be impurity was 2 nm, i.e., much smaller than d. By selective doping of both the central areas of wells and barriers or only in the central areas of wells, a situation was realized where the upper Hubbard band (A^+ center) or the lower Hubbard band (A^0 center) were filled in equilibrium [4]. This case is of interest because transport via the upper Hubbard band may exhibit some special features that are caused by the large radii of the states. We found that for the A^0 impurity band C is smaller than its theoretical value ($Co = 6.2$) by 4 times, and for A^+ impurity band by 30 times and depends strongly on impurity concentration. So the coefficient in the Efros–Shklovski law proves to be much smaller than follows from the Efros–Shklovski theory and depends on proximity to the metal–insulator transition. This may be related to both κ divergence and correlated hopping. We exclude the κ divergence because the barrier width (25 nm) was much larger than observed localization radii.

2 Experiment The method of growing and doping multilayer structures was described in a previous paper [4]. The temperature dependence of the ohmic resistance for samples with A^0 centers and with different concentration of A^+ centers are shown in Fig. 1 and Fig. 2. Samples with A^0 centers had very large resistance at low temperatures, so we have measured only one sample with high acceptor concentration ($N = 10^{18}$ cm^{-3}). One can see that the low-temperature part of resistance for both A^0 and A^+ impurity bands can be described by the E–S law, so the parameter T_0 can be estimated. Besides, pre-exponential factors estimated for these all samples were found to be close to $\rho_0 = h/e^2 \sim 10^4 \ \Omega$.

Sample parameters are given in Table 1. Hole concentrations were calculated from the Hall coefficient at 300 K for the 15 nm well, localization radii were obtained from the positive quadratic magnetoresis-

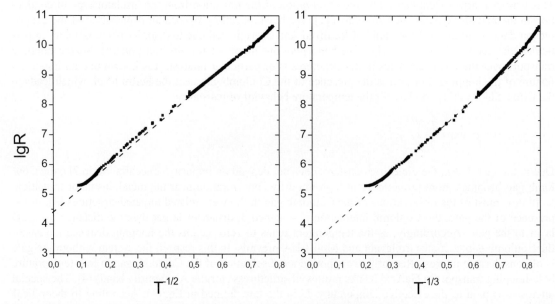

Fig. 1 Resistivity (per one well)–temperature dependence for samples with A^0 centers ($N = 10^{18}$ cm^{-3}) in two scales (Mott and E–S). It is seen that E–S gives the better fit to the experimental data.

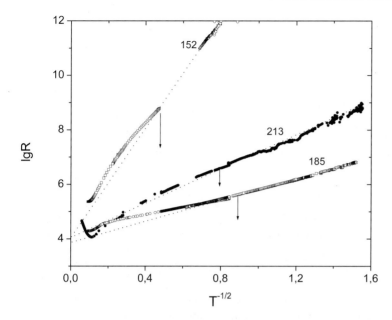

Fig. 2 Resistivity–temperature dependence for samples with different concentration of A^+ centers. Arrows indicate the crossover temperatures T_c.

Table 1

N sample	N_A^0, N_A^+ (cm^{-3})	T_0, K	a from MR (nm)	C
185 (A^+)	10^{18}	20	13	0.2
213 (A^+)	3×10^{17}	60	12	0.6
152 (A^+)	1×10^{17}	500	10	3.5
200 (A^0)	10^{18}	300	6	1.5

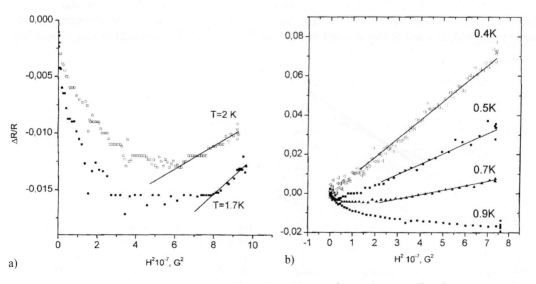

a)

b)

Fig. 3 Positive quadratic magnetoresistance: (a) for samples with A^0 centers ($N = 10^{18}$cm^{-3}), (b) for samples with A^+ centers ($N = 10^{18}$ cm^{-3}).

tance (Figs. 3a, b):

$$\ln\left(\frac{R(H)}{R(0)}\right) = 0.002 \left(\frac{T_0}{T}\right)^{3/2} \left(\frac{eHa^2}{ch}\right)^2 . \qquad (3)$$

Coefficient C has been evaluated with the help of Eq. (2) and with the assumption that $\kappa = 12$.

One can see that C decreases when the impurity concentration increases in the A^+ band and at the lowest concentration it remains much smaller than the theoretical value 6.2. For the A^0 band we have only one sample and we expected that in this case C should be close to the theoretical value, but this is not the case and C is still lower than 6.2.

3 Discussion The Efros–Shklovskii theory for single-electron hopping in 2D gives a universal value $C = 6.2$. In addition, it gives the temperature-dependent non-universal pre-exponential factor $\rho_0 \sim T/\gamma e^2$, where γ is a typical phonon frequency. On the other hand numerical simulations of a Coulomb glass with multielectron hopping have given a lower value of $C = 0.62$ [6]. Note that the electron-assisted hopping is expected to have a universal prefactor $\rho_0 = h/e^2$ [5]. Analysis of our data gives such a universal value of ρ_0.

In Fig. 4 the resistance–temperature dependence for sample N185 is shown with the temperature dependence of $\rho_0 \sim T/\gamma e^2$. The corresponding factor A when extrapolated to high temperatures proves to be smaller than h/e^2, which can not be true for phonon-assisting hopping. Therefore, the situation with the temperature-independent universal $\rho_0 = h/e^2$ seems to be more realistic.

As for the coefficient C for GaAs/AlGaAs systems, different values of C were obtained experimentally (from 2.5 and 0.6) [3], but no dependence of these values on carrier or impurity concentration was reported. According to our results, for A^0 impurity band the value of C is smaller than its theoretical value ($C_0 = 6.2$) by 4 times, and for the A^+ impurity band by 30 times; in the latter case a strong dependence of C on impurity concentration was found. Thus the coefficient in the Efros–Shklovski law proves to be non-universal.

As was mentioned above, both the "universal" prefactor and the small values of C can be related to electron-assisted (correlated) hopping. To the best of our knowledge the only attempt to develop an analytical model of correlated hopping within a semi-phenomenological approach was made in recent theoretical work by Kozub et al. [7]. Following the ideas of [7] one comes to the E–S law and universal prefactor for 2D; however, the effective temperature in the E–S law appears to be different from the estimate given by Efros and Shklovskii and is given as $T_0 \sim C_1 e^2/\kappa a$. Thus instead of being purely nu-

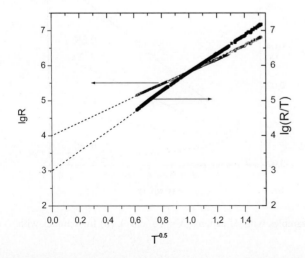

Fig. 4 Temperature dependence of the resistance (sample N185) on two scales: not taking into account and taking into account the temperature dependence of ρ_0.

meric the coefficient C_1 appears to be:

$$C_1 \sim (\kappa^{1/2}/e)\,(ag_f)^{-1/2},\tag{4}$$

where $g_f = N/W$ is the Fermi level density of states, N is impurity concentration, W is width of impurity band. Thus, at first, C_1 decreases with an increase of concentration, and then it decreases with an increase of a, which takes place in the vicinity of the metal–insulator transition.

These considerations can also remove the apparent contradictions between the theoretical prediction that the Coulomb gap value $\Delta_c = \pi/2(e^4/\kappa^2)\,g_f$ in 2D should increase with an increase of concentration and experimental data where such an increase was not observed [8]. For the 2D-crossover temperature from the Mott law to the E–S law $T_c = (e^6/\kappa^3)\,g_f^2 a$ (following an assumption that an effective energy band for Mott hopping is equal to Δ_c) should increase with an increase of g_f. However, we have seen the opposite behavior for our samples with A^+ centers, Fig. 2. Using the model in [7] one obtains for the crossover temperature $T_c = (e^3/\kappa^{3/2})\,(g_f/a)^{1/2}$; thus in the vicinity of metal–insulator transition where a diverges, the crossover temperature decreases, which is in agreement with the experiment.

Summarizing, we have shown that in 2D doped GaAs/AlGaAs non-gated structures hopping in the impurity band is over the Coulomb-gap states and has a character of assisted (correlated) hopping.

Acknowledgement   We acknowledge financial support by RFFI grant N 03-02-17516.

References

[1]   E. Abrahams, S. V. Kravchenko, and M. P. Sarachik, Rev. Mod. Phys. **73**, 251 (2001).
[2]   V. I. Kozub and N. V. Agrinskaya, Phys. Rev. B **64**, 245103 (2001).
[3]   F. W. van Keuls, X. L. Hu, H. W. Jiang, and A. J. Dahm, Phys. Rev. B **56**, 1161 (1997).
[4]   N. V. Agrinskaya, V. I. Kozub, V. M. Ustinov, A. V. Chernyaev, and D. V. Shamshur, JETP Lett. **76**, 360 (2002).
[5]   I. L. Aleiner, D. G. Polyakov, and B. I. Shklovskii, in Proc. 22 Int. Conf. on the Physics of Semicond. Vancouver 1994, p. 787, World Scientific, Singapore (1994).
[6]   A. Perez-Garrdo, M. Ortunio, E. Guevas, J. Ruiz, and M. Pollak, Phys. Rev. B **55**, 8630 (1997).
[7]   V. I. Kozub, S. D. Baranovskii, and I. Shlimak, Solid State Commun. **113**, 587 (2000).
[8]   S. I. Khondaker, I. Shlimak, J. T. Nicholls, M. Pepper, and D. A. Ricthie, Phys. Rev. B **59**, 4580 (1999).

phys. stat. sol. (c) **1**, No. 1, 126–130 (2004) / **DOI** 10.1002/pssc.200303612

Hopping transport through an array of Luttinger liquid stubs

A. L. Chudnovskiy[*]

I. Institut für Theoretische Physik, Universität Hamburg, Jungiusstr. 9, 20355 Hamburg, Germany

Received 1 September 2003, accepted 4 September 2003
Published online 28 November 2003

PACS 71.10.Pm, 72.20.Ee, 73.21.Hb

We consider a thermally activated transport across and array of parallel one-dimensional quantum wires of finite length (quantum stubs). The disorder enters as a random tunneling between the nearest-neighbor stubs as well as a random shift of the bottom of the energy band in each stub. Whereas one-particle wave functions are localized across the array, the plasmons are delocalized, which affects the variable-range hopping. A perturbative analytical expression for the low-temperature resistance across the array is obtained for a particular choice of plasmon dispersion.

Variable-range hopping (VRH) has been the subject of numerous experimental and theoretical investigations [1, 2], as in represents a major transport mechanism in disordered systems that are insulating at zero temperatures. Whereas the single particle wave functions of an insulator are localized, the localization of many particle excitations, such as plasmons, in presence of interactions does not obviously follow the localization of single particle wave functions, which reflects itself in VRH. In this paper we report the results on plasmon-mediated VRH in a model of an insulating system with delocalized plasmons.

We calculate the variable range hopping resistance across an array of parallel periodically placed quantum wires of finite length L and interwire spacing a (quantum stubs). Neighbor stubs are separated by tunnel barriers of random height. We consider a spinless case. The disorder enters the model both as a random tunneling between the nearest-neighbor stubs, and as a random shift of the bottom of the energy spectrum of each stub. Because the length of each stub L and the spacing between the neighbor stubs a are nonrandom, the interactions within each stub as well as the interactions between the different stubs are not affected by the disorder. In the result, the plasmons in the array are delocalized, whereas the one particle wave-functions are localized in the direction across the array. In the limit of infinite length $L \to \infty$ of each stub, the plasmons in the array form a sliding Luttinger liquid phase, and that the plasmon spectrum is continuous [3]. At finite length of the wires the plasmon spectrum represents a set of plasmon bands with a continuously varying wave vector p describing the propagation across the array, and a discrete number of a band n that relates to discrete plasmon levels in each stub.

To calculate the resistance across the array at low finite temperatures, we adopt an approach of Ref. [4], assuming that the resistance of the array is determined by few "breaks", that is the junctions between the two stubs with highest resistance. Let us parametrise the tunneling matrix element between the two stubs in the form $t_{i,i+1} = \exp\left(-|y_{i,i+1}|/a\right)$. The parameter y can be associated with an effective distance between the two stubs. This effective distance is random, its distribution follows from the distribution of the heights of potential barriers. Since a "break", being a junction with expo-

[*] Corresponding author: e-mail: achudnov@physik.uni-hamburg.de, Phone: +49 40 42838 2433 Fax: +49 40 42838 6798

nentially large resistance, is not shorted by other resistances connected in parallel, the resistance of a break can be written in the form $R_1 = R_0 \exp\left[2|y_{i,i+1}|/a + f(E_a, T)\right]$. Function $f(E, T)$ accounts for the effect of thermally activated plasmons in the resistance of a break. The random addition energy E_a includes the random shifts of the bottoms of the energy specrum of each stub. The probability density for the resistance $\rho(u) = R/R_0 = e^u$ is proportional to e^{-gA} (we keep the notations of [4]), where A is the area in the (y, E_a) phase space that results in the resistance e^u. That area is given by the integration over the phase space with the additional condition $2|y_{i,i+1}|/a + f(E_a, T) = u$. The resistance is calculated as

$$R = R_0 L_y \int\limits_0^\infty \mathrm{d}u\, \rho(u)\, \mathrm{e}^u = R_0 L_y \int\limits_0^\infty \mathrm{e}^{u - gA(u)}, \tag{1}$$

where L_y is the length across the array. According to the discussion above, the basic quantity needed to estimate the resistance of the disordered array is the resistance of a break between the two nearest neighbor stubs that we denote as R_1. In what follows we calculate the conductance σ_1 between the two nearest neighbor stubs in the array. Then, since the resistance of a break is much larger than the resistance of the neighbor regions, it can be evaluated as $R_1 = 1/\sigma_1$.

We calculate the tunneling current between the two nearest neighbor stubs with numbers $n = 0$ and $n = 1$. We assume that the tunneling occurs through a single pinhole connecting the two stubs. Without loss of generality we take the coordinate of the pinhole in the stub as x. A fermion field in a given stub can be represented as a sum of the right- and left-moving chiral components. We assume that the size of the tunneling region (a pinhole) is still larger than $1/(2k_F)$, where k_F denotes the Fermi wave-vector in a stub. Then the tunneling between the modes with different chiralities is suppressed, and the major contribution to the current is given by the tunneling between the fermionic modes with equal chiralities. The probability of a single hop is characterized by the correlation function

$$X_1(\tau) = \sum_{\chi = R, L} \langle T_\tau(\Psi_{0\chi}(0, \tau)\, \Psi_{1\chi}^\dagger(0, \tau)\, \Psi_{1\chi}(0, 0)\, \Psi_{0\chi}^\dagger(0, 0)) \rangle, \tag{2}$$

where $\chi = R, L$ denotes the chirality. The current associated with the single-particle nearest neighbor hopping is given by [5, 6]

$$I_1(V) = -2e\Im X^R(\omega = -eV) = -e \int\limits_{-\infty}^\infty \mathrm{d}t\, \mathrm{e}^{i\omega t}[X_-(\tau = it + 0) - X_+(\tau = it + 0)], \tag{3}$$

where $X_+(\tau) = X_1(\tau > 0)$ and $X_-(\tau) = X_1(\tau < 0)$. Due to the Luttinger liquid character of each stub, a chiral fermion operator $\hat{\Psi}_{n\chi}(x)$ with coordinate x in a stub with number n can be expressed using the bosonization rules as follows

$$\Psi_{n\chi}(x) \sim \left(\frac{2\pi}{L}\right)^{1/2} \hat{F}_{n\chi}\, \mathrm{e}^{-i\phi_{n\chi}(x)} = \left(\frac{2\pi}{L}\right)^{1/2} \hat{F}_{n\chi} \exp\left[\int\limits_{-1}^1 \mathrm{d}p\, \phi_{p\chi}(x)\, \mathrm{e}^{-i\pi pn}\right], \tag{4}$$

where $\phi_{p\chi}(x)$ is a chiral bosonic field, and $\hat{F}_{n\chi}$ is a Klein factor. In the last equality a discrete Fourier transform with respect to the number of the stub has been used. Using (4), the correlator (2) splits into the product of the correlator of Klein factors and the correlator of bosonic exponents. The correlator of the Klein factors is calculated to give

$$\langle (\hat{F}_{0,\chi}(\tau)\, \hat{F}_{1,\chi}^\dagger(\tau)\, \hat{F}_{1,\chi}(0)\, \hat{F}_{0,\chi'}^\dagger(0)) \rangle = \frac{1}{a^2} \exp\left[(E(N_0, N_1) - E(N_0 + 1, N_1 - 1))\, \tau\right]. \tag{5}$$

Here $E(N_0, N_1)$ denotes the energy of the ground state, and $E(N_0 + 1, N_1 - 1)$ denotes the energy of the states with changed occupation numbers of the stubs 0 and 1. The difference $E_a = E(N_0 + 1, N_1 - 1) - E(N_0, N_1)$ is the additive charging energy for transfer of charge from the stub 1 to the stub 0.

© 2004 WILEY-VCH Verlag GmbH & Co. KGaA, Weinheim

The dynamics of plasmons in the array is governed by the action of a sliding Luttinger liquid [3]

$$S = \int\limits_{-1}^{1} dp \int\limits_{0}^{\beta} d\tau \int\limits_{-L/2}^{L/2} dx \, \frac{1}{2K_p} \left\{ \frac{1}{u_p} |\partial_\tau \Theta_p|^2 + u_p |\partial_x \Theta_p|^2 \right\}, \tag{6}$$

where the Luttinger liquid constant K_p and the plasmon velocity u_p can be expressed through the interactions within each stub and between the stubs. Due to the finite length of the stub L, the x-components of plasmon wave-vectors are discrete $\left(k_n = \frac{\pi n}{L} \right.$ for zero boundary conditions at the ends of each stub $\left. \right)$. The interaction between the stubs leads to the formation of plasmon bands from each of the discrete energy levels of a single stub. The plasmon dispersion within each band is characterized by the dependence of the plasmon velocity u_p on the wave vector p across the array.

To calculate the correlator of bosonic exponents

$$X_b^{\chi\chi}(\tau) = \langle T_\tau (e^{-i\phi_{0\chi}(0,\tau)} \, e^{i\phi_{1\chi}(0,\tau)} \, e^{-i\phi_{1\chi}(0,0)} \, e^{-i\phi_{0\chi}(0,0)}) \rangle, \tag{7}$$

that enters (3), we represent the chiral field $\phi_{p\chi}(x)$ as a combination of two dual bosonic fields $\Theta_p(x)$ and $\Phi_p(x)$, $\phi_{p,R/L}(x) = (\Theta_p(x) \pm \Phi_p(x))/2$. The fields $\Theta_p(x,\tau)$, $\Phi_p(x,\tau)$ are related to free bosonic fields $\tilde{\Theta}_p(x,\tau)$, $\tilde{\Phi}_p(x,\tau)$ by a rescaling $\Theta_p = \tilde{\Theta}_p \sqrt{K_p}$, $\Phi_p = \tilde{\Phi}_p / \sqrt{K_p}$ (the coordinates are suppressed) with complementary rescaling of velocity $u_p \to u_p/K_p$ [5]. Rescaling the fields $\Theta_p(x,\tau)$, $\Phi_p(x,\tau)$, and using the finite-temperature correlator of free bosonic fields, one finds that the correlators of bosonic exponents for left- and right-moving fields are equal and can be written in the form ($\tau > 0$)

$$X_{b\pm}(\tau) = \exp \left[-\int\limits_0^1 dp \, \kappa_p (1 - 2\cos(\pi p) + \cos(2\pi p)) \left\{ \sum_{m=1}^{\infty} \ln \left[1 - \exp \left(-\frac{\pi u_p}{2L} (m\beta - \tau) \right) \right] \right. \right.$$

$$\left. \left. + \ln \left[1 - \exp \left(-\frac{\pi u_p}{2L} (m\beta + \tau) \right) \right] - 2 \ln \left[1 - \exp \left(\frac{\pi u_p}{2L} m\beta \right) \right] + \ln \sinh \left(\pm \frac{\pi u_p}{2L} \tau \right) \right\} \right], \tag{8}$$

where $\kappa_p = K_p + 1/K_p$.

Analytical evaluation of the functions $X_{b\pm}(\tau)$ for general dispersion u_p, coupling constant κ_p, and at any temperature seems to be impossible. To obtain analytical expressions, we make the following simplifying assumptions. The coupling constant is assumed to be p-independent, $\kappa_p = \kappa$. Furthermore, we consider a simplified form for plasmon dispersion $u_p = v_0 + v_1 p$. That dispersion law corresponds to a plasmon band of the width $2v_1$ ($-1 \leq p \leq 1$). We remind that p is the wave number for the propagation of plasmons across the array. Under the above approximations, it is possible to find analytical expressions for the thermally activated nearest neighbor hopping in the low-temperature limit. Notice, that in approximative evaluations it is much more important to keep the p-dependence of the velocity u_p that reflects the formation of plasmon bands than the p-dependence of the coupling constant κ_p. In the lowest order in p-dependent terms, the latter just leads to the averaging of the single Luttinger liquid result over the coupling constant.

In Eq. (9) the term $\ln \sinh \left(\pm \frac{\pi u_p}{2L} \tau \right)$ gives the zero temperature result, whereas the temperature correction is contained in the sum over m. Consider, for example, the term $\sum_{m=1}^{\infty} \ln \left[1 - \exp \left(-\frac{\pi u_p}{2L} (m\beta - \tau) \right) \right]$, the analysis of other terms being similar. Expanding the logarithm, we obtain

$$\sum_{m=1}^{\infty} \ln \left[1 - \exp \left(-\frac{\pi u_p}{2L} (m\beta - \tau) \right) \right] = -\sum_{n=1}^{\infty} \frac{1}{n} \exp \left[\frac{\pi u_p}{2L} n\tau \right] \sum_{m=1}^{\infty} \exp \left[-\frac{\pi u_p}{2L} nm\beta \right]. \tag{9}$$

One can see from (10) that at low temperatures (large β) the terms with large m in the sum over m are exponentially suppressed, independently of n. The leading order at very low temperatures is given by the term with $m = 1, n = 1$. Therefore, in the low-temperature limit, we keep in (8) the terms with $m = n = 1$ only. Substituting (5) and (8) in (3), leaving only the leading terms at low temperature, and differentiating over V at $V = 0$, we obtain the conductance between the nearest-neighbor stubs in the form

$$\sigma_1 = -e \left. \frac{dI(V)}{dV} \right|_{V=0} \approx 8\pi^2 e \left(\frac{\kappa L}{v_1} \right)^2 \left(1 - 3 \frac{\kappa L}{\beta v_1} e^{-\beta \pi w/L} \right) e^{-\beta(E_c + 2\pi w/L)} , \tag{10}$$

where $w = v_0 + v_1$. Eq. (11) represents the final result of the calculation of conductance between the two neighbor stubs. It contains the lowest order correction from the plasmon band in form of the multiplicative factor $\left(1 - 3 \frac{\kappa LT}{v_1} e^{-\beta \pi w/L} \right)$. Using (11) the function $f(E_a, T)$ can be approximated as

$$f(E_a, T) = \beta E_a + \varphi(T) \approx \beta \left(E_a + \frac{2\pi w}{L} \right) + 3 \frac{\kappa L}{\beta v_1} e^{-\beta \pi w/L} \tag{11}$$

To proceed with calculation of resistance, we note that the values of E_a have to be restricted from above by some energy of order of the interlevel distance in a single stub $\pi w/L$. Imposing the restriction $|E_a| \leq \frac{\pi w}{L}$, we obtain for the area in the phase space

$$A = \begin{cases} \dfrac{aT}{2}(u - \varphi)^2, & u < \beta \dfrac{\pi w}{L} + \varphi \\[2ex] a \dfrac{\pi w}{L}(u - \varphi) - \dfrac{aT}{2} \left(\dfrac{\pi w}{L} \right)^2, & u > \beta \dfrac{\pi w}{L} + \varphi \end{cases} \tag{12}$$

The expression for the resistance becomes

$$\frac{R}{R_0 L_y} = \int\limits_0^{\varphi + \beta \pi w/L} du \, e^{u - \frac{ga}{2\beta}(u-\varphi)^2} + \int\limits_{\varphi + \beta \pi w/L}^{\infty} du \exp \left[u \left(1 - \frac{\pi w g a}{L} \right) \right] \exp \left[\frac{\pi w g a}{L} \left(\varphi + \beta \frac{\pi w}{L} \right) \right] . \tag{13}$$

The second term in (14) converges only for $\frac{1}{ga} < \frac{\pi w}{L}$, which expresses the condition that the interlevel distance between the plasmon modes of a single isolated stub is larger than the average energy separation between the localized energy levels in the array. The condition of convergence in (14) implies that the saddle point of the first integral in (14) falls within the integration limits. Then one can evaluate the first integral by the steepest descent method. Finally, one obtains the resistance in the form

$$R \approx R_0 L_y \left[1 + 3 \frac{\kappa L}{v_1} T e^{-\frac{\pi w}{LT}} \right] \left\{ \sqrt{\frac{2\pi}{gaT}} \exp \left[\frac{1}{2gaT} + \frac{2\pi w}{LT} \right] + \frac{1}{\frac{\pi w}{L} ga - 1} \exp \left[\frac{\pi w}{LT} \left(3 - ga \frac{\pi w}{2L} \right) \right] \right\} . \tag{14}$$

Comparing the obtained result with the resistance of the noninteracting random chain $R \propto \sqrt{\frac{2\pi}{gaT}} \exp \left[\frac{1}{2gaT} \right]$ one notices two major differences. Firstly, the spectrum of a single stub introduces a natural restriction on the minimal density of the localized states $ga > \frac{L}{\pi w}$. Secondly, at low temperatures, the finite width of the plasmon bands (given by v_1) and the interactions (characterized by κ) enter as a prefactor $\left[1 + 3 \frac{\kappa L}{v_1} T e^{-\frac{\pi w}{LT}} \right]$. We conclude, that for a dispersion linear in p, the effect of interaction depends on the sign of v_1, leading either to an enhancement (for $v_1 > 0$) or to a reduction (for $v_1 > 0$) of the resistance.

Acknowledgements The author is grateful to M. Raikh and to D. Pfannkuche for illuminating discussions. Financial support from Sonderforschungsbereich 508 is gratefully acknowledged.

References

[1] T. Vuletic, B. Korin-Hamzic et al., Phys. Rev. B **67**, 184521 (2003); C. Cui, T. A. Tyson et al., Phys. Rev. B **67**, 104107 (2003).
[2] B. I. Shklovskii, Phys. Rev. B **67**, 045201 (2003); M. M. Fogler, S. Teber and B. I. Shklovskii, cond-mat/0307299; M. Foygel, R. D. Morris, and A. G. Petukhov, Phys. Rev. B **67**, 134205 (2003).
[3] R. Mukhopadhyay, C. L. Kane, and T. C. Lubensky, Phys. Rev. B **64**, 045120 (2001).
[4] M. E. Raikh, I. M. Ruzin, Sov. Phys. JETP **68**, 642 (1989).
[5] A. O. Gogolin, A. A. Nersesyan, A. M. Tsvelik, Bosonization in Strongly Correlated Systems, Cambridge University Press, 1998.
[6] G. Mahan, Many-Particle Physics, New York, 1981.

phys. stat. sol. (c) **1**, No. 1, 131–135 (2004) / **DOI** 10.1002/pssc.200303620

Non-exponential localization in two-dimensional disordered systems

Masaki Goda[*, 1], **Takahiro Aoyagi**[1], and **Yutaka Ishizuki**[2]

[1] Faculty of Engineering, and [2]Graduate School of Science and Technology,
  Niigata University, Niigata 950-2181, Japan

Received 1 September 2003, revised 4 October 2003, accepted 6 October 2003
Published online 28 November 2003

PACS 71.23.-An

Vanishing of exponential localization in two-dimensional disordered systems is examined by the combination of an analytical calculation and of a numerical examination. We extend our analytical theory to be applicable to the ordinary tight-binding system. And, with the aid of a numerical calculation, the absence of exponential localization in two-dimension is strongly suggested in a weakly disordered system.

1 Introduction

Localization of wave functions in disordered systems has been one of the important problems to be solved since the middle of the last century. One-parameter scaling theory of localization presented in the late 1970's by Abraham et al. seemed to give an overview how the localization and delocalization are realized in increasing the system-size [1]. The theory claimed that almost all eigenstates are exponentially localized in one- and two-dimension (1- and 2-D). However, it is also well known that rigorous argument has been made only for the case of 1-D [2, 3].

In 2-D, there is a controversy. Two-parameter scaling theory by Caveh describes a "mobility edge" separating the exponential localization and the power-law localization [4]. Some questions have been given on the validity of the one-parameter scaling theory [5, 6].

One of the authors (M. Goda) has presented, in collaboration with M. Ya Azbel, an analytical theory of non-exponential localization in 2-D. This describes statistical properties of wave propagation in a model strip [7, 8, 9]. The semi-infinite strip with the width S consists of a free space and randomly distributed delta-function potentials on some lattice points. We may call it model A.

The theory is based on studying the product of the transfer matrices $T_L = \prod_{\ell=1}^{L} t_\ell$. We have pointed out that the S-th largest (that is the smallest positive) Lyapunov exponent of the Hermitian matrix $M_L = T_L^+ T_L$ along the strip (of the length L) vanishes with increasing S in a weakly disordered system. Because the absolute square of the transmitted wave $|\psi_L|^2$ is described by M_L as $<\psi_\ell | \psi_\ell> = <\psi_0 | M_L | \psi_0>$, where ψ_0 is the incident incoming wave, the vanishing of the smallest positive Lyapunov exponent of M_L describes the vanishing of the exponential localization. The theory is analytic except for a crucial assumption that

$$< \lambda_S(M_L) > \cong \lambda_S(< M_L >), \tag{1}$$

* Corresponding author: e-mail: mgoda@cc.niigata-u.ac.jp, Phone: +81 25 262 6762, Fax: +81 25 262 7744

where $\lambda_r(M_L)$ $(1 \leq r \leq S)$ is the r-th largest eigenvalue (that is, $\lambda_S(M_L)$ is the smallest one bigger than or equal to unity) of M_L and $< \dots >$ denotes to take an ensemble average. At a glance, the assumption (1) looks dangerous because we have many experiences on how random phase works. However, we should also remind that the random phase of the wave functions in random systems does not play any role in M_L, since it corresponds to the absolute square of the wave function $< \psi_L | \psi_L >$. The equality of Eq. (1) takes place when $(M_L)^{1/2L}$ has a convergent property with increasing L. It is thus important to confirm the convergent property of the Hermitian matrix $(M_L)^{1/2L}$ to justify the equality (1) and confirm the existence of the non-exponential localization in 2-D.

2   Localization of eigenstates of a tight-binding electronic system

We extend the above theory to cover the well known case of the tight-binding electronic system on a strip of square lattice with only diagonal disorder. We find that the essential structure of the theory does not change at all. We may call the tight-binding system model B.

Let ψ_ℓ be the wave function at a region assigned by ℓ ($0 \leq \ell \leq L$; positive integer) along the strip. Then the transfer matrix is defined as $\psi_\ell = t_\ell \, \psi_{\ell-1}$. We expand ψ_ℓ by a complete set of bases $\{\phi_\ell^{(\alpha,\mu)}\}$ of plane wave, where $\phi_\ell^{(\alpha,\mu)}(s) = \exp\{i\alpha k_\mu \ell\}\exp\{iK_\mu s\}$ $(\alpha = +1,-1)$ $(s = 1 \sim S)$. Two wave numbers αk_μ and $K_\mu = 2\pi\mu/S$ $(\mu = 0, \pm 1, \pm 2,,, \pm(S-1)/2)$ are those along the strip and perpendicular to the αk_μ axis, respectively. Energy conservation law requires $k_\mu^2 + K_\mu^2 = k^2$, where k is the wave number of the incident incoming wave. Then $\psi_\ell = \sum_a c_\ell(a)\phi_\ell^a$ $(a = (\alpha,\mu))$, $c_\ell = t_\ell \, c_{\ell-1}$, $t_\ell = \Theta_\ell W$ and

$$\Theta_{a'}^a = \delta_{a,a'} - i\alpha \frac{V_\ell(\mu'-\mu)}{2\sin k_\mu}, \tag{2}$$

where V_ℓ is the Fourier transform of the diagonal potentials perpendicular to αk_μ axis. Diagonal matrix $W_{a'}^a = \delta_{a,a'}\exp\{i\alpha k_\mu\}$ describes the phase shift. The only essential difference in model B is the term $\sin k_\mu$ in the denominator of Eq. (2) instead of k_μ in the model A. The localization length ξ is now defined as

$$<\xi^{-1}> = \lim_{L\to\infty} \frac{<\log\lambda_S(M_L)>}{2L} = \lim_{L\to\infty} < L_S(M_L) >, \tag{3}$$

where $L_r(M_L)$ is the r-th largest Lyapunov-exponent (L-exponent) of M_L. Then our assumption (1) guarantees that if $\lim_{L\to\infty} L_S(< M_L >)$ vanishes the localization length diverges. Our theory describes the vanishing of $\lim_{L\to\infty} L_S(< M_L >)$ on an energy region corresponding to that of a virtual crystal, in a weakly disordered regime. Once the problem becomes to find the eigenvalues of $(< M_L >)^{1/L} = < M_L >$, further calculation is made analytically. The eigenvalue equation is formally written as,

$$\sum_\mu \Pi_\mu\{(\lambda - q_\nu)\delta_{\mu,\nu} - \varepsilon u_\nu^\mu\} = 0 \quad \varepsilon \propto 1/S, \tag{4}$$

where $\{\Pi_\mu\}$ is the eigenmatrix. The term q_ν is diagonal with respect to the "channel" ν and describes a virtual crystal. The "inter- and intra-channel" interaction is described by εu_ν^μ with the coefficient

$\varepsilon \propto 1/S$. This suggests the virtual crystal to survive in the 2-D limit where $S \to \infty$. Actually, it is rigorously shown by investigating the shift of the eigenvalues, by the term εu_ν^μ, from the unit circle of the complex energy plane, by using a generalized Rayleigh's theorem [10].

3 Numerical examination to confirm the assumption (1)

We show a numerical evidence of justifying (1). It consists of showing (a) the convergent property of the distribution of each L-exponent of T_L (and hence of M_L) with increasing L, and of showing (b) two statistical inequalities. Each of them is sufficient and essential for proving our conclusion of the existence of non-exponential localization in two-dimension.

We generate $10^3 \sim 10^4$ samples of random strip of square lattice with the width $S(= 1 \sim 10)$ and of the length 10^3. The nearest neighbor interaction t is set to be unity and the diagonal elements are taken as identical random variables with a uniform distribution over the period $[-\varepsilon, \varepsilon]$. Because we have to examine the case of weak disorder, we take $\varepsilon = 0.0001$. The numerical calculation is made by double precision and the transfer matrices $\{t_\ell\}$ are multiplied up to $L = 10^3$ by keeping the product of $2S$ eigenvalues of T_L to be unity within the accuracy of single precision.

We find that the $r-$th L-exponent $L_r \equiv \ln\{|\lambda_r|\}/L$, where λ_r is the $r-$th eigenvalue, of T_L $(r = 1 \sim S)$ behaves like $L_r = 1/L^\eta + L_r(\infty)$ and the standard deviation σ_r^2 of the distribution of L_r over the samples converges like $\sigma_r^2 \propto 1/L^\alpha$. The convergent property $\sigma_S^2(\ell)$ is exemplified for $S = 10$ in Fig. 1, and the powers of convergence $\alpha_1(S)$ and $\alpha_S(S)$ are shown in Fig. 2, as a function of $1/S$. The $1/S$ plot suggests the convergent property of L_1 and L_S in the limit $S \to \infty$.

The Fig. 3 and 4 compares, respectively, the L-exponent $L_S(L)$ of T_L and of $< T_L >$ for the cases $S = 5$ and $S = 10$. These show clearly how the random phase of the wave function works for the L-exponent L_r to shrink. But on the contrary, Fig. 5 and 6, showing the corresponding L-exponent of M_L, give us an absolutely different feature. The smallest positive L-exponent of the ensemble averaged M_L ($< M_L >$) is always equal or bigger than the ensemble averaged one of M_L. That is,

$$< L_S(M_L) > \le L_S(< M_L >). \tag{5}$$

We also confirm another inequality

$$< (\lambda_S(M_L))^{1/2L} > \cong (\lambda_S(< M_L >))^{1/2L}, \tag{6}$$

which is numerically similar to Eq. (5) because we examine the case of weak disorder.

4 Summary

We examined the localization problem in 2-D by the combination of analytical calculation and numerical examination. We extended our analytical theory to be applicable to model B and numerically found the inequalities (5) and (6). Expecting Eqs. (5) and (6) are valid for $S \to \infty$, Eq. (5) directly guarantees the absence of exponential localization in 2-D when the L-exponent L_S of $< M_L >$ vanishes in the limit $S \to \infty$.

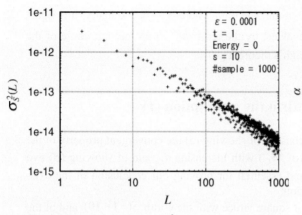

Fig. 1　Standard deviation $\sigma^2(L)$ of the L-exponent L_S of T_L.

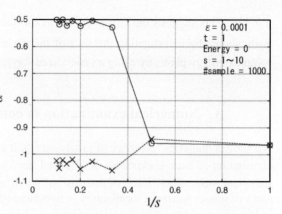

Fig. 2　$1/S$ plot of the exponent of power α of $\sigma^2(L) = AL^\alpha$ for the maximum L-exponent (○) and the minimum L-exponent (×).

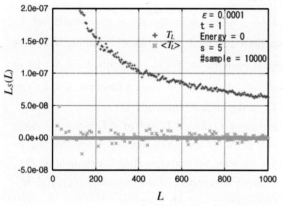

Fig. 3　Mean L-exponent $L_S(L)$ of T_L (+) and that of $< T_L >$ (×) for $S = 5$.

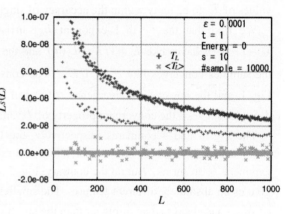

Fig. 4　Mean L-exponent $L_S(L)$ of T_L (+) and that of $< T_L >$ (×) for $S = 10$.

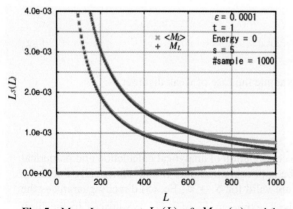

Fig. 5　Mean L-exponent $L_S(L)$ of M_L (+) and that of $< M_L >$ (×) for $S = 5$.

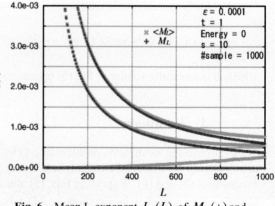

Fig. 6　Mean L-exponent $L_S(L)$ of M_L (+) and that of $< M_L >$ (×) for $S = 10$.

References

[1] E. Abrahams, P. W. Anderson, D. C. Liccialdello, and T. V. Ramakrishnan, Phys. Rev. Lett. **42**, 673 (1979).
[2] K. Ishii, Suppl. Prog. Theor. Phys. **53**, 77 (1973).
[3] S. Kotani, Taniguchi Symp. SA, Katata, Ed. by K. Ito, 225 (Kinokuniya, 1984).
[4] M. Kaveh, J. Phys. C **17**, L79(1985).
[5] B. L. Altshuler, V. E. Kravtsov, and I. V. Lerner, in *Anderson Localization*, Ed. by T. Ando and H. Fukuyama, Springer Proceeding in Physics, Vol. **28**, p. 300 (Springer, 1988).
[6] N. Tit, N. Kumar, J. W. Halley, and H. Shore, Phys. Rev. B **47**, 15988 (1993).
[7] M. Ya. Azbel, Phys. Rev. B **26**, 4735 (1982).
[8] M. Goda, M. Ya. Azbel, and H. Yamada, Int. J. Mod. Phys. B **13**, 2705 (1999) .
[9] M. Goda, M. Ya. Azbel, and H. Yamada, Physica B **296**, 66 (2001).
[10] See J. H. Wilkinson, *The Algebraic Eigenvalue Problem* (Oxford, 1965).

phys. stat. sol. (c) **1**, No. 1, 136–139 (2004) / **DOI** 10.1002/pssc.200303633

Quantum effects in Mott's variable-range hopping

J. Prior[*], **A. M. Somoza**, and **M. Ortuño**

Departamento de Física, Universidad de Murcia, Murcia 30.071, Spain

Received 1 September 2003, revised 23 September 2003, accepted 23 September 2003
Published online 28 November 2003

PACS 72.10.Bg, 72.20.Ee, 72.80.Ng

We have simulated numerically Mott's variable–range hopping for one–dimensional systems taking into account the exact form of the states in the calculation of the spatial factors contributing to the resistance between states. DC hopping conductance is obtained by a percolation procedure in which each node corresponds to a states and their conections are governed by their hopping resistances. Our results show important differences with respect to the standard calculations, based on the asumption of an exponential dependence with distance of the effective resistances between states.

1 Introduction

Mott's variable–range hopping refers to the conduction mechanism between localized states in the absence of interactions at very low temperatures so that a compromise between jumping distances and energy penalties must be reached. The aim of this work is to discuss the effects of quantum fluctuations on Mott's variable–range hopping in finite one–dimensional samples. Up to now, quantitative calculations on this regime assume hydrogen-like wavefunctions which results in an exponential overlap with an exponent equal to $2r/\xi$, where ξ is the localization length, which is assumed to have a unique well-define value.

In the hopping regime, following Miller and Abrahams [1], one replaces the transport problem by a random resistor network in which sites i and j are connected by the resistance

$$R_{i,j} = R_0 \exp\left\{\frac{2r_{ij}}{\xi}\right\} \exp\left\{\frac{E_{i,j}}{k_B T}\right\} \tag{1}$$

where r_{ij} is the distance between sites and $E_{i,j}$ is their energy difference if they are on different sides of the Fermi level, E_F, and $\max\{|E_F - E_i|, |E_F - E_j|\}$ otherwise. A naive extension of Mott's argument to one–dimension predicts a conductance of the form:

$$\sigma = \sigma_0 \exp\left\{-(T/T_0)^{1/2}\right\} \tag{2}$$

with T_0 given by $T_0 = 1/(k\xi N(E_F))$, where $N(E_F)$ is the density of states at the Fermi level and k is Boltzmann's constant. As already noted by Kurkijarvi [2], the special nature of one–dimensional systems leads to corrections on the previous formula which depend on the size of the system. Taking these effects into account, Lee [3] and Serota et al. [4] obtained, for not too large samples, the following self-consistent equation for the average, over disorder realizations, of the logarithm of the resist-

---

[*] Corresponding author: e-mail: jprior@um.es, Phone: +34 968 367 383, Fax: +34 968 364 148

ance $\langle \ln R \rangle$

$$\left(\frac{T_0}{T}\right)\left\{\ln\frac{2L}{\xi}+\ln\left[\langle\ln R\rangle\left(\frac{T}{T_0}\right)\right]\right\}=\langle\ln R\rangle^2 . \tag{3}$$

where L is the length of the sample. Raikh and Ruzin [5] undertook a detailed analytical treatment of the model and obtained the distribution of sample resistances.

2 Model

We consider a one–dimensional sample of size L described by the standard Anderson-Hubbard Hamiltonian

$$H = \sum_i \epsilon_i a_i^\dagger a_i + t \sum_i a_{i+1}^\dagger a_i + \text{h.c.}, \tag{4}$$

where the operator $a_i^\dagger$ (a_i) creates (destroys) an electron at site i of a regular lattice and ϵ_i is the energy of this site chosen randomly between $(-W/2, W/2)$ with uniform probability. The hopping matrix element t is taken equal to -1 and the lattice constant equal to 1, which sets the energy and length scales, respectively. In order to reduce edge effects, we use periodic boundary conditions, i.e., $a_{L+1}^\dagger = a_1^\dagger$, $a_0^\dagger = a_L^\dagger$. The system size varies from $L = 100$ to 2000 and we consider two disorder strenghts, $W = 2$, which corresponds to a localization length $\xi = 25$, and $W = 10$, for which $\xi = 1.4$, at $T = 0$.

In real disordered systems, the wavefunctions decay with distance in a very complicated way, and so the overlaps between them are not of a simple exponential form. In this work, we want to use realistic data for the wavefunctions and study their implications in the final resistance of the sample. We first calculate the states of the system ϕ_α and then the effective bond resistances between them which are proportional to the exponential energy factor appearing in Eq. (1) and to the matrix element [6]

$$D_{\alpha,\beta} \equiv |\langle\phi_\alpha| e^{iqx} |\phi_\beta\rangle|^{-2}, \tag{5}$$

representing the coupling of the states α and β by a phonon of momentum q. To conserve energy, the phonon momentum must be proportional to the energy difference between the states. These matrix elements enter directly in the calculation of the bond resistances between states and, in the standard approach, they are approximated by the exponential function of distances appearing in Eq. (1). We found that once the states are not extremely localized, they turn out to be very different from the exponential form assumed in Eq. (1). With the exact evaluation of $D_{\alpha,\beta}$, we are considering the full distribution of values of the localization length, and so we are taking into account the effects of quantum fluctuations on hopping. In the following we will refer to the results based on an exponential form of the resistance as "standard" and to our results as "quantum".

A percolation algorithm calculates the sample resistance from the effective resistances between states, for both the standard and the quantum models. In the standard model, the localization length is obtained through a linear fit of the logarithm of the matrix elements $D_{\alpha,\beta}$ as a function of the distance between the centers of the states α and β. We have checked that this procedure is better than using the localization length extracted from the conductivity at $T = 0$.

3 Numerical results

Figure 1 shows $\langle \ln R \rangle$ as a funtion of $T^{-1/2}$ for the standard model, for sample sizes ranging between 250 and 2000. In all the data presented in this work the disorder strength is $W = 2$, which is closer to the experimental situation [7], but similar conclusions are obtained for the case $W = 10$. Mott's law would correspond to a straight line of constant slope in Fig. 1. Only the largest samples show a linear behavior and, even in this case, the slope depends on the sample size. This is in qualitative agreement

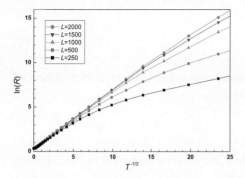

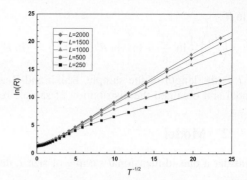

Fig. 1 $\langle \ln(R) \rangle$ for several lengths as a function of $T^{-1/2}$ for the standard model.

Fig. 2 $\langle \ln(R) \rangle$ for several lengths as a function of $T^{-1/2}$ for the quantum model.

with Eq. (3). If we neglect the second term on the left–hand side of this equation, we obtain Mott's law with a size dependent logarithmic correction to the characteristic temperature. A better approximation is obtained by inserting the resulting expression in the second term of Eq. (3) [4]

$$\langle \ln R \rangle \sim \left(\frac{T_0}{T}\right)^{1/2} \left(\ln\left\{ \frac{2L}{\xi}\left(\frac{T_0}{T}\right)^{1/2} \left[\ln\frac{2L}{\xi}\right]^{1/2} \right\} \right)^{1/2}. \tag{6}$$

This equation incorporates a logarithmic deviation with respect to Mott's law, which can be clearly appreciated in Fig. 1 for the shorter samples. However, a full quantitative agreement between numerical results and existing theories is not posible, even for this standard model. Care must be taken when deducing values of the localization length from experimental results of Mott's variable–range hopping in one–dimension.

The results of $\langle \ln R \rangle$ versus $T^{-1/2}$ for the quantum model are shown in Fig. 2. The parameters are the same as in Fig. 1. Note that the vertical scales of both figures are different. The trends for both models are similar, but the results for the quantum model are more resistive than those for the standard one. This is a consequence of the one–dimensional geometry of the system, which results in a decrease of the conductance with the inclusion of fluctuations in the individual resistances, since a region with large resistance cannot be avoided in order to percolate in these systems.

In order to compare better the two models, we plot $\langle \ln R \rangle$ as a function of $T^{-1/2}$ for the standard model (solid dots) and for the quantum model (empty circles) for the case $L = 2000$ in Fig. 3. The error bars in this figure correspond to the standard deviation of the data. We have considered 100 samples in this calculation, and so the error in the mean is 10 times smaller than the standard deviation.

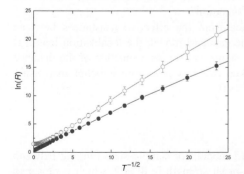

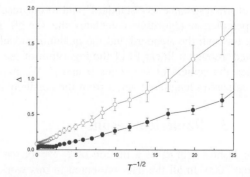

Fig. 3 $\langle \ln(R) \rangle$ as a function of $T^{-1/2}$ for the standard (solid dots) and the quantum (empty circles) calculations of Mott's law.

Fig. 4 Fluctuations of the hopping conductance for the standard (solid dots) and the quantum (empty circles) models.

Serota et al. [4] predicted that the fluctuations Δ of the logarithm of the resistance should grow with decreasing temperature according to the expression

$$\Delta \equiv \langle (\ln R - \langle \ln R \rangle)^2 \rangle^{1/2} = \left(\frac{T_0}{T} \right)^{1/2} \left[\ln \frac{2L}{\xi} \right]^{-1/2} . \tag{7}$$

Figure 4 shows the results for Δ as a function of $T^{-1/2}$ for the standard model (solid dots) and for the quantum one (empty circles). These values are the same as the error bars in Fig. 3. The estimated errors for Δ are shown as error bars in Fig. 4. We can note the agreement of the standard model with the previous prediction, Eq. (7). The results for the quantum model follow a similar trend to those for the standard one, although they are almost double in magnitude. Quantum fluctuations are of the order of the geometrical fluctuations, previously considered in the hopping regime.

4 Summary and conclusions

We have studied Mott's variable–range–hopping law in one–dimensional systems incorporating the effects of quantum fluctuations. We fully diagonalized the Hamiltonian, Eq. (4), and evaluated the matrix element $D_{\alpha,\beta}$, representing the phonon coupling between states, to obtain the effective hopping resistances between states. We used these values in a clasical percolation calculation of the DC conductance. We found that quantum fluctuations play an important role in hopping and should be taken into account in any calculation of these problems.

Acknowledgements The authors would like to acknowledge financial support from the Spanish DGCYT, project numbers BFM2000–1059 and BFM2000–1319 and a grant (JP).

References

[1] A. Miller and E. Abrahams, Phys. Rev. **120**, 745 (1960).
[2] J. Kurkijarvi, Phys. Rev. B **8**, 922 (1973).
[3] P. A. Lee, Phys. Rev. Lett. **53**, 2042 (1984).
[4] R. A. Serota, R. K. Kalia, and P. A. Lee, Phys. Rev. B **33**, 8441 (1986).
[5] M. E. Raikh and I. M. Ruzin, in: Mesoscopic Phenomena in Solids, edited by B. I. Altshuler, P. A. Lee, and R. A. Webb (Elsevier Science, Amsterdam, 1991).
[6] N. F. Mott and E. A. Davies, Electronic Processes in Non-crystalline Materials (Clarendom Press, Oxford, 1979).
[7] F. Ladieu, D. Mailly, and M. Sanquer, J. Phys. I France. **3**, 2321 (1993).

phys. stat. sol. (c) **1**, No. 1, 140–143 (2004) / **DOI** 10.1002/pssc.200303637

Magnetic field dependence of the magnetic length for localized electrons in QHE

M. Saglam[*,1] and **B. Boyacioglu**[2]

[1] Ankara University, Department of Physics, Faculty of Sciences, 06100 Tandogan, Ankara, Turkey
[2] Ankara University, Dikimevi Vocational School of Health, 06590-Cebeci, Ankara, Turkey

Received 1 September 2003, revised 9 October 2003, accepted 10 October 2003
Published online 28 November 2003

PACS 73.43.Cd

Magnetic field dependence of the magnetic length for localized electrons in Quantum Hall Effect (QHE) systems is calculated. It is found that in the localization region the magnetic length becomes inversely propotional to the magnetic field. Therefore in the localization region the shrinkage of the cyclotron radius becomes faster compared to the delocalized region where we have the conventional magnetic length $\left(\dfrac{\hbar c}{eB}\right)^{1/2}$ which is inversely proportional to the square root of the magnetic field.

1 Introduction

The physics of 2D electron systems subjected to very high magnetic fields attracted more attention after the discovery of the Integer Quantum Hall Effect (IQHE) and the Fractional Quantum Hall Effect (FQHE) [1, 2]. The Quantum Hall Effect and the localization are inseparable concepts. More than two decades past the discovery of the QHE, the nature of the transitions between adjacent Hall plateaus is still subject to continuing research and localized electrons are responsible for the QHE plateaus [3–9]. Since the classic calculation of Yafet, Keyes and Adams [10] demonstrating the shrinkage of shallow donor wave functions in large magnetic fields, it has been widely recognized that a magnetic field could be used to tune a metallic sample through the metal-insulator (MI) transition. Several experimental groups (Keyes et al. [11], Ishida et al. [12] and Robert et al. [13]) have succesfully proved the idea of using a large magnetic field to tune a metallic sample through the (MI) transition. Aoki [2] showed the localization by a computer experiment. Shklovskii and Efros [15] discussed the effective radius of electron motion for an hydrogenlike atom in a magnetic field by using an envelope wave function approximation. The aim of this study is to calculate the effective magnetic lenght (cyclotron radius of ground state Landau level) as a function of the magnetic field for localized electrons in QHE systems on the basis of Bockstedte et al. [14] calculations. Bockstedte et al. [14] investigated a phase diagram for the localization-delocalization transition of a QHE system in a perpendicular magnetic field B with a numerical method and found that below a critical concentration ($n_{\mathrm{crit}} = [0.246 \pm 0.004] \, a^{-2}$) all states would be localized, here a is the impurity radius and n is the areal concentration of the interface electrons. In the present work we focus our attention on the effective magnetic length that they derived. We show that in the localization region the magnetic length becomes inversely propotional to the magnetic field. Therefore in the localization region the shrinkage of the cyclotron radius becomes faster compared to the delocalized region where we have the conventional magnetic length $\left(\dfrac{\hbar c}{eB}\right)^{1/2}$ which is inversely propotional to the square root of the magnetic field.

---
[*] Corresponding author: e-mail: Mesude.Saglam@science.ankara.edu.tr

2 Formalism

The system we consider consist of shallow impurities distributed randomly on a two-dimensional (2-d) lattice. Let the donor be located at the center of a cylindrical coordinate system (ρ, ϕ, z) in which the z axis is directed along the magnetic field B. The vector potential A is chosen in the form [15]:

$$A = \tfrac{1}{2}[B \times r].$$ (1)

In this symetrical gauge the ground state wave function F is found to be the solution of the Schrödinger equation

$$-\frac{\hbar^2}{2m^*}\left\{\frac{1}{\rho}\frac{\partial}{\partial\rho}\left[\rho\frac{\partial F}{\partial\rho}\right] + \frac{\partial^2 F}{\partial z^2}\right\} + \frac{\hbar^2\rho^2}{8m^*\lambda^4}F + V(z)F = EF$$ (2)

where $V(z) = \tfrac{1}{2}kz^2$ is potential of donor located at the centre and λ is the magnetic length given by

$$\lambda = \left(\frac{\hbar c}{eB}\right)^{1/2}.$$ (3)

The effect of the magnetic field has been reduced to the appearence of the second term, which works like a spring pulling an electron to the z axis $\left(\text{the force constant being } K^* = \dfrac{\hbar^2}{4m^*\lambda^4}\right)$.

In the absence of the harmonic impurity potential, $V(z)$, for $z = 0$, the solutions of Eq. (2) are simply the harmonic oscillator wave functions

$$|\rho_l\rangle = F_l(\rho) = \left(\frac{\alpha}{\sqrt{\pi}\,2^l l!}\right)^{1/2} H_l(\alpha\rho)\,e^{-(\alpha^2\rho^2)/2}$$ (4)

where $|\rho_l\rangle$ is Dirac notation of $F_l(\rho)$, $\alpha = \left(\dfrac{m^*K^*}{\hbar^2}\right)^{1/2}$ and $H_l(\alpha\rho)$ is the Hermite polynominal of order l ($l = 0, 1, 2, 3\ldots$). If we calculated the ground state wave function we find

$$F_0(\rho) = \frac{1}{(\sqrt{2\pi}\,\lambda)^{1/2}}\,e^{\frac{-\rho^2}{4\lambda^2}}$$ (5)

In the presence of the harmonic impurity potential, $V(z) = \tfrac{1}{2}kz^2$, the ground state wave function takes the form

$$F_0(r) = F_0(\rho)\xi_0(z) = \frac{1}{(\sqrt{\pi}\bar{\lambda})^{1/2}}e^{\frac{-\rho^2}{4\lambda^2}}\xi_0(z)$$ (6)

where $F_0(\rho)$ ($\rho = \sqrt{x^2 + y^2}$) is ground state wave function with the eigenvalue $E_0 = \tfrac{1}{2}\hbar\omega_c$ in the absence of $V(z)$ and $\xi_0(z)$ is the ground state wave function of the one-dimensional harmonic oscillator in z-direction. Here $\bar{\lambda}$ is the effective magnetic length which is given by [14]

$$\bar{\lambda} = a\left(1 + \frac{a^4}{4\lambda^4}\right)^{-1/2}$$ (7)

where a is the zero-magnetic field impurity radius and $\lambda = \sqrt{\dfrac{\hbar c}{eB}}$ is conventional magnetic length which is inversely proportional to the square root of B.

Next we look at the localization condition in two dimensions. It was shown by Bockstedte et al. [14] that below a critical concentration ($n_{\text{crit}} = [0.246 \mp 0.004]\,a^{-2}$) all states will be localized. In their calculation it is assumed that all the interface electrons are in the lowest Landau level, so the filling factor ν is equal to unity which corresponds to the condition $D = n$. Here n is the total number of interface electron and D is the degeneracy which is related to the magnetic field by the relation

$$D = \left(\frac{L}{2\pi}\right)^2\frac{2\pi eB}{\hbar c}.$$ (8)

The condition $D = n$ defines a normalizing field, B_{nor} which is given by

$$B_{nor} = \frac{nhc}{e} \tag{9}$$

where $n = \frac{N}{L^2}$ is the concentration. The magnetic lenght corresponding to B_{nor} will be λ_{nor} which is given by

$$\lambda_{nor} = \sqrt{\frac{\hbar c}{eB_{nor}}} = \frac{1}{\sqrt{2\pi n}} . \tag{10}$$

A typical concentration for QHE systems is between $2 \times 10^{11} - 10^{12}$ cm^{-1} which produce λ_{nor} smaller than a (for GaAs $a \approx 100$Å). Therefore for $B > B_{nor}$ magnetic length is smaller than the Bohr radius ($\lambda < a$).

Let us recall the localization condition of Bockstedte et al. [14] which says at low concentrations, $n < n_{crit} = [0.246 \mp 0.004]\, a^{-2}$ all states will be localized. Since for $B > B_{nor}$ we have $\lambda < a$ in the Bockstedte localization condition a should be replaced by λ. Namely, when $n\lambda^2 < (0.246 \mp 0.004)$ all states will be localized. From Eq. (10) for $B > B_{nor}$, we can write $n\lambda^2 < \frac{1}{2\pi} < (0.246 \mp 0.004)$. Therefore for $B > B_{nor}$ all states will be localized and since λ is smaller than a Eq. (7) reduces to

$$\bar{\lambda} = a\left(1 + \frac{a^4}{4\lambda^4}\right)^{-1/2} = \frac{2\lambda^2}{a} = \frac{2\hbar c}{aeB} \sim \frac{1}{B} . \tag{11}$$

Therefore in the localization region the effective magnetic length becomes inversely proportional to the magnetic field. That means once the localization starts the shrinkage of the cyclotron radius for ground state and hence the shrinkage of the wave function becomes faster compared to the delocalized region where have the conventional magnetic length $\lambda = \left(\frac{\hbar c}{eB}\right)^{1/2}$ which is inversely proportional to the square root of magnetic field.

3   Results and discussion

Magnetic field dependence of the magnetic length for localized electron in two dimensional systems is calculated. We find that in the localization region the magnetic length becomes inversely proportional to magnetic field. We believe that the present result will bring some explanation for understanding the plateaus in QHE. As it is well known Coulomb energy plays an important role in localization, therefore it can not be ignored for localized electrons [7–9]. With the above obtained result we can show that the total energy (Landau energy + Coulomb energy) does not change with the magnetic field for localized electrons. To show this we write total energy for localized electron in the ground state of Landau level:

$$E = \frac{1}{2}\, \hbar\omega_c - \frac{e^2}{\bar{\lambda}} + \text{constant} \tag{12}$$

where   is the dielectric constant and $\frac{e^2}{r}$ is replaced by second term because the effective radius is replaced by $\bar{\lambda}$. If we substitute the value of $\bar{\lambda}$ from Eq. (11) , $a = \frac{\hbar^2}{m^*e^2}$ and $\omega_c = \frac{eB}{m^*c}$ in Eq. (12) we see that the magnetic field dependence disappears and the total energy becomes constant which corresponds to the plateaus in QHE.

As a result of the present investigation, we show that in the localized region the magnetic lenght becomes inversely propotional to the magnetic field. Therefore the total energy does not change with the magnetic field which is related to the plateaus in QHE.

References

[1] T. Ando, A. B. Fowler, and F. Stern, Rev. Mod. Phys. **54**, 437 (1982).

[2] H. Aoki, Rep. Prog. Phys. **50**, 655 (1987).

[3] H. Aoki and T. Ando, Phys. Rev. Lett. **54**, 831 (1985).

[4] A. M. M. Pruisken, Phys. Rev. Lett. **61**, 1297 (1988).

[5] J. T. Chalker and G. J. Daniell, Phys. Rev. Lett. **61**, 593 (1988).

[6] B. Huckestein and B. Kramer, Phys. Rev. Lett. **64**, 1437 (1990).

[7] D. G. Polyakov and B. I. Shklovskii, Phys. Rev. B **48**, 11167 (1993).

[8] S. L. Sondhi, S. M. Girvin, J. P. Carini, and D. Shahar, Rev. Mod. Phys. **69**, 315 (1997).

[9] F. Hohls, U. Zeitler, R. J. Haug, R. Meisels, K. Dybko, and F. Kucher, Physica E **16**, 10 (2003).

[10] Y. Yafet, R. W. Keyes, and E. N. Adams, J. Phys. Chem. Solids **1**, 137 (1956).

[11] W. Keyes and R. J. Sladek, J. Phys. Chem. Solids **1**, 143 (1956).

[12] S. Ishida and E. Otsuka, J. Phys. Soc. Japan **42**, 542 (1977).

[13] J. L. Robert, A. Raymond, R. L. Aulombard, and C. Bousquet, Phil. Mag. B **42**, 1003 (1980).

[14] M. Bockstedte and S. F. Fisher, J. Phys.: Condens Matter **5**, 6043 (1993).

[15] B. I. Shklovskii and A. L. Efros, Electronic Properties of Doped Semiconductors, Springer series in Solid State sciences 45, (1984).

phys. stat. sol. (c) **1**, No. 1, 144–147 (2004) / **DOI** 10.1002/pssc.200303609

Charge carrier properties below and above the metal–insulator transition in conjugated polymers – recent results

H. B. Brom[*, 1], **H. C. F. Martens**[1], **I. G. Romijn**[1], **I. N. Hulea**[1, 2], **H. J. Hupkes**[1], **W. F. Pasveer**[2, 3], **M. A. J. Michels**[2, 3], **A. K. Mukherjee**[4], and **R. Menon**[4]

[1] Kamerlingh Onnes Laboratory, P.O. Box 9504, 2300 RA Leiden, The Netherlands
[2] Dutch Polymer Institute, P.O. Box 902, 5600 AX Eindhoven, The Netherlands
[3] Eindhoven University of Technology, P.O. Box 513, 5600 MB Eindhoven, The Netherlands
[4] Department of Physics, Indian Institute of Science, Bangalore, 560012 India

Received 1 September 2003, accepted 3 September 2003
Published online 28 November 2003

PACS 71.30.+h, 72.80.Le, 77.22.Ch

The strong doping dependence of the dc conductivity in conjugated polymers below the metal–insulator transition (MIT), illustrated for PPV, is argued to be due to an increase in the density of states *and* the growth in size of the regions, in which the charge is delocalized. Obviously in the metallic states these regions have to overlap. Phase sensitive dielectric spectroscopy data show that when this overlap is achieved less than 1% of the donated charges (at least in polypyrrole) participates in metallic transport. The data up to the MIT allow a tentative picture of the density of states.

1 Introduction Chemical doping of conjugated polymers, like polyacetylene, polypyrrole (PPy) and poly(*p*-phenylene vinylene) (PPV) makes them conducting and even let them pass through a metal–insulator transition (MIT) to become metallic. Often the conduction processes below the metal–insulator transition are described by (variable range) hopping and above the MIT as Drude-like. Here we show that if one has to explain the strong doping-concentration dependence of the dc conductance below the MIT, the variable-range hopping-mechanism between point-like sites has to be modified to allow for the growing size of the localized regions and indicate how such a modification can be justified theoretically [1]. For a characterization of the metallic state close to the MIT we use dielectric data on PPy from 10^{-4} to 4 eV down to 4.2 K [2], to argue that only 1% of the carriers are delocalized. The subtle metallic features and the anomalies in carrier dynamics are modelled by coherent and incoherent transport between short conjugated segments. The data up to the MIT can be reconciled to a Gaussian density of states.

2 Experimental (OC₁C₁₀-PPV) has been doped in solution with $FeCl_3$ [1]. The doping of polypyrrole with PF_6 is done electrochemically [2, 3]. The dc-data are always taken in the ohmic regime. The dielectric scans up to 1 THz (4 meV) were made in transmission with phase sensitive equipment. Between 1 THz and the UV the samples were measured in reflectance [2].

3 Results and discussion Figure 1 shows the characteristic temperature (*T*) and concentration (*c*) dependence of the dc conductivity of $FeCl_3$ doped PPV. The conductivity (*σ*) has the familiar exponential dependence on $T^{-1/4}$ at low temperatures, characteristic for Mott variable range hopping [4] and at fixed *T* an 8 orders of magnitude increase with one order in concentration. To gain insight into the physics behind these features, we made the choice to stay within the framework of variable range hopping, while

---

* Corresponding author: e-mail: brom@phys.leidenuniv.nl, Phone: +31 71 5275425, Fax: +31 71 5275404

allowing for a finite (c-dependent) volume V_0 of the transport sites (i.e. delocalized regions), and an energy dependence of the density of states $g(E)$. The squared wave function outside V_0 decays within a (localization) length $1/\alpha$. For a carrier at the Fermi level E_F the hopping probability depends exponentially on the hopping distance R between the delocalized regions and the activation energy ($E - E_F$). The density of final sites N, which can be reached with activation energy less than $E - E_F$, is given by $N(E,E_F) = \int g(E') \, dE'$, where the integration is made from E_F to E. Mott's criterion states that the conductivity is governed by those hops for which E and R are such that about one state lies within a volume V, defined by $VN \approx 1$, which makes R a unique function of V and the radius A of V_0 (for simplicity V_0 will be assumed to be spherical). The conductivity can now be written as $\sigma = \sigma_0 \exp[-\alpha R - \beta(E-E_F)]$ with $\beta = 1/k_B T$ and σ_0 a prefactor. By performing some mathematics [1] an explicit expression for the c dependence of the conductivity can be derived, which at sufficiently low temperatures can be simplified to $\sigma = \sigma_0 \exp[\alpha A(c)] \exp[-(T_0(c)/T)^{1/4}]$, Mott's variable-range-hopping formula with the important addition of the factor $\exp[\alpha A(c)]$. The density of states can be calculated from the value of $k_B T_0 \propto 1/g$ [5] or from numerical fits (see Fig. 1) based on the equations obtained with the model. The values for $g(E)$ agree closely and are shown in Fig. 2. At a given temperature the growth of the delocalized regions (the radius A increases by a factor 4 in the measured concentration range) and the increase in the density of states appear to be equally important for the increase in conductivity.

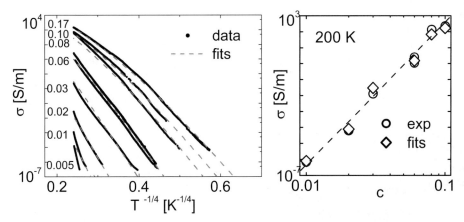

Fig. 1 DC-conductivity (σ) versus $T^{-1/4}$ (left) for various c and c-dependence of σ at 200 K (right) in FeCl$_3$ doped PPV. The concentration levels are expressed per monomer. The fits shown by the dashed lines (left) and diamonds (right) are made with the model discussed in the text. The dashed line in the right figure corresponds to $\sigma \propto c^8$ [1].

Theoretically [6, 7] and experimentally [8] arguments have been given for the density of states to be Gaussian. Although the data are consistent with such a profile, see Fig. 2, the range is clearly insufficient. At present we are studying PPV in an electrochemically gated transistor [9], where by changing the gate voltage the doping level could be varied by two orders of magnitude, and the density of states is directly obtained as function of energy. Preliminary results [10] confirm the Gaussian fit presented in Fig. 2 (in view of these new results, the previous published values of $g(E)$ [2] have been adjusted by a factor of 3).

By further increasing the doping level the delocalized regions will begin to overlap and we expect to reach the metal–insulator transition. We leave aside the interesting questions around such a quantum phase transition [11] and concentrate on the number of charges, which are really delocalized in the metallic state. For an answer we investigated the metallic state of polypyrrole by reflectance measurements up to 4 eV. As an extension of previous studies of other groups [12–15] phase sensitive transmission

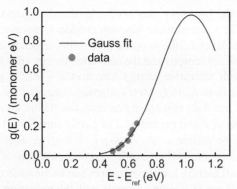

Fig. 2 Reconstructed density of states per monomer as function of energy. The Gaussian fit is mainly inspired by new (preliminary) data obtained in an electrochemically-gated transistor [10].

measurements (below 4 meV) complemented these data and appeared to be essential in the Kramers-Kronig analysis. The new data show convincingly that more than 99% of the charges remain localized (responsible for the features above 0.1 eV). The remaining (1% or less) charges contribute to the transport by hopping and tunnelling, see Fig. 3, which leads to the low energy structure and the negative dielectric constant. The optical data are well described by a model that includes both coherent tunnelling and incoherent hopping between delocalized regions. The unusually long values of the scattering time found previously [16] and low density of delocalized charges stem from the small coherent tunnel rate [2].

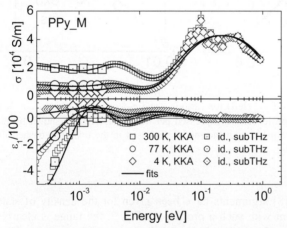

Fig. 3 Dielectric data on the most metallic polypyrrole sample (PPy_M). The subTHz (below 4 meV) phase sensitive measurements, shown here for three temperatures, have been performed down to 4.2 K [2]. The reflection data above 0.01 eV are temperature independent [12] (data shown are taken at room temperature; KKA stands for Kramers-Kronig analysis).

Acknowledgements   This work forms part of the research program of the Dutch Polymer Institute and is also financially supported by FOM-NWO. We like to acknowledge Paul Blom (RUG) and Eric Meulenkamp (Philips) for useful discussions about the density of states.

References

[1] H. C. F. Martens, I. N. Hulea, I. Romijn, H. B. Brom, W. F. Pasveer, and M. A. J. Michels, Phys. Rev. B **67**, 121203(R) (2003).

[2] I. G. Romijn, H. J. Hupkes, H. C. F. Martens, H. B. Brom, A. K. Mukherjee, and R. Menon, Phys. Rev. Lett. **90**, 176602 (2003).

[3] C. O. Yoon et al., Phys. Rev. B **49**, 10851 (1994).

[4] N. F. Mott, Philos. Mag. **19**, 835 (1969).

[5] H. Böttger and V. V. Bryksin, Hopping Conduction in Solids (Akademie Verlag, Berlin, 1993).

[6] H. Bässler, Phys. stat. sol. (b) **175**, 15 (1993).

[7] S. D. Baranovskii, H. Cordes, F. Hensel, and G. Leising, Phys. Rev. B **62**, 7934 (2000).

[8] H. C. F. Martens, P. W. M. Blom, and H. F. M. Schoo, Phys. Rev. B **61**, 7489 (2000).

[9] E. A. Meulenkamp, J. Phys. Chem. B **103**, 7831 (1999); A. J. Roest, J. J. Kelly, D. Vanmaekelbergh, and E. A. Meulenkamp, Phys. Rev. Lett. **89**, 036801 (2002).

[10] I. N. Hulea et al., unpublished.

[11] S. Sachdev, Quantum Phase Transitions (Cambridge University Press, Cambridge 1999).

[12] R. S. Kohlman et al., Phys. Rev. Lett. **74**, 773 (1995); **77**, 2766 (1996); **78**, 3915 (1997).

[13] K. Lee et al., Phys. Rev. B **52**, 4779 (1995), Adv. Mater. **10**, 456 (1998).

[14] A. B. Kaiser, Adv. Mater. **13**, 927 (2001), Rep. Prog. Phys. **64**, 1 (2001).

[15] V. N. Prigodin and A. J. Epstein, Synth. Met. **125**, 43 (2002).

[16] H. C. F. Martens et al., Phys. Rev. B **63**, 073203 (2001); **64** R201102 (2001); **65**, 079901 (2002).

phys. stat. sol. (c) **1**, No. 1, 148–151 (2004) / **DOI** 10.1002/pssc.200303611

Inter-chain and intra-chain hopping transport in conducting polymers

S. Capaccioli [*, 1], **M. Lucchesi**[1], **D. Prevosto**[1], and **P. A. Rolla**[1]

[1] INFM and Department of Physics, University of Pisa, via F.Buonarroti 2, 56127 Pisa, Italy

Received 2 September 2003, revised 24 September 2003, accepted 24 September 2003
Published online 28 November 2003

PACS 71.55.Jv, 72.20.Ee, 72.80.Le, 73.61.Ph, 77.22 Gm

The d.c. conductivity σ_{dc} and the a.c. electric response of poly(3-decylpyrrole) prepared with different dopants and synthesis conditions were studied over a broad temperature (75–300 K) and frequency (100 Hz–40 MHz) range, both in bulk state (films) and in solution. Concerning films, the temperature dependence of σ_{dc} followed Mott's law, with the hopping parameters strongly doping dependent. A.c. conductivity σ_{ac} was strictly coupled to d.c. transport, so that a master curve resulted from a plot of normalised complex permittivity versus a critical frequency proportional to σ_{dc}. Similar results to the bulk case were observed for non-dilute solutions, whereas for very dilute solutions σ_{dc} scaled with temperature as the reciprocal of solvent viscosity. Moreover, a steeper bilogarithmic slope of σ_{ac} versus frequency was found. Inter-chain charge transport in dilute solutions can be attributed to dissociated counterions diffusing through the solvent, whereas intra-chain hopping can eventually contribute only at high frequencies.

1 Introduction

The conductivity of conjugated polymers is increased by doping, which introduces topological defects (polarons and bipolarons) along the polymer chains [1]. The overall system is kept neutral by counterions, pinned to the macromolecular charged defects. Charge transport occurs in these systems by carriers both moving along the macromolecular chains (intra-molecular conduction) and hopping between different chains (inter-molecular conduction) [1]. Recent results evidenced the occurrence in many systems of "transverse" bipolarons, enhancing the inter-chain transport, with a transition from 1-D (anisotropic) to isotropic 3-D hopping [2]. Strong packing and disorder of polymer chains, typical of most solid samples, cannot allow to distinguish between inter- and intra-chain processes: their electrical properties are common to many disordered materials, like amorphous semiconductors [3]. Anyway, keeping macromolecules apart, as it occurs for very dilute solutions, could be a good method to obtain non-interacting chains, as Nowak and co-workers showed [4]. So the behaviour of intra-chain hopping would finally emerge. This goal led us to analyse d.c. and a.c. electrical conductivity of poly(3n-decylpyrrole) (P3DP), soluble in weakly polar solvents due to its long alkylic chain, thus allowing to prepare solutions and thin films easily. Electrical properties of diluted solutions and bulk films have been measured and compared.

2 Experimental

P3DP films were obtained via the galvanostatic electrochemical oxidation of the 3-decylpyrrole dissolved in propylene carbonate with different galvanic current densities I_g and different doping electrolytes [3]. Very diluted solutions of chemically polymerised P3DP (doped Cl$^-$) at different polymer weight concentrations (C = 0.01–1.0 %wt) were prepared in tetrachloromethane CCl$_4$, a non polar liquid

---

* Corresponding author: e-mail: capacci@df.unipi.it, Phone: +39 050 2214322, Fax: +39 050 2214333

with low static permittivity to produce a weak electrostatic interactions between the dissolved molecules and the solvent. Particle size in solutions was assessed by QELS (Quasi Elastic Light Scattering) and SANS (Small Angle Neutron Scattering): chain clusters formed cylinders of radius (1.5 ± 0.2) nm and length (22 ± 2) nm, corresponding to aggregates of chains of nearly 50 monomers. The percolation threshold for sticks of such aspect ratio is expected to be C = 0.2% in weight. The complex admittance of a parallel golden plate capacitor, filled with disk shaped films or with P3DP/CCl$_4$ solutions, was measured in the temperature (77–320 K) and frequency range (100 Hz–40 MHz) using HP4194A Impedance Analyzer.

3 Results and discussion

3.1 Electrical response of films

The temperature dependence of the d.c. conductivity for all the solid samples (Fig. 1a) was well described by the Mott's law (variable range hopping VRH model [5, 6]):

$$\sigma_{dc}(T) = \frac{A}{T^b} \exp\left(-\left(\frac{T_0}{T}\right)^{\gamma}\right) \qquad (1)$$

with the hopping parameters strongly dependent on dopants and on synthesis conditions: in particular, $\gamma = 0.25$ (typical for 3-D VRH) for all the samples, apart from $\gamma = 0.5$ for toulensulfonate (TsO$^-$) doped samples. The value $\gamma = 0.5$ is often associated to 3-D hopping in systems with parabolic density of state near E_F or to 1-D hopping with constant $N(E_F)$. The latter sounds more probable, as the huge size of (TsO$^-$) may hinder the charge transfer between adjacent chains, producing 1-D hopping. T_0 depends on the localisation and density of the states: for the most conducting sample (BF4$^-$ doped, $I_g = 0.1$ mA/cm^2) T_0 was $5 \cdot 10^6$ K, whereas for the less conducting sample (TsO$^-$ doped, $I_g = 0.05$ mA/cm^2) T_0 was $4.7 \cdot 10^4$ K.

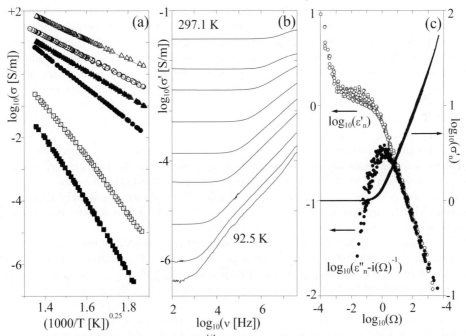

Fig. 1 (a) D.c. conductivity vs. $(1000/T)^{1/4}$ of different P3DP films (solid and open symbols are related respectively to $I_g = 0.05$ and 0.1 mA/cm^2; squares, triangles and circles are related to TsO$^-$, BF$_4^-$, ClO$_4^-$ dopants; lines are from VRH fits); (b) Real part of conductivity for P3DP(TsO$^-$) vs. frequency for T = (297.1 K; 247.4 K; 219.0 K; 185.7 K; 157.6 K; 138.7 K; 115.8 K; 99.6 K; 92.5 K) from upper to lower spectrum; (c) Bilogarithmic plot of normalised permittivity (real and imaginary part, left y-axis) and conductivity (real part, right y-axis) vs. normalised frequency $\Omega = \omega/\omega_c$ for P3DP(TsO$^-$) (same temperatures as in (b)).

Concerning the response to electrical alternating field, Fig. 1b shows that conductivity increases with increasing frequency, from a constant value equal to d.c. conductivity to a region where its frequency dependence is power-law like. Moreover, real and imaginary part of permittivity (after subtracting σ_{dc} contribution) reveal spectra associated to a relaxation process (loss peak) to be ascribed to the displacement of hopping charges. Dielectric strength of this relaxation $\Delta\varepsilon = \varepsilon'(\omega)-\varepsilon_\infty$ is associated to such displacements. In disordered systems dominated by hopping and where the transport occurs among more than two localised states during an electric field period, there is a critical frequency ω_c, proportional to the critical hopping rate and then to σ_{dc}, often located near the maximum of loss peak, defining an universal scale for frequency $\Omega = \omega/\omega_c$ and $\omega_c \propto \sigma_{dc}/(\varepsilon_0\Delta\varepsilon)$ [3, 6, 7]. If we plot conductivity and permittivity, normalised to their dc.values ($\varepsilon_n = \varepsilon(\omega)/\Delta\varepsilon$ and $\sigma_n = \sigma(\omega)/\sigma_{dc}$), versus Ω, the electrical responses at different temperatures collapse on a master curve (Fig. 1c), with the following asymptotic limits:

$$\Omega \ll 1, \sigma_n(\Omega) = 1 + i\alpha_1\Omega + \alpha_2\Omega^2; \ \Omega \gg 1, \sigma_n(\Omega) \cong 1 + K_1\Omega^{s'} + iK_2\Omega^{s''} \tag{2}$$

the exponents s' and s'' of the power law valid at high Ω are both smaller than 1 (s' in Fig. 1c is ranging between 0.5 and 0.7), decreasing with T and slightly increasing with Ω.

3.2 Electrical response of solutions

The a.c. electric response of P3DP solutions was very similar to that measured in the bulk: Fig. 2a shows the validity of the master curve after the scaling procedure with ω_c. The regime of power law is evident ($0.72 < s' < 0.78$) and even the loss peak could be found after subtracting d.c. contribution to ε'', but it was located at lower Ω (~0.1) than in the bulk case. The exponent s' decreased with T, as predicted by a number of hopping models [6, 7], and it was slightly bigger than in the bulk. For concentration below the threshold of $C_0 = 0.2\%$wt., a reduction of contribution of inter-chain hopping was expected, and consequently a dramatic reduction of σ_{dc}. On the contrary, even below C_0, a d.c. conduction was present, as it can be noted in Fig. 2b and in Fig. 2c. In particular, it was no evidence of percolation threshold at C_0 (inset Fig. 2c). On the other hand, the exponent s' was independent of C ($s' \sim 0.7$) for a wide range of high concentration, but it had a strong increase just near C_0 tending to one in the very dilute regime. This quasi-linear increase of $\sigma(\omega)$ causes a frequency independent imaginary permittivity (so-called "flat losses"), typical for conducting impurities in insulating matrix. More information can be obtained by studying the temperature dependence of conductivity (Fig. 2b) both in the dilute (C = 0.13%wt) and concentrated (C = 0.66%) regime. The d.c. conductivity of both systems had an Arrhenius temperature dependence $\sigma_{dc} = \sigma_0 \exp{(E_{act}/k_BT)}$. The activation energy was $E_{act} = 48$ meV for the dilute solution (C = 0.13%wt) and $E_{act} = 63$ meV for C = 0.66%. Moreover, when the solvent crystallised (at T = 247 K), the conductivity of dilute solution vanished, becoming lower than the background conductivity due to the stray contribution of the cell spacers (measured with the cell filled by the pure solvent). High frequency a.c. conductivity slightly increased with increasing T and it was only weakly affected by solvent solidification. This anomalous behaviour can be explained by taking into account two different microscopic mechanisms active to transport charge in dilute solutions.

(i) At low frequency, i.e. for long range charge transport (investigated in the electrical field period), long compared to the dimension of the chain aggregates dissolved in the solution, hops between localised sites are very unlikely. The hopping process, when it occurs, is mainly between nearest neighbours. In this regime, ionic conductivity becomes important: in fact, there is a finite probability of dissociation for counterions and polycations and the diffusion of ionic charge carriers bridges charge transport between separate chains or clusters. As ionic mobility is inversely proportional to solvent viscosity, it is not strange that E_{act} for dilute solution was 48 meV, very close to the CCl_4 viscosity activation energy (45 meV). Same explanation holds also for the vanishing of σ_{dc} when the solvent was frozen. For C=0.66% E_{act} was bigger than in the dilute regime, but it was comparable with values measured in the bulk state of P3DP (60–200 meV) [3]. Summarising, in this regime charge transport occurs by means of hopping transitions or, alternatively, by ionic diffusion through the solvent.

(ii) At high frequency and very short lengths, the transport occurs only by hopping in the single chain or in a small cluster and it can involve even a number of sites, but located into the cluster. For this reason high frequency conductivity is weakly affected by temperature. It is noteworthy that the behaviour of the exponent s′, slightly decreasing with temperature and tending to 1 at low concentration, has been reported in several systems where conducting impurities are embedded in an insulating matrix, like carbon nanotubes in polymers, or blends of insulating and conducting polymers [8] and it was explained in the framework of hopping theories. In particular, Böttger and Bryksin [6] evaluated an expression for the ac hopping conductivity for 1-D system, where $\sigma(\omega) \propto (j\omega)^s$ with exponent $s = \alpha d/(\alpha d+1)$, where α^{-1} was the localisation length and d the intersite separation. Thus, choosing $\alpha d \gg 1$, an exponent $s \sim 1$ is expected.

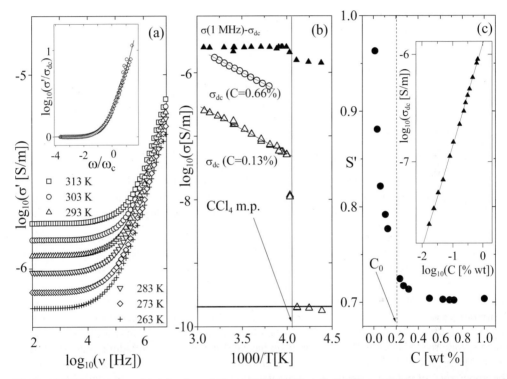

Fig.2 a) Real part of conductivity for solution of P3DP(Cl⁻) at C=0.66% and (inset) its master curve after rescaling according to critical frequency and σ_{dc}; b) d.c. (open symbols) and a.c. conductivity (solid symbols) at concentrations below (triangle) and above (circle) percolation threshold: solid lines are Arrhenius fitting curves, vertical line indicates solvent melting point, horizontal one the background stray conductivity; c) exponent s′ of the power law of a.c. conductivity vs. concentration (dotted vertical line shows percolation threshold) at fixed temperature (T=278 K) and (inset) the linear trend of d.c. conductivity vs. concentration (bilogarithmic slope 0.95±0.02).

References

[1] *Handbook of conducting polymers* (2nd edition), ed. by T.A. Skotheim et al., Dekker Inc., New York (1998).

[2] L. Zuppiroli, M.N. Bussac, S. Paschen, S. Chauvet, and L. Forro, Phys. Rev. B **50**, 5196 (1994).

[3] S. Capaccioli, M. Lucchesi, P.A. Rolla, and G. Ruggeri, J. Phys.: Condens.Matter **10**, 5595 (1998).

[4] M.J. Nowak, D. Spiegel, S. Hotta, A.J. Heeger, and P.A. Pincus, Synth. Met. **28**, 399 (1989).

[5] N.F. Mott and E.A. Davis, Electronic Processes in non crystalline materials, Clarendon Press, London (1971).

[6] H. Böttger and V.V. Bryskin,Hopping conduction in solids, VHC, Berlin (1985).

[7] A.R. Long, Hopping conductivity in the intermediate frequency regime, p. 201 in: "Hopping Transport in Solids", ed. by M. Pollak and Shlovskii, North Holland, Amsterdam (1991)

[8] R. Singh et al., J. Mater. Sci. **33**, 2067 (1998); S. Barrau et al., Macromolecules **36**, 5187 (2003); P. Dutta, S. Biswas, M. Ghosh, S.K. De, and S. Chatterjee, Synth. Met. **122**, 455 (2001).

phys. stat. sol. (c) **1**, No. 1, 152–155 (2004) / **DOI** 10.1002/pssc.200303618

Hopping polaron transport in disordered organic solids

I. I. Fishchuk[*, 1]**, A. Kadashchuk**[2]**, H. Bässler**[3]**, and S. Nešpůrek**[4]

[1] Institute for Nuclear Research, National Academy of Sciences of Ukraine, Prospect Nauky 47, 03680 Kyiv, Ukraine

[2] Institute of Physics, National Academy of Sciences of Ukraine, Prospect Nauky 46, 03028 Kyiv, Ukraine

[3] Institute of Physical, Nuclear and Macromolecular Chemistry and Material Science Center, Philipps-Universität Marburg, Hans-Meerwein-Strasse, 35032 Marburg, Germany

[4] Institute of Macromolecular Chemistry, Academy of Sciences of the Czech Republic, Heyrovský Sq. 2,162 06 Prague 6, Czech Republic

Received 1 September 2003, accepted 4 September 2003
Published online 28 November 2003

PACS 71.38.–k, 72.20.Jv, 72.80.Ng

An effective medium theory based on Marcus jump rate equation is formulated to describe nondispersive hopping polaron charge transport in a disordered organic material. The theory was found be in well agreement with relevant experiments and computer simulation data. It is shown that the Poole-Frenkel-type field dependence of mobility $\ln \mu \propto \sqrt{E}$ occurs when the energetic correlation effects have been taken into account. Our model suggests a possible test for distinguishing between polaron and polaron-free transport in disordered solids.

1 Introduction In the past decade, many results concerning the charge carrier transport in disordered organic materials were published and described by a formalism based on disorder elaborated by Bässler and co-workers [1, 2]. Recently the formalism has been extended by including intersite energetic correlations [3]. Despite of the success of the disorder formalism, there were extensive discussions concerning the importance of polaron effects for charge transport in organic disordered solids [4, 5]. Purely polaron models are less used because of their principle limitation related to the magnitude of physical parameters such as the polaron activation energy and transfer integral. The recent computer simulations by Parris et al. [4] have demonstrated that the problem related to physical parameters can be solved assuming the small-polaron transport occurs in the presence of correlated energetic disorder.

Recently we applied the effective medium approach (EMA) to describe charge-carrier transport in weakly disordered polaron-free three-dimensional (3D) organic media [6] as well as in disordered media containing traps [7]. In this work we present an analytical EMA theory to describe the charge transport in disordered materials where polaron formation takes place.

2 Nondispersive hopping transport

2.1 General theory A self-consistent theory based on the effective medium approach was recently formulated [7] to describe nondispersive charge carrier transport in disordered 3D organic systems for arbitrary electric fields $\boldsymbol{E} = \{E, \ 0, \ 0\}$ and has been applied to polaron-free materials. In general case, the effective hopping drift mobility μ_e can be obtained from the following expression

---

[*] Corresponding author: e-mail: ifishch@kinr.kiev.ua, Phone: +380 44 265 3969, Fax: +380 44 265 4463

$$\mu_{\rm e} = a \frac{W_{\rm e}^+ - W_{\rm e}^-}{E}, \tag{1}$$

where a is an average distance between neighboring hopping sites. $W_{\rm e}^+$ and $W_{\rm e}^-$ are the effective jump rates in the direction along and against the electric field, respectively. In two-site cluster approximation $W_{\rm e}^+$ and $W_{\rm e}^-$ can be determined [6] as the following formulas:

$$W_{\rm e}^+ = \frac{\left\langle \dfrac{W_{12}^+}{W_{12}^+ + W_{21}^-} \right\rangle}{\left\langle \dfrac{1}{W_{12}^+ + W_{21}^-} \right\rangle} \quad , \quad W_{\rm e}^- = \frac{\left\langle \dfrac{W_{21}^-}{W_{12}^+ + W_{21}^-} \right\rangle}{\left\langle \dfrac{1}{W_{12}^+ + W_{21}^-} \right\rangle}, \tag{2}$$

where W_{12}^+ and W_{21}^- are the jump rates between two adjacent sites along and opposite to the electric field direction, respectively. The angular brackets denote energetic configuration averaging. Here positional disorder was neglected.

For disordered materials with a weak electron-phonon coupling and/or low-temperature limit when polaron effects are unimportant, the Miller-Abrahams (M-A) jump rate [9] can be used

$$W_{\rm ij} = W_0 \exp\left[-\frac{|\varepsilon_{\rm j} - \varepsilon_{\rm i}| + (\varepsilon_{\rm j} - \varepsilon_{\rm i})}{2k_{\rm B}T} \right], \tag{3}$$

where W_0 is the prefactor, ε_k is the energy of site k.

Polaron effects might be important in the high-coupling and/or high-temperature limit. In such case, the nonadiabatic small polaron hopping given by Marcus jump rate [9] could be used

$$W_{\rm ij} = \frac{J^2}{\hbar} \sqrt{\frac{\pi}{4E_a k_{\rm B}T}} \; \exp\left(-\frac{E_a}{k_{\rm B}T}\right) \exp\left[\frac{\varepsilon_{\rm j} - \varepsilon_{\rm i}}{2k_{\rm B}T} - \frac{(\varepsilon_{\rm j} - \varepsilon_{\rm i})^2}{16E_a k_{\rm B}T} \right], \tag{4}$$

where J is the transfer integral and E_a is the small polaron activation energy.

Under the applied electric field in the Eqs. (3), (4), the value of $\varepsilon_{\rm j} - \varepsilon_{\rm i}$ should be substituted by $\varepsilon_{\rm j} - \varepsilon_{\rm i} + e(r_{\rm j} - r_{\rm i}) \; E$, where e is the unit charge, r_k is the spatial position of site k. Values $W_{\rm e}^+$ and $W_{\rm e}^-$ can be calculated by substituting Eqs. (3), (4) into Eq. (2) in the presence of electric field. To perform the configuration averaging in Eq. (2) for $W_{\rm e}^+$, one should choose the distribution functions for the starting state ε_1 and the target state ε_2. As we demonstrated before [6–8], in thermodynamic equilibrium, the target and starting states are described by the density-of-states (DOS) distribution $P(\varepsilon_2)$, and the asymptotic occupational density-of-states (ODOS) distribution of carriers after establishing of thermal equilibrium, $P(\varepsilon_1)$, respectively. Thus, for calculation of $W_{\rm e}^+$ we choose the Gaussian DOS distribution with the width σ for $P(\varepsilon_2)$ and shifted by $\varepsilon_0 = -\sigma^2/k_{\rm B}T$ the Gaussian ODOS distribution for $P(\varepsilon_1)$. To calculate $W_{\rm e}^-$ one should use ε_1 and ε_2 instead of ε_2 and ε_1, respectively, in the distribution functions.

2.2 Hopping transport in a polaron-free system First, for the sake of completeness, let us start with the M-A jump rate given by Eq. (3). For the considered case of $x = \sigma/k_{\rm B}T \gg 1$ we can use concept of the percolation energy level $\varepsilon_{\rm p}$ which, when only energetic disorder is considered, is located at

$\varepsilon_p \cong -\sigma/2$ [7]. Therefore, calculating W_e^+ one can put in Eq. (6) $P(\varepsilon_2) = \delta(\varepsilon_2 - \varepsilon_p)$. Performing calculations in the limiting case of zero-field $(E \to 0)$ one gets from Eq. (1) the relation

$$\mu_e = \mu_0 \, \exp\left[-\frac{1}{2}\left(\frac{\sigma}{k_B T}\right)^2\right],$$ (5)

where $\mu_0 = ea^2 W_0/k_B T$. Further, we will limit our consideration by the electric field range important for experiments, i.e. we assume $1/x \ll eaE/\sigma \ll x$. One could take into account the energetic correlation effects using the method described in Ref. [7]. Doing so, eaE/σ should be substituted by $\sqrt{x/2}\sqrt{eaE/\sigma}$ in obtained expression. Then one gets

$$\mu_e = \mu_0 \, \exp\left\{-\frac{1}{2}\left(\frac{\sigma}{k_B T}\right)^2 + \frac{1}{\sqrt{2}}\left[\left(\frac{\sigma}{k_B T}\right)^{3/2} - \left(\frac{\sigma}{k_B T}\right)^{1/2}\right]\sqrt{\frac{eaE}{\sigma}}\right\}.$$ (6)

The obtained Eq. (6) agrees well with both results of computer simulations [4] of charge transport in materials devoid of polaron effects and with available experimental data of the dependence of charge carrier mobility on electric field which obey the well-known Poole-Frenkel-type dependence $\ln \mu \propto \sqrt{E}$.

2.3 Hopping transport in a polaron system   Let us consider polaron transport using the Marcus jump rate given by Eq. (4) in disordered systems. Performing calculation under the same approximation as used above, in the limiting case of zero-field $(E \to 0)$ we get the relation

$$\mu_e = \mu_1 \, \exp\left[-\frac{E_a}{k_B T} - \frac{1}{8q^2}\left(\frac{\sigma}{k_B T}\right)^2\right],$$ (8)

where $\mu_1 = ea^2 W_1/2q k_B T$, $W_1 = (J^2/\hbar)\sqrt{\pi/4E_a k_B T}$, $q = \sqrt{1 - xy/8}$, $y = \sigma/E_a$.

In the case when $\sigma^2/8E_a k_B T \ll 1$ one can derive the apparent effective Arrhenius activation energy of the polaron mobility as

$$E_{eff} = -k_B\left[d \ln \mu_e /d(1/T)\right] = E_a + \frac{1}{4}\frac{\sigma^2}{k_B T}.$$ (9)

This equation differs somewhat from the conventional expression $E_{eff} = E_a + (8/9)\sigma^2/k_B T$ [1], which was used for estimating of the material parameters E_a and σ from the temperature dependence of the mobility. Generally, Eq. (8) derived from the Marcus jump rate seems to be the most appropriate approach to estimate material parameters E_a and σ. But in this case the value E_{eff} is not a linear function of $1/T$. Therefore, Eq. (9) is inappropriate for estimating the polaron energy E_a by extrapolating E_{eff} to the infinite temperature. For that we must use a more complicated method [10].

Now we will limit ourselves to the field range important for experiments. Calculation after taking into account the energetic correlation effects led to the following expression:

$$\mu_e = \mu_1 \, \exp\left\{-\frac{E_a}{k_B T} - \frac{1}{8q^2}\left(\frac{\sigma}{k_B T}\right)^2 + \frac{1}{2\sqrt{2}q^2}\left[\left(\frac{\sigma}{k_B T}\right)^{3/2} - \left(\frac{\sigma}{k_B T}\right)^{1/2}\right]\sqrt{\frac{eaE}{\sigma}}\right\}.$$ (10)

The obtained Eq. (10) agrees well with computer simulation results [5] and relevant experimental data on field dependence of polaron transport in disordered 3D organic solids, which corresponds to the Poole-Frenkel type of the dependence $\ln \mu \propto \sqrt{E}$.

3 Conclusion   As it follows from the presented results (Eq. (8)), observation of the temperature dependence $\ln \mu \propto 1/T^2$ for zero-field charge mobility does not necessary mean the absence of polaron formation. Therefore, generally speaking, the temperature dependence cannot be considered as a crucial test for the existence of polaron effects. Analysis of the present theory suggests an important test which, in principle, could be used for distinguishing between polaron and polaron-free transport. It appears that slopes of the electric field dependence of the mobility (Poole-Frenkel (PF) factors), defined as $\alpha_0 = \partial \ln\left(\mu_e / \mu_0\right) \big/ \partial \sqrt{E}$, varies almost linearly with $T^{-3/2}$ in the case of polaron-free transport when the M-A formalism is applicable. Alternatively, the dependence of the PF factor in representation $\alpha_1 = \partial \ln\left(\mu_e / \mu_1\right) \big/ \partial \sqrt{E}$ vs. $T^{-3/2}$ notably deviates significantly from a straight line in the case of polaron model. Thus, the presence of such deviation should imply the presence of polarons in the system.

Acknowledgements   This research was supported by NATO Grant No. PST.CLG 978952, by the project "Optodynamics" at the Philipps University, and by grants No. A1050901 and AV0Z4050913 from the Grant Agency of the Academy of Sciences of the Czech Republic.

References

[1] H. Bässler, phys. stat. sol. (b) **175**, 15 (1993) (and references therein).
[2] H. Bässler, in: Semiconducting Polymers: Chemistry, Physics and Engineering, eds. G. Hadziioannou and P. F. van Hutten (Wiley-VCH, Weinheim, 2000), p. 365.
[3] S. V. Novikov, D. H. Dunlap, V. M. Kenkre, P. E. Paris, and A. V. Vannikov, Phys. Rev. Lett. **81**, 4472 (1998).
[4] P. E. Parris, V. M. Kenkre, and D. H. Dunlap, Phys. Rev. Lett. **87**, 126601 (2001).
[5] I. I. Fishchuk, Philos. Mag. B **81**, 561 (2001).
[6] I. I. Fishchuk, D. Hertel, H. Bässler, and A. Kadashchuk, Phys. Rev. B **65**, 125201 (2002).
[7] I. I. Fishchuk, A. Kadashchuk, H. Bässler, and D. S. Weiss, Phys. Rev. B **66**, 205208 (2002).
[8] A. Miller and E. Abrahams, Phys. Rev. **120**, 745 (1960).
[9] R. A. Marcus, Rev. Mod. Phys. **65**, 599 (1993).
[10] I. I. Fishchuk, A. Kadashchuk, H. Bässler, and S. Nešpůrek, Phys. Rev. B **67**, 224303 (2003).

phys. stat. sol. (c) **1**, No. 1, 156–159 (2004) / **DOI** 10.1002/pssc.200303623

High conductivity of defect doped polymers in metal–polymer–metal systems

A. N. Ionov[*, 1], **V. M. Svetlichnyi**[2], and **R. Rentzsch**[3]

[1] Ioffe Physico-Technical Institute, 194021 St. Petersburg, Russia
[2] Institute of Macromolecular Compounds, RAS, 199004 St. Petersburg, Russia
[3] Institut für Experimentalphysik, Freie Universität Berlin, 14195 Berlin, Germany

Received 1 September 2003, revised 22 September 2003, accepted 2 October 2003
Published online 28 November 2003

PACS 71.55.Jv, 73.40.Ns, 73.61.Ph, 74.45.+c

We have observed that amorphous films of polyimide PI (R-Oph$_2$O) and–co-polymer exhibit high conductivity in Metal–Polymer–Metal (M–P–M) structures and supercurrents in Superconductor–Polymer–Superconductor (S-P-S) structures if the polymer films were prepared at $T < 150$ °C. In the crystalline state, films of PI (R-Oph$_2$O) also become highly conductive in M–P–M structures and have supercurrents in S–P–S structures. We consider that the conductivity effect is connected in some way with the effect of polymer electrification.

1 Introduction It is well known that doped π-conjugated bulk polymers such as Polyacetylene show an electrical conductivity like that in a disordered solid [1]. The contributions to the electrical transport are intramolecular transport, intermolecular transport or between polymer fibers. At a small doping level the temperature dependence of conductivity corresponds to hopping transport where the activation energy depends on the doping concentration. At an intermediate doping level the transport obeys the variable range hopping laws at low temperatures similar to the behavior of doped semiconductors with an impurity concentration corresponding to the dielectric side of the metal–insulator transition, whereas at high doping levels the conductivity may be of metallic type with a maximum absolute value of conductivity up to 10^5 S/cm.

In regard to non-conjugated polymers there is the common opinion that all of them have good dielectric properties only at electric fields less than the breakdown field (E_c). However that is true for bulk polymers only. The first observation to our knowledge that undoped, non-conjugated thin polymer films have a high conductivity was published in [2]. In this paper polypropylene films of very high molecular weight and a small thickness of about 80 nm were studied. They were placed between a planar electrode and a needle under pressure, required for a good electrical contact. The effect of high conductivity at $E \ll E_c$ was observed at separate points of the surface. Later, the high-conductivity effect was also observed in thin films of atactic polypropylene [3, 4], in polyimide films prepared by Langmuir–Blodgett technique [5] and in poly(3,3′–phthalidylidene-4,4′-biphenylene)} [5]. There, it was also confirmed that the conductivity occurs through many channels orientated perpendicular to the polymer surfaces and connecting the two metallic electrodes. This fact explained the strong anisotropy of conductivity and it was the essential difference from doped conjugated polymers. It was also shown that the cross-section of the conductive channels grows with increasing current. The smallest observed value of the cross-section was less than 10^{-10} cm^2. The estimated value of the polymer conductivity was higher than 10^{12} S/cm at liquid helium temperatures. Moreover, if the electrodes undergo a transition into the superconducting state, a supercurrent flows through the polymer film [4, 6, 7]. In a small magnetic field, when H was

---

[*] Corresponding author: e-mail: ionov@tuch.ioffe.rssi.ru

applied perpendicularly to the current direction, Josephson voltage oscillations have been obtained at $I \geq I_c$ which behave like a great number of parallel-connected Josephson contacts. From the period of the $U(H)$ oscillations the distance between neighboring conducting channels was found to be 5–20 μm [8]. The characteristic scale of the oscillations increases with increasing current. This can be understood if the distance between the channels decreases with increasing number of parallel Josephson contacts with increasing of current as the cross-section of the channels increases. This behavior is opposite to that obtained in weak links of ordinary superconductors or high-temperature superconductors where the number of Josephson junctions and their dimensions do not depend on the current. The aim of the present paper is to find other polymer materials with abnormally high conductivity phenomena as well as to find mechanisms for understanding these phenomena.

2 Experimental proceedure   The polyimide [PI(R-Oph$_2$O)] - Poly-[4,4′-*bis*(4″-N-phenoxy)diphenyl]imide of 1,3-*bis*(3′,4-dicarboxyphenoxy)benzene (Fig. 1) – has been prepared by a two stage synthesis.

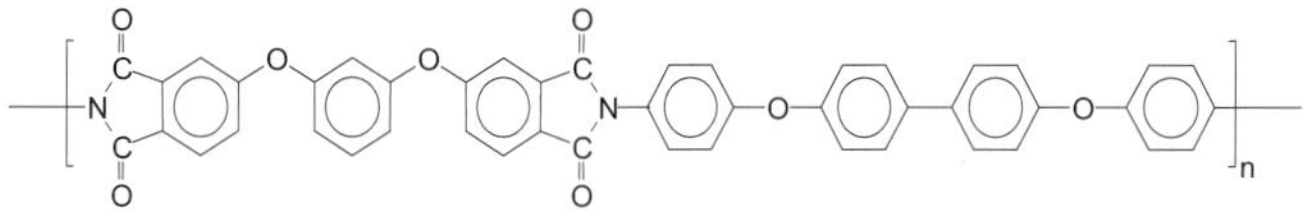

Fig. 1

At first synthesis step a polyimide precursor was obtained as a 20% mass solution in N-methyl-2-pyrrolidone. Then a droplet of 20% mass solution was diluted down to a 5% (weight) in N-methyl-2-pyrrolidone and deposited on electrodes. We used; i) polished metallic bulk electrodes (Nb, Sn); ii) gold stripes with width 1–2 mm and about 0.05μm thickness on a glass or polished silicon substrate_ for sandwich geometry; iii) gold stripes with the gaps in the range 0.8–2.8 μm on GaAs substrates for planar geometry. After a heat treatment at 80–100 °C for 1 hour; the major part of the organic solvent was removed and polymer films with a thickness of 0.6–1.3 μm were obtained. X-ray investigation showed that in this case PI (R-Oph$_2$O) is in the amorphous state, i.e. the orientation of macromolecules is chaotic. In a second synthesis step additional heat treatment up to 320 °C gave rise to imidized polyimide precursors. In this case at heating PI (R-Oph$_2$O) from 80 to 280 °C at a chosen heating speed, a partially crystalline state was produced [9]. The degree of crystallinity as obtained from X-ray investigations was about 30–40%. Further heating from 280 to 320 °C resulted again in transformation to the amorphous state.

2.   Copolymer [co-poly[4,4′-*bis*(4″-N-phenoxy) diphenyl-sulfone–α,ω-*bis* (γ-amino propyl)oligodimethylsiloxane]imide of 1,3-*bis*(3′,4-dicarboxyphenoxy)benzene] (Fig. 2) has been synthesized in accordance with the method of a single-stage high temperature imidization in solution in N-methylpyrrolidone - toluene mixture which was described in [10].

The degree of imidization controlled by IR spectroscopy was not less than 97 %. The structure of copolyimide is shown in figure 2:

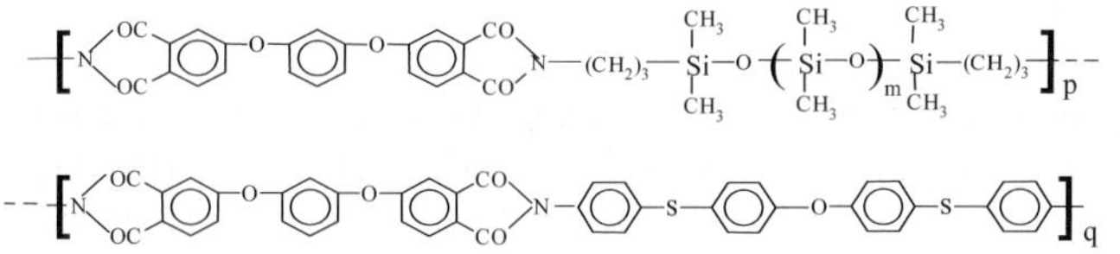

Fig. 2

where $m = 18$, $p = q = 40$.

The films were prepared by deposition of a droplet of 5 % wt. solution of copolyimide in N-methyl-2-pyrrolidone on electrodes (Nb, Sn). The films were heated in air for 1 h at a constant temperature of 370–400 K to remove the major part of the polymer solvent.

The thickness of polymer films in cases (1) and (2) were determined by an interference microscope MII-4.

3 Experimental results   The surfaces of the polymers were analyzed by an atomic force microscope (AFM). Figure 3 presents the AFM topography of partially crystalline PI (R-Oph$_2$O). As seen from the figure the structure is inhomogeneous with many pinholes. The density of pinholes is about 10^8 cm^2. Here the darker grey tone corresponds to deeper points of the polymer film. It is also seen from the figure that the pinhole direction is at some angle with the polymer surface. The inner size of isolated craters obtained by AFM is in the range 50–500 nm. AFM images of amorphous PI (R-Oph$_2$O) films obtained on heating to 320 °C indicate them to be more homogeneous, without pinholes.

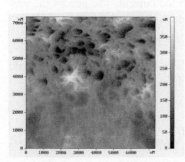

Fig. 3   AFM image of partially crystalline PI(R-Oph$_2$O).

The width of the valleys is only about 1–2μm and their depth is about 10–15 nm. Similar surface topographs were obtained for amorphous PI (R-Oph$_2$O) and copolymer heated in a range not above 80–100 °C.

4 Conductivity   For the usual reasons the metallic conductivity in both polymer structures should be absent. Firstly there are no carriers for current in polymer chains. Secondly even in the case of conjugated polymers and after high doping; hopping conductivity is observed. However, we observed a metallic conductivity when films prepared under the specified procedure (see above) were placed between metallic electrodes (Fig. 4).

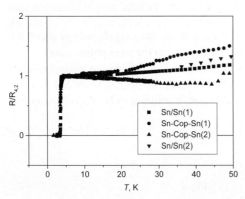

Fig. 4 Typical temperature dependence of the normalized resistance in a Metal–Polymer–Metal sandwich structure with a copolymer thickness about 1 μm and in a Metal / Metal structures.

The effect of conductivity _ disappeared above 180–190 °C for PI (R-Oph$_2$O) when the polymer is in an amorphous state, then it_ appears again in a temperature range of 270–290 °C, just at the transition from the amorphous to partially crystalline state. Above 300–320 °C the conductivity again disappears at the transition from the crystalline to the amorphous state.

For copolymer the conductivity effect exists in the amorphous state up to 190–210 °C only. We have done mass-spectroscopy analyses and found that it was in the temperature range 180–210 °C that intensive evaporation process of rest solvent was started. From this fact one may conclude that organic solvents took an active part in forming of conducting channels. However, for PI (R-Oph$_2$O) at crystalline state organic solvent cannot play any role for forming conducting states because of almost full evaporation of the solvent at 280 °C. A common reason should exist which can explain conducting state in both polymers.

Our experiment showed that in the crystalline state the conductivity effect is very sensitive to an external pressure. For example at relatively small pressure the resistance can be some tens of Ohms at room temperature. In this case there is no supercurrent in Superconductor–Polymer–Superconductor (S–P–S) sandwich structures. Moreover the resistance can increase with decreasing temperature as _was observed in [11]. However, with increasing pressure the resistance increases down to 10^{-1}–10^{-3} Ω. In this case the supercurrent in S–P–S structures was observed, which can be explained by proximity effect. The absolute value of the external pressure is too small to change the intrinsic structure of macromolecules or supramolecular arrangement. We consider that the main role of pressure is to change the contact area between polymer and free electrode touching the polymer surface. The effect of pressure was more significant for a polymer in the crystalline state as we consider due to greater roughness of the surface. In this case due to the electrification effect [12] the charge density inside polymer film accepted from electrodes by pinhole's traps changes as a consequence of changing the contact area. We did not observe conductivity when one of electrodes had square about 30 nm^2 for film thickness of about one micron.

As mentioned above the conductivity effect depends on the annealing temperature. In the amorphous state one of the reasons may be the presence of rest solvent which possibly modifies polymer chains and produces centers for trapping of electrons or holes from the metallic electrodes. We will call such polymer chains "defect polymer chains". We consider that in this case there is a critical density of accepted charge from electrodes which excites macromolecules and as a result some self-organization of macromolecules is produced which leads to the high conductivity effect in some region from one electrode to another at room temperature. When the temperature of annealing is increased, the polymer channel loses the ability to accept charge from electrodes at the same film thickness and the conductivity effect disappears.

In conclusion apart from doped conjugated polymers where hopping conductivity is observed we found that thin films of some polymers placed between metallic electrodes have metallic conductivity up to high temperatures. There is supercurrent through the polymer film if electrodes become superconductors. We believe that effect of "electrification" (charging of the polymer by metallic electrodes in an electrical circuit) plays a significant role in forming a metallic state in polymer channels.

Acknowledgements This work was supported in part by the program "Low-Dimensional Quantum Structures".

References

[1] A. Heeger, S. Kivelson, J. Schrieffer, and W. Su, Rev. Mod. Phys. **60**, 781 (1988).

[2] N.S. Enikopopyan, S.G. Grusdeva, N.M. Galashina, E.I. Shkljarova, and L.N. Grigorov, Dokl. AN USSR **283**, N6, 1404 (1985).

[3] N.S. Enikolopyan, L.N. Grigorov, and S.G.Smirnova, Pis'ma Zh. Eksper. Teor. Fiz. **49**, 326 (1989) [JETP Lett. **49**, 371 (1989)]

[4] V.M. Arkhangorodskii, A.N. Ionov, V.M. Tuchkevich, and I.S. Shlimak, Pis'ma Zh. Eksper. Teor. Fiz. **51**, 56 (1990) [JETP Lett. **51**, 67 (1990)].

[5] O.A. Scaldin, A.Yu. Zherebov, A.N. Lachinov, A.N. Chuvyrov, and V.A. Delev, Pis'ma Zh. Eksper. Teor. Fiz. **51**, 141 (1990).

[6] A.N. Ionov, V.A. Zakrevskii, and I.M. Lazebnik, Pis'ma Zh. Tekh. Fiz. **25**, 36 (1999) [Tech. Phys. Lett. **25**, 691 (1999)].

[7] A.N. Ionov and V.A. Zakrevskii, Pis'ma Zh. Tekh. Fiz. **26**, 34 (2000) [Tech. Phys. Lett. **26**, 910 (2000)].

[8] A.N. Ionov, A.N. Lachinov, and R. Rentzsch, Pis'ma Zh. Tekh. Fiz. **28**, 69 (2002) [Tech. Phys. Lett. **28**, 608 (2002)].

[9] V.E. Yudin, V.M. Svetlichnyi, G.N. Gubanova, A.L. Didenko, T.E. Sukhanova, V.V. Kudriavtsev, S. Ratner, and G. Marom, J. Appl. Polymer Sci. **83**, 2873 (2002).

[10] J. Yilgor and B.C Johnson, Amer. Chem. oc. Polymer Preprints **27**, 54 (1986).

[11] A.M. El'yashevich, A.N. Ionov, M.M. Rivkin, and V.M. Tuchkevich, Fiz. Tverd. Tela **34**, 3457 (1992) [Sov. Phys. Solid. State **34**, 1850 (1993)].

[12] J. Lowell and A.C. Rose-Innes, Adv. Phys. **29**, 947 (1980).

phys. stat. sol. (c) **1**, No. 1, 160–163 (2004) / **DOI** 10.1002/pssc.200303630

Hopping transport of charge carriers in nanocomposite materials

Sergey V. Novikov

A. N. Frumkin Institute of Electrochemistry, Leninsky prosp. 31, 119071 Moscow, Russia

Received 1 September 2003, revised 3 October 2003, accepted 6 October 2003
Published online 28 November 2003

PACS 72.20.Ee, 72.80.Le, 73.61.Ph

Charge transport in polymers containing J-aggregates of cyanine dyes serving as fast transport channels is considered. Shape of photocurrent transient depends on initial spatial distribution of carriers. If carriers are homogeneously generated at the electrode surface, then at the initial stage current raises reflecting carrier collection to fast channels and corresponding rate constant is determined by channel density.

1 Introduction   Under certain conditions organic polymers (polyimides and others) doped with cyanine dyes demonstrate formation of nanosized dye aggregates in the bulk of the polymer matrix [1, 2]. These so called J-aggregates [3] have spatially ordered structure and usually contain many molecules of a dye. Resulting composite polymeric materials demonstrate intense electroluminescence with narrow emission bands (attributed to J-aggregates) that could be exploited in electronic organic devices [2].

There are reliable experimental indications that J-aggregates should serve as traps for electrons and holes, and yet hopping mobilities of both kinds of carriers are much greater (by order of magnitude or even more) in nanocomposite material in comparison with undoped polymer [2, 4]; polymer matrices doped with non-aggregated dyes do not demonstrate increase in carrier mobility. This observation suggests that in relatively thin (with thickness $l \simeq 100-200$ nm) transport layers aggregates provide channels connecting the opposite electrodes so carriers could travel from one electrode to another without need to enter into the polymer matrix itself. Motion of carriers inside quasi-crystalline channels with reduced energetic disorder naturally explains higher mobilities of carriers in composite materials. Recently formation of sparse mesh of very long gently curved fibrils (with length 200–300 nm and longer) consisting of individual threads of J-aggregates with thickness of about 2.3 nm or, at larger concentration, of bundles of interwoven threads (with diameter about 10 nm) has been directly visualized in aqueous solutions of cyanine dyes [5]. Typical mesh scale is about 100 nm or greater.

Thus, it seems that the simplest yet reasonable model of the composite material should be an isolated quasi-crystalline cylindrical channel with radius b that connects two electrodes and is surrounded by disordered polymer matrix. We used the lattice model of dipolar glass [6] to describe disordered polar polymer serving as a source of residual energetic disorder $U(r)$ inside the channel. Sites of the channel carry no dipole moment and all energy fluctuations at these sites are due to contributions of the surrounding dipolar matrix (see Fig. 1). In the case of non-zero moment, aggregate's molecules still do not contribute to the energetic disorder and provide only spatially periodic addition to energy landscape. To model difference in positions of energy levels in the dipolar matrix and J-aggregate a constant value Δ was subtracted from the energy of every site inside the channel. Every basic cell

---

* e-mail: novikov@gagarinclub.ru, Phone: +7 095 952 24 28, Fax: +7 095 952 08 46

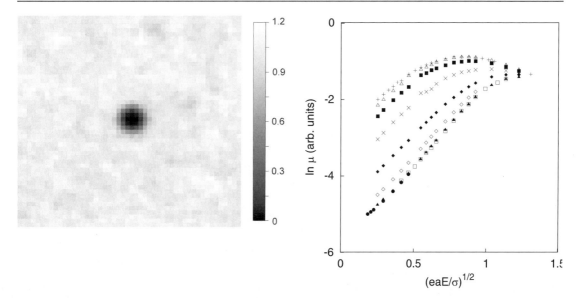

Fig. 1 Fast channel with $b = 3a$ in dipolar matrix. Distribution of variance $\sigma_l^2 = \langle U^2 \rangle_l / \sigma^2$ across a cross-section of the basic lattice sample is shown (average is calculated along lines parallel to channel direction), here σ^2 is energy variance in undoped dipolar glass; typically, $\sigma \simeq 0.1$ eV. Minimal value of σ_l^2 is about 0.1.

Fig. 2 Mobility field dependence for MC simulation, $\sigma/kT = 3.9$: dipolar glass ($\square$); composite with $b = 3a$ for different values of Δ/kT: 33.3 (+), 16.7 ($\triangle$), 10 ($\blacksquare$), 8.3 ($\times$), 6.7 ($\blacklozenge$), 5 ($\diamond$), 0 ($\bullet$), correspondingly; dipolar glass with "channel" arranged by setting traps with $\Delta/kT = 16.7$ in cylinder domain ($\blacktriangle$).

with linear size L contains one channel, so the surface density of channels is $1/L^2$. In our simulations we used $L = 50a$, where $a \approx 1$ nm is lattice scale.

2 Monte Carlo simulation of TOF currents

Monte Carlo (MC) simulation of time-of-flight (TOF) current transients demonstrates that in nanocomposite material carrier mobility μ becomes greater and for weak electric field E the usual Poole-Frenkel dependence

$$\mu \propto \exp\left(\gamma \sqrt{E}\right) \tag{1}$$

takes place with coefficient γ having the same magnitude as in the case of undoped matrix (see Fig. 2). Simulation reveals a characteristic feature of the transient in the case of homogeneous carrier generation at the electrode surface: for small time current raises with time (after initial sharp decrease, associated with energetic relaxation) which reflects collection of carriers to fast transport channels (see Fig. 3).

Simulation data suggest that two conditions have to be fulfilled for realization of fast transport: first, aggregates should have quasi-crystalline structure that provides reduced energetic disorder inside channels; second, transport energy levels in aggregates have to be located at lower energies in comparison to the transport levels in surrounding matrix, thus serving as effective traps for carriers. Effective trapping is absolutely necessary for realization of fast transport, thus Δ should be large enough (see Figs. 2 and 3, curves for $\Delta/kT = 5$ and $\Delta/kT = 16.7$). Without trapping carriers quickly leave fast channels and get trapped by deeper energetic wells in the surrounding disordered matrix, effective carrier collection does not take place, and photocurrent transient has an ordinary form (Fig. 3, curve for $\Delta/kT = 5$). In materials discussed in [2, 4] Δ is large enough ($\Delta \simeq 0.5$ eV), while preliminary experimental data indicate that formation of J-aggregates does not change mobility of holes in composite materials with small Δ [7].

If carriers are generated in channels, then, naturally, initial raise of the current does not take place. Yet trapping of carriers in channels is still a necessary condition for fast transport: if Δ is too small, carriers quickly diffuse out of channels. In that case initial sharp decrease of current becomes pro-

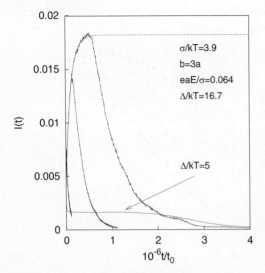

Fig. 3 Current transients for transport layers with different thickness: 500, 5,000, and 20,000 lattice planes, correspondingly; $1/t_0$ is the hopping frequency. Broken line shows the fit of the transient for 20,000 planes for Eq. (7). The bottom line is the transient for $\Delta/kT = 5$ and 5,000 planes.

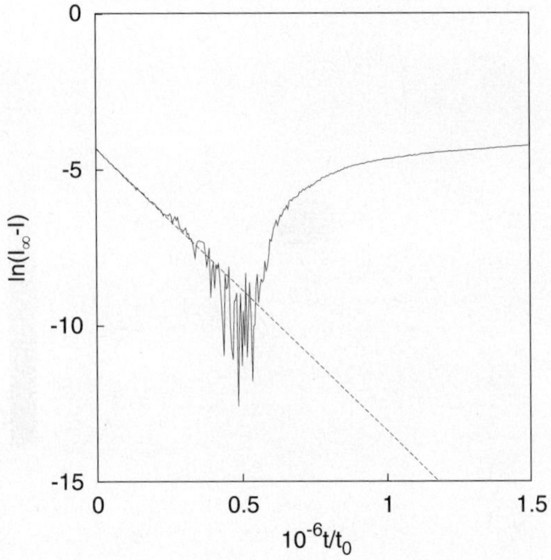

Fig. 4 Fitting of the transient (Fig. 3, 20,000 planes) for Eq. (7). Best fit (broken line): $k_0 = 9.04 \times 10^{-6} t_0^{-1}$.

longed reflecting spatial relaxation of carriers. Here fast transport could be realized only in very thin transport layers, in thicker layers mobility curve $\mu(E)$ does not differ from the corresponding curve for the undoped material.

3 Current relaxation at the initial stage Initial current raise in the case of homogeneous photogeneration may be reasonably well described by a simple model that considers the carrier motion as a superposition of the drift along field direction and 2D transverse diffusion with diffusion coefficient D to the channel serving as a sink (for simplicity we consider cylinder domain with radius $R \approx L/2$)

$$\frac{\partial c}{\partial t} = \frac{D}{r} \frac{\partial}{\partial r} \left(r \frac{\partial c}{\partial r} \right) \tag{2}$$

with boundary conditions $c(b,t) = 0$, $\left. \frac{\partial c}{\partial r} \right|_{r=R} = 0$, and initial condition $c(r,0) = c_0$, $R > r > b$. Usual procedure gives solution with Bessel functions

$$c(r,t) = \sum_{n=0}^{\infty} A_n \exp(-k_n t) \left[J_0(\lambda_n r) - \frac{J_0(\lambda_n b)}{Y_0(\lambda_n b)} Y_0(\lambda_n r) \right], \qquad \lambda_n = \sqrt{k_n/D}, \tag{3}$$

with eigenvalues λ_n satisfying equation

$$Y_0(z) J_1(\zeta z) - J_0(z) Y_1(\zeta z) = 0, \qquad z = \lambda b, \qquad \zeta = R/b. \tag{4}$$

For $\zeta \gg 1$, the only case where the model of isolated transport channel may be applied, the smallest root of Eq. (4) is

$$z_0 \approx \frac{1}{\zeta} \sqrt{\frac{2}{\ln \zeta - \gamma_E}}, \tag{5}$$

where $\gamma_E = 0.5772\ldots$ is Euler constant, and for large n an asymptotic equation is valid

$$z_n \approx \frac{\pi}{\zeta - 1} \left(n + \frac{1}{2} \right), \qquad n \gg 1. \tag{6}$$

In fact, Eq. (5) has accuracy of 5% even for $\zeta = 3$, and Eq. (6) has the same accuracy even for $n = 1$.

If time is not too small, then only the first term (with smallest eigenvalue) in the sum (3) is important (if $\zeta \gg 1$, then $k_1/k_0 \gg 1$), so total number of carriers in the matrix is $N_m(t) \propto N_m^0 \exp(-k_0 t)$ and for current we have

$$I(t) \propto \left(N_c^0 + N_m^0\right) \mu_c - N_m^0(\mu_c - \mu_m) \exp(-k_0 t), \tag{7}$$

here N_c^0 and N_m^0 are initial numbers of carriers in channels and matrix, and μ_c and μ_m are their mobilities, correspondingly. Analyzing $I(t)$, we could find k_0 (see Figs. 3 and 4).

This scenario suggests that field dependence of k_0 should reflect that of D and the ratio k_0/D does not depend on E. Computer simulation carried out in close resemblance to that described in [6] proved that this is indeed the case: for $\sigma/kT = 3.9$ and $R/b = 8.33$ $k_0 a^2/D \approx 1.6 \times 10^{-3}$, while Eq. (5) gives $k_0 a^2/D \approx 2 \times 10^{-3}$. Typically, k_0^{fit} and k_0^{calc} differ by factor of 2–3; we cannot expect better agreement with our crude approach. If N_c^0/N_m^0 and μ_c/μ_m can be estimated, then k_0 could be calculated from the TOF data for thin layers, where current raises linearly with time $I(t) \propto N_m^0 \mu_m + N_c^0 \mu_c + N_m^0(\mu_c - \mu_m)k_0 t$. Exponential term in Eq. (7) develops only if transit time for channels t_{TOF} is large enough, namely $k_0 t_{\text{TOF}} 1$, this leads to a simple estimation for transport layer thickness

$$l \frac{\mu_c}{\mu_m} \gtrsim \frac{eER^2}{kT}. \tag{8}$$

For $\mu_c/\mu_m = 10$ and $E = 10^5$ V/cm at room temperature we have $l \gtrsim 1.5\mu$ for $R = 20$ nm.

The most important result of this consideration is the possibility to estimate typical distance $2R$ between conducting channels knowing k_0 and D. Simulation shows that D may be reasonably well estimated by the Einstein equation $D = kT\mu/e$ from the TOF data. It would be extremely interesting to compare TOF data for channel density with direct visualization of the composite layer surface with the use of AFM: how many visible channels are in fact conducting? Then we could estimate typical current density in a channel and consider an important problem of thermal stability of nanocomposite devices.

4 Conclusion Extensive MC simulation of charge carrier transport in nanocomposite polymer materials has been carried out. Data show that the key feature of the TOF transients in materials with well isolated channels for the case of homogeneous (or, even better, out of channels) carrier generation is an initial raise of photocurrent which reflects collection of carriers to fast channels. At this stage current temporal dependence follows simple exponential law, and the rate constant provides direct information about surface density of conducting channels. Experimental observation of this particular behavior could serve as a reliable test of the model. Proper initial spatial distribution of carriers may be achieved by choosing a suitable photoexcitation wavelength.

Acknowledgements Partial finacial support from the ISTC grant 2207 and RFBR grants 02-03-33052 and 03-03-33067 is appreciated. I am greatly indebted to A. V. Vannikov, E. I. Mal'tsev, A. D. Grishina, and A. R. Tameev for numerous discussions. This work was partially supported by the University of New Mexico and utilized the UNM-Alliance LosLobos Supercluster at the Albuquerque High Performance Computing Center.

References

[1] F. C. Spano and S. Mukamel, Phys. Rev. A **40**, 5783 (1989).
[2] E. I. Mal tsev, D. A. Lypenko, B. I. Shapiro, M. A. Brusentseva, G. H. W. Milburn, J. Wright, A. Hendriksen, V. I. Berendyaev, B. V. Kotov, and A. V. Vannikov, Appl. Phys. Lett. **75**, 1896 (1999).
[3] E. E. Jelley, Nature **138**, 1009 (1936).
[4] E. I. Maltsev, D. A. Lypenko, V. V. Bobinkin, B. I. Shapiro, A. R. Tameev, J. Wright, V. I. Berendyaev, B. V. Kotov, and A. V. Vannikov, Mol. Cryst. and Liq. Cryst. **361**, 217 (2001).
[5] H. von Berlepsch and C. Böttcher, J. Phys. Chem. B **106**, 3146 (2002).
[6] S. V. Novikov, D. H. Dunlap, V. M. Kenkre, P. E. Parris, and A. V. Vannikov, Phys. Rev. Lett. **81**, 4472 (1998).
[7] A. D. Grishina and A. R. Tameev, private communication.

phys. stat. sol. (c) **1**, No. 1, 164–167 (2003) / **DOI** 10.1002/pssc.200303631

Temperature and field dependence of the mobility in 1D for a Gaussian density of states

W.F. Pasveer*, **P.A. Bobbert,** and **M.A.J. Michels**

Group Polymer Physics, Eindhoven Polymer Laboratories and Dutch Polymer Institute,
Technische Universiteit Eindhoven, P.O. Box 513, 5600 MB Eindhoven, The Netherlands

Received 1 September 2003, accepted 3 September 2003
Published online 28 November 2003

PACS 72.10.-d,72.80.Le.

The temperature and field-dependent mobility of a charge carrier in a gaussian density of states has been
analyzed, based on a numerically exact solution of the Master equation. In this way we get a microscopic
insight into the origin of the mobility and find some new features pointing to relevance of the Fermi level
and of variable-range hopping to sites further away than nearest ones.

1 Introduction It is by now a well-established fact that the mobility in many disordered organic solids,
such as conjugated polymers, deviates from normal Arrhenius-type behaviour. Modelling in the case of
Gaussian disorder gave $\mu \sim \exp[-(\sigma/kT)^2]$, while $\mu \sim \exp[\sqrt{E}]$ was found for the dependence on electric
field[1]; here parameters were chosen such that the results can effectively be treated as nearest neighbor
hopping (NN). In the present letter we want to investigate the influence of hopping to more distant sites
(see also[2]) and we want to examine the dependence of the mobility on Fermi level and electric field. For
the sake of simplicity we focus on 1D systems only; the results can be extended to higher dimensions,
but in recent years systems with effective 1D hopping such as columnar discotic liquid-crystalline glasses
also turn out to be of interest, due to their potential for technological applications in i.e. light-emitting
diodes.[3, 4, 5]

2 Theory and methodology The incoherent motion of a charge carrier for which doubly occupied
electronic states are not allowed can be described by the Master equation.

$$\frac{\partial n_i}{\partial t} = -\sum_{j \neq i}[W_{ij}\, n_i\, (1 - n_j) - W_{ji}\, n_j\, (1 - n_i)], \tag{1}$$

where n_i is the occupational density of site i with position vector $\mathbf{R}_i$ and energy ε_i, and W_{ij} is the transition
rate hops from site i to j. For the case of interest, W_{ij} is suggested by Miller and Abrahams[6] as

$$W_{ij} = \begin{cases} \nu_0 \exp(-\alpha \mid \mathbf{R}_{ij} \mid) \exp[-(\varepsilon_j - \varepsilon_i)/kT], & \varepsilon_j > \varepsilon_i \\ \nu_0 \exp(-\alpha \mid \mathbf{R}_{ij} \mid), & \varepsilon_j < \varepsilon_i \end{cases} \tag{2}$$

Here ν_0 is an intrinsic rate and ε includes a contribution $-q\mathbf{E}\cdot\mathbf{R}_{ij}$ from the external field $\mathbf{E}$; q is the particle
charge. In the above-stated form the Master equation is non-linear and is difficult to handle. However,
following Butcher[7] we linearize with respect to $\mathbf{E}$. We substitute $n_i = n_i^0 + n_i'$ and $W_{ij} = W_{ij}^0 + W_{ij}'$,

* Corresponding author: e-mail: w.f.pasveer@tue.nl, Phone: +31 40 247 5688, Fax: +31 40 244 5253

with n_i^0 is the equilibrium Fermi-Dirac distribution and $W_{ij}' = (\partial W_{ij}/\partial E)_{E \to 0}$. Note that we resort to the stationary Master equation because the interest is in the steady-state mobility μ^*. In our numerical calculations, we first generate randomly distributed energies on each site of an array of lattice constant R_0 and size $N = 10000$, by sampling a Gaussian of width σ and zero mean. Charges are introduced by setting the chemical potential in the FD distribution. Then we apply an electric field and solve the linearized equation for $n_i'/[n_i^0 (1 - n_i^0)]$. This scaling with $n_i^0 (1 - n_i^0)$ turned out to be very important to cover the low-temperature regime. In case we also want to study the electric- field dependence, we do not take into account the factors $(1 - n_i)$ and instead of the FD distribution we use a Boltzmann distribution. The mobility μ^* can be calculated from the current between all sites (i, j), either including the factors $(1 - n_i)$ or not:

$$\mu^* \mathbf{E} = \mathbf{v} = \frac{L R_0}{< n_i^0 >} \sum_i \sum_j [\mathbf{R}_{ij} W_{ij} \, n_i (1 - n_j) + \mathbf{R}_{ji} W_{ji} \, n_j (1 - n_i)]. \tag{3}$$

3 Numerical results

3.1 Nearest-neighbor vs. variable-range hopping

For convenience we introduce units $q = R_0 = \nu_0 = 1$, and for all calculations we take $\alpha = 1$. We consider a reduced energetic-disorder parameter σ/kT[1] with values between 1 and 8.

In Fig. 1 we show the results for the mobility as a function of temperature in the cases of VRH and of NNH only. We have taken an example with the Fermi level at $\varepsilon = 0$. It is clearly observed from Fig. 1 that

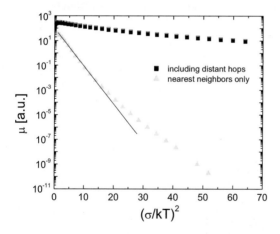

Fig. 1 Mobility as a function of temperature for NNH only and for VRH. Averaging over different disorder realizations is performed to get error bars smaller than the symbol size.

for only NNH in the regime $(\sigma/kT)^2 < 15$, as used in literature, we recover $\mu^* \sim \exp[-(\sigma/kT)^2]$, but for larger values of $(\sigma/kT)^2$ it deviates. However if we include VRH the magnitude of the mobility changes drastically. To obtain a detailed picture of what happens in the (R, ε) space, we write our simulation results in the form

$$\mu^*(\beta) = \int d\mathbf{R} d\varepsilon \mu(R, \varepsilon; \beta), \tag{4}$$

with the distribution $\mu(R, \varepsilon; \beta)$ following from the summand in (4). Fig. 2 shows a typical example of the distribution in 1D, $(\sigma/kT = 7.5)$; the figure has a mirror image for $\varepsilon < 0$, understandable from energy conservation. Fig. 2 clearly shows a peak at $(R_p, 0)$ giving the maximum contribution to the mobility.

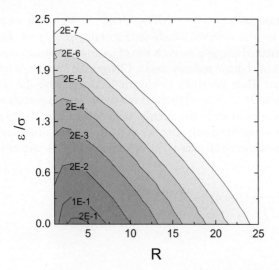

Fig. 2 Distribution of hops (R, ε) and their contribution to the mobility at a fixed temperature $\sigma/kT = 7.5$, where $\mu_{max} = 0.23$

From the position of the peak and the fact that there is a large non-negligible (R, ε) domain, determined by hopping distances much larger than R_p, we have to conclude that the influence of hopping to neighbors far beyond the nearest ones is very important at low temperatures.

3.2 Influence of the Fermi level   The results for mobility as a function of temperature for different positions of the Fermi level is shown in Fig. 3. First note that for all positions of ε_F we have $\mu^* \sim \sigma/kT$ in

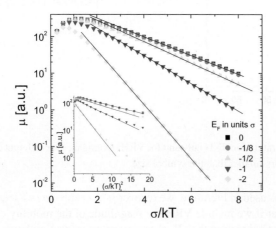

Fig. 3 Mobility as a function of temperature for different positions of the Fermi level. Averaging over different disorder realizations is performed to get error bars smaller than the symbol size.

the low-temperature regime. Second, for low ε_F the mobility drops down very quickly. Looking at the inset we see that the mobility plotted vs. $(\sigma/kT)^2$ initially follows a straight line, but for lower temperatures it starts to curve, as is also observed in the data shown in figure 11 of Hartenstein *et al.*[8]; after the curvature point the regime linear in σ/kT sets in.

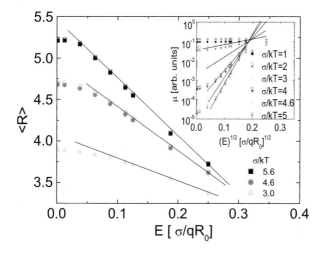

Fig. 4 Influence of electric field on $< R >$ for different values of σ/kT.

3.3 Field dependence A characteristic behavior for the field dependence of the mobility is shown in the inset of Fig. 4. The data are plotted vs. $\sqrt{E}$, the so-called Poole-Fenkel representation (PF). From this figure it is clear that that the mobility only shows PF behavior (i.e. a straight line when plotted vs. $\sqrt{E}$) in a limited range. In the low-field region the mobility tends to increase parabolically as expected[9], next we observe the PF behavior and finally the mobility will decrease as E^{-1} for all temperatures because the drift velocity saturates, a direct consequence of the Miller-Abrahams hopping rates. Having seen in the previous subsection that the effect of hopping to more distant sites than nearest ones play an important role at low temperatures, let us see the influence of the field on top. For this we calculate the field dependence of the average hopping distance, $< R >$, weighed with $\mu(R, \varepsilon; \beta)$ for some fixed values of σ/kT. The outcome is visualized in Fig. 4. One sees that with increasing E the hopping distance decreases. This is most pronounced for low temperatures. So again our conclusion must be, hops to distant neighbors have to be taken into account, because they certainly have a non-negligible contribution to the mobility, although VRH becomes less important at high fields.

Acknowledgements This work forms part of the research programme of the Dutch Polymer Institute.

References

[1] H. Bässler, Phys. Stat. Sol. (b) **175**, 15 (1993).
[2] K. Kohary et al., Phys. Rev. B **63**, 94202 (2001).
[3] W.R. Caseri, et al., Adv. Mat. **15**, 125 (2003).
[4] I. Bleyl, C. Erdelen, H.-W. Schmidt, D. Haarer, Philos. Mag. B **79**, 463 (1999).
[5] L.K. Gallos, B. Movaghar, and L.D.A. Siebbeles, Phys. Rev. B **67**, 165417 (2003).
[6] A. Miller and E. Abrahams, Phys. Rev. **120**, 745 (1960).
[7] P.N. Butcher, in *Linear and Nonlinear Electronic Transport in Solids*, edited by J.T. Devreese and V.E. Van Doren (Plenum, New York, 1976), 341/381.
[8] B. Hartenstein, H. Bässler, S. Heun, P. Borsenberger, M. Van der Auwerer, F.C. De Schryver, Chem. Phys. **191**, 321 (1995).
[9] S.V. Novikov, D.H. Dunlap, V.M. Kenkre, P.E. Parris, and A.V. Vannikov, Phys. Rev. Lett. **81**, 4472 (1998).

phys. stat. sol. (c) **1**, No. 1, 168–171 (2004) / **DOI** 10.1002/pssc.200303636

Concentration dependence of the hopping mobility in disordered organic solids

O. Rubel[1], **S. D. Baranovskii**[*1], **P. Thomas**[1], and **S. Yamasaki**[2]

[1] Department of Physics and Material Sciences Center, Philipps-University Marburg, D-35032 Marburg, Germany
[2] AIST (ASRC), AIST Tsukuba-Central 4, 1-1-1 Higashi, Tsukuba, Ibaraki 305-8562, Japan

Received 1 September 2003, accepted 3 September 2003
Published online 28 November 2003

PACS 72.20.Ee, 72.80.Le, 72.80.Ng

Traditionally the dependance of the drift mobility, μ, on the concentration of localized states, N, in disordered organic solids is plotted in the form $\mu \propto \exp[-C(N\alpha^3)^{-p}]$ with $p = 1/3$ and constant C. This representation cannot be correct, because transport in disordered organic solids is essentially a variable-range-hopping process with a weaker dependence $\mu(N)$. We study this dependence theoretically and show that both parameters p and C strongly depend on temperature and hence they are not universal. Only at very high temperatures the formula with $p = 1/3$ is valid. The result is significant in particular for a correct diagnostics of the localization length α from the measured dependence $\mu(N)$.

1 Introduction

Organic photoconductors, such as conjugated and molecularly doped polymers and organic glasses represent a rapidly evolving research area due to their current and potential applications in various electronic devices [1, 2, 3]. Transport properties of charge carriers attract most attention of the researchers. Correspondingly, the most studied physical quantity is the drift mobility, μ, of charge carriers. It has already become a tradition in the field of organic solids to represent the dependence of the drift mobility μ on the concentration of localized states N in the form

$$\mu \propto \exp[-C(N\alpha^3)^{-p}], \tag{1}$$

with constant C and $p = 1/3$ [4, 5, 6]. In this formula α is the localization length of carriers in the localized states. This representation, however, cannot be correct, because the hopping drift mobility in organic disordered solids evidences also a strong temperature dependence and hence the transport process is essentially a variable-range-hopping process (VRH). While the dependence described by Eq. (1) with constant C and $p = 1/3$ should be expected for the nearest-neighbour hopping (NNH) [7], the VRH should demonstrate a weaker concentration dependence and furthermore the exponent should also depend on temperature leading in the form of Eq. (1) to the temperature-dependent coefficient C. For example, in the case of a uniform energy distribution of localized states the concentration dependence of the the drift mobility is described by the famous Mott's law with $p = 1/4$ and $C \propto T^{-1/4}$ [7, 8].

Straightforward computer simulations have shown that the experimentally observed temperature dependence of the drift mobility [3], $\mu \propto \exp[-(T_0/T)^2]$, evidences the energy distribution of localized states

---

* Corresponding author: e-mail: baranovs@staff.uni-marburg.de, Phone: +49 6421 2825582, Fax: +49 6421 2827076

(DOS) in organic disordered media to be Gaussian [2]:

$$g(\epsilon) = \frac{N}{\sigma\sqrt{2\pi}} \exp\left(-\frac{\epsilon^2}{2\sigma^2}\right). \tag{2}$$

Here σ is the characteristic scale of the energy distribution.

It is our aim in this report to obtain the concentration dependence of the drift mobility for the VRH transport of charge carriers via localized states with a Gaussian energy distribution and a random spatial distribution. Using the theoretical approach based on the concept of the transport energy [9] we show below that, as expected for the VRH process, the dependence $\mu(N)$ is weaker than $\ln(\mu) \propto N^{-1/3}$ and, more important, it essentially depends on temperature. Eq. (1) with $p \approx 1/3$ appears only approximately valid for high temperatures and/or for extremely low concentrations of localized states.

2 Theory

The jump rate of a carrier from an occupied localized state i to an empty localized state j is a product of a prefactor ν_0, a carrier wave-function overlap factor, and a Boltzman factor for jumps upward in energy [10]:

$$\nu_{ij} = \nu_0 \exp\left(-\frac{2r_{ij}}{\alpha}\right) \exp\left(-\frac{\epsilon_j - \epsilon_i + |\epsilon_j - \epsilon_i|}{2kT}\right). \tag{3}$$

Here ϵ_i and ϵ_j are the energies of the localized states on sites i and j, respectively; r_{ij} is the distance between the sites; ν_0 is the attempt-to-escape frequency; k is the Boltzmann constant. The most appropriate way to calculate the drift mobility μ of charge carriers in a disordered system with an exponentially broad distribution of local transition probabilities given by Eq. (3) is to average the inverse hopping rates over the real transitions performed by a mobile charge carrier [9]. Hence the carrier drift mobility can be evaluated as

$$\mu \sim (e/kT)R^2(\epsilon_t)\langle t\rangle^{-1}, \tag{4}$$

where e denotes the elementary charge, $\langle t\rangle$ is the average hopping time and $R(\epsilon_t)$ is the typical length of hopping transitions to the transport energy, ϵ_t, from lower energy states determined by the concentration of localized states below ϵ_t. Neglecting the concentration dependence of the prefactor $(e/kT)R^2(\epsilon_t)$ in Eq. (4) compared to the exponential term in $\langle t\rangle$ one obtains for the drift mobility

$$\ln\left\{\mu/[eR^2(\epsilon_t)/kT]\right\} = -2\left[\frac{4\sqrt{\pi}}{3B}N\alpha^3 \int_{-\infty}^{x_t/\sqrt{2}} \exp(-t^2)\,dt\right]^{-1/3} - \frac{x_t}{kT/\sigma} - \frac{1}{2(kT/\sigma)^2}. \tag{5}$$

The numerical coefficient $B = 2.7$ is introduced into Eq. (5) in order to warrant the existence of an infinite percolation path over the states with energies below ϵ_t. The transport energy as the function of the temperature and the concentration is evaluated as a solution of the equation [9]

$$\exp\left(\frac{x^2}{2}\right)\left[\int_{-\infty}^{x/\sqrt{2}} \exp(-t^2)\,dt\right]^{4/3} = \frac{kT}{\sigma}\left(9\sqrt{2\pi}N\alpha^3\right)^{-1/3}. \tag{6}$$

It is Eq. (5) that determines the dependence of the carrier drift mobility on parameters kT/σ and $N\alpha^3$. From the above equations it is clear that only these two parameters are essential for the results.

In the present work we study the dependence of the drift mobility, μ, on $N\alpha^3$ keeping kT/σ as a parameter. Although Eq. (5) does not provide a pure power law for the dependence of $\ln(\mu)$ on $N\alpha^3$,

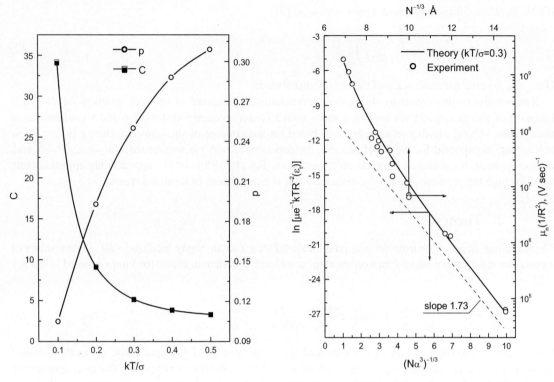

Fig. 1 Temperature dependence of the parameters C and p in Eq. (1).

Fig. 2 Concentration dependence of the drift mobility evaluated from Eq. (5) for $kT/\sigma = 0.3$ and observed in experiment [4].

we will always try to represent the concentration dependence of the carrier drift mobility in the form of Eq. (1) aiming at a comparison of our theoretical results with experimental data. According to numerous experimental studies of the temperature dependence of the drift mobility in various organic disordered solids, particularly in doped polymers, the width σ of the DOS energy distribution in such systems ranges from 0.065 to 0.150 eV depending on the kind of dopant, dopant concentration, and host polymers [2, 3]. The temperature ranges in the experimental studies from 200 K to 350 K [3]. Hence the values of the parameter kT/σ in real experiments range between 0.1 and 0.5.

In Fig. 1 we show the values of parameters p and C that provide the best fit for the solution of Eq. (5) in the form of Eq. (1). These data were calculated for $N\alpha^3$ in the range $0.001 < N\alpha^3 < 1.0$. It is well seen in Fig. 1 that for low enough temperatures the values of the parameter p are essentially less than $1/3$. This is to be expected for the VRH transport mechanism based on the interplay between the spatial and energy factors in the exponent of the transition probability given by Eq. (3) [7, 8]. Energy dependent terms in Eq. (3) diminish the effect of the spatially dependent terms in the expression for the transition probability and they lead to a weaker dependence of the carrier mobility on $N\alpha^3$ compared to the NNH regime with $p = 1/3$, in which only spatial terms determine the mobility of charge carriers. The values of p in Fig. 1 obtained for a Gaussian DOS differ from the value $p = 1/4$ in the Mott's law valid for a constant, energy-independent DOS. With rising temperature, the value of p increases and it becomes close to $1/3$ at $kT/\sigma \approx 0.5$. For higher temperatures one should expect the validity of the nearest-neighbour hopping regime with $p \approx 1/3$. The concentration dependence of the drift mobility in the form of Eq. (1) with parameters p and C given in Fig. 1 is the main result of our report.

3 Discussion

The data plotted in Fig. 1 raise the question: how was it possible to plot numerous experimental data in the form of Eq. (1) with $p = 1/3$ for kT/σ essentially smaller than 0.5? In order to answer this question, we plot in Fig. 2 the values of $\ln[\mu e^{-1} kTR^{-2}(\epsilon_t)]$, obtained by solving Eq. (5), and experimental data for the drift mobility in TNF/PVK [4] as a function of $N^{-1/3}$. In such plots, it is not possible to fit the data by a straight line in the whole range of concentrations. However, it is possible in a restricted concentration range and it has been done so far in all interpretations of experimental data in organic disordered media [4, 5]. One of the consequences of such plots was the apparent dependence of the localization length α on the concentration of impurities N [4, 5]. Our theoretical results in Fig. 2 demonstrate the same curvature of the $\ln[\mu e^{-1} kTR^{-2}(\epsilon_t)]$ versus $(N\alpha^3)^{-1/3}$ plots as the experimental data and they provide the fit for experimental data with a constant localization length α.

A strong temperature dependence of the slope parameter C, which is well seen in Fig. 1, evidences that the expression for the drift mobility μ cannot be factorized as $\mu(kT/\sigma, N\alpha^3) = \mu_0 \chi(kT/\sigma)\varphi(N\alpha^3)$. This is the essence of the VRH transport mechanism and it is in contrary to numerous statements in the scientific literature, which claim such a factorization possible (see, for example, [6]). Such a factorization is approximately possible only in the extreme case corresponding to the strong inequality $(N\alpha^3)^{-1/3} \gg \sigma/kT$ [7]. In such a case energy dependent terms in Eq. (3) (of the order σ/kT) do not contribute much to the jump probability compared to the spatially dependent terms (of the order $(N\alpha^3)^{-1/3}$). Under such circumstances the transport path of charge carriers is determined solely by geometrical factors. Such transport mechanism is known as the nearest-neighbor hopping. The drift mobility in the NNH is at best described by the percolation theory [7] and it is given by Eq. (1) with $p = 1/3$ and $C \approx 1.73$ [7]. Indeed a slope coefficient of our theoretical curve in Fig. 2 is very close to 1.73 for the limit of low concentrations $N\alpha^3$.

4 Conclusions

A theory for the dependence of charge carrier drift mobility in the hopping regime on the concentration of localized states is suggested for disordered organic solids. Theoretical consideration in the framework of a consistent concept of the transport energy in application to the VRH transport gives the dependence $\mu(N)$ in the form of Eq. (5). The solution of this equation can be approximately represented in the form of Eq. (1) with parameters p and C given in Fig. 1. The results of the theory agree with experimental data better than those of the simplistic approach with $p = 1/3$ and constant C.

Acknowledgements Financial support of the Deutsche Forschungsgemeinschaft and of the Founds der Chemischen Industrie is gratefully acknowledged. Autors also like to thank the Optodynamic Center of the Philipps-University Marburg and the European Graduate College "Electron-Electron Interactions in Solids".

References

[1] G. Hadziioannou and P. van Hutten, editors, *Semiconducting Polymers*, Wiley, Weinheim, 2000.
[2] H. Baessler, Phys. Status Solidi B **175**, 15 (1993).
[3] P. M. Borsenberger, E. H. Magin, M. van der Auweraer, and F. C. de Schryver, Phys. Status Solidi A **140**, 9 (1993).
[4] W. Gill, in *Proceedings of the Fifth International Conference on Amorphous and Liquid Semiconductors*, edited by J. Stuke and W. Brenig, page 901, London: Taylor and Francis, 1974.
[5] S. J. S. Lemus and J. Hirsch, Philos. Mag. **53**, 25 (1986).
[6] A. Nemeth-Buchin, C. Juhasz, V. Arkhipov, and H. Baessler, Philos. Mag. Lett. **74**, 295 (1996).
[7] B. I. Shklovskii and A. L. Efros, *Electronic Properties of Doped Semiconductors*, Springer, Heidelberg, 1984.
[8] N. F. Mott, J. Non-Crystal. Solids **1**, 1 (1968).
[9] S. D. Baranovskii, H. Cordes, F. Hensel, and G. Leising, Phys. Rev. B **62**, 7934 (2000).
[10] A. Miller and E. Abrahams, Phys. Rev. **120**, 745 (1960).

phys. stat. sol. (c) **1**, No. 1, 172–175 (2004) / **DOI** 10.1002/pssc.200303645

Nonlocal electron–phonon coupling: influence on the nature of polarons

V. M. Stojanović[*], **P. A. Bobbert**, and **M. A. J. Michels**

Group Polymer Physics, Eindhoven Polymer Laboratories and Dutch Polymer Institute, Technische Universiteit Eindhoven, P.O. Box 513, 5600 MB Eindhoven, The Netherlands

Received 1 September 2003, accepted 4 September 2003
Published online 28 November 2003

PACS 63.20.Kr, 71.38.−k

We present a variational approach to an extended Holstein model, comprising both local and nonlocal electron–phonon coupling. The approach is based on the minimization of a Bogoliubov bound to the free energy of the coupled electron-phonon system, and is implemented for a one-dimensional nearest-neighbor model, with Einstein phonons. The ambivalent character of nonlocal coupling, which both promotes and hinders transport, is clearly observed. A salient feature of our results is that the local and nonlocal couplings can compensate each other, leading to a supression of polaronic effects.

1 Introduction   A common starting point in the study of the electron (or exciton)-phonon (henceforth e–ph) interaction in non-polar materials is the Holstein Hamiltonian, which accounts for a dependence of the electronic on-site energies upon the lattice degrees of freedom (local e–ph coupling). The evidence is accumulating, yet, that an effective dependence of electronic transfer integrals upon phonon states, termed nonlocal e–ph coupling, could also be of importance in certain classes of organic materials, such as the molecular crystals of polyacenes and organic solid state excimers [1]. Given the fact that the relevant electronic and phononic energies in the polyacenes are comparable, the approaches that lean on a classical treatment of the lattice dynamics [2] seem to be out of place. In order to reveal at least qualitative features of both zero- and finite-temperature polaron structure, we set out to analyse an extended Holstein model, starting from a nonlocal generalization of the Lang-Firsov transformation. We exploit the method of temperature-dependent, variationally optimized canonical transformations [3], based upon the Bogoliubov bound to the free energy, and thereby generalize the work of Yarkony and Silbey [4], pertaining to local coupling only.

2 Model   The system under study consists of an excess electron (hole, exciton) interacting with non-polar optical phonons, and is described by an extended Holstein Hamiltonian, encompassing an electronic part, a phonon part, and the interaction (e–ph) part ($\hbar = 1$ in what follows):

$$H = \sum_{m,n} \varepsilon_{mn} a_m^\dagger a_n + \sum_q \omega_q (b_q^\dagger b_q + \tfrac{1}{2}) + \sum_{q,m,n} \omega_q g_{mn}^q (b_q + b_{-q}^\dagger) a_m^\dagger a_n \, . \tag{1}$$

Here $a_m^\dagger$ and $b_q^\dagger$ create an electron in the Wannier state at site m, and a phonon with wave vector q and frequency ω_q, respectively. ε_{mn} is a compact notation for the electronic on-site energies ($\varepsilon_{mm} \equiv \varepsilon$) and transfer integrals ($\varepsilon_{mn}, m \neq n$). Besides, we assume the form $g_{mn}^q = g_{mn}(\mathrm{e}^{-iq \cdot R_m + \mathrm{e}^{-iq \cdot R_n}})/2\sqrt{N}$, where g_{mn} depends only on $|R_m - R_n|$, and where N is the number of sites. This form obeys the

---

[*] Corresponding author: e-mail: v.s.stojanovic@tue.nl, Phone: +31 40 247 56 88, Fax: +31 40 244 52 53

property $(g_{mn}^q)^* = g_{nm}^{-q}$ required for the hermiticity of H_{e-ph}, and satisfies the translational symmetry. In order to account for both local and nonlocal e–ph coupling, we utilize a nonlocal unitary transformation $H \rightarrow \tilde{H} = e^S H e^{-S}$ of the displacement-operator type, the generator of which has the form

$$S = \sum_{m,n,q} A_{mn}^q (b_q^\dagger - b_{-q}) a_m^\dagger a_n .$$ (2)

The transformation parameters A_{mn}^q have to meet the same conditions as the q-dependent e–ph coupling constants, in order to satisfy the anti-hermiticity of S ($S^\dagger = -S$) and translational invariance. While, as can easily be shown, the latter property guarantees the invariance ($e^S \, P \, e^{-S} = P$) of the total crystal momentum $P = \sum_k k a_k^\dagger a_k + \sum_q q b_q^\dagger b_q$, we assume the form $A_{mn}^q = \lambda g_{mn}^q$, wherein λ is the as yet undetermined (implicitly temperature-dependent) variational parameter, representing a measure of the "phonon-dressing" of the excitation. The generator of our canonical transformation can thus be rewritten in the form $S = \sum_{m,n} C_{mn} a_m^\dagger a_n$, with $C_{mn} = \lambda \sum_q g_{mn}^q (b_q^\dagger - b_{-q})$. It yields the relations

$$e^S a_m e^{-S} = \sum_n (e^{-C})_{mn} a_n ; \qquad e^S b_q e^{-S} = b_q + \sum_{m,n} \left[(e^C b_q e^{-C})_{mn} - b_q \delta_{mn} \right] a_m^\dagger a_n .$$ (3)

In the dilute limit, upon introducing the notation $\tilde{f}_{mn} = (e^C f e^{-C})_{mn}$, the transformed Hamiltonian takes up the form $\tilde{H} = \sum_{m,n} \tilde{V}_{mn} a_m^\dagger a_n + \sum_q \omega_q (b_q^\dagger b_q + 1/2)$, where the use of the transformation tsf in the sense of a temperature-dependent optimal transformation[3] is necessitated by the presence of the phonon operators in $\tilde{V}_{mn}(\{b_q^\dagger; b_q\}) = \tilde{\varepsilon}_{mn} + \sum_q \omega_q (b_q^\dagger \tilde{g}^q + \tilde{g}^q \tilde{b}_{-q} + b_q^\dagger \tilde{b}_q - b_q^\dagger b_q)_{mn}$.

3 Free-energy minimization

If the Hamiltonian $H = H_0 + V$ is transformed by means of a unitary transformation $U = e^S$, so that $\tilde{H} = e^S H e^{-S} = \tilde{H}_0 + \tilde{V}$, its corresponding free energy $F = -\beta^{-1} \ln [\text{Tr} (e^{-\beta H})]$ obeys the inequality $F \leq -\beta^{-1} \ln \text{Tr} (e^{-\beta \tilde{H}_0}) + \langle \tilde{V} \rangle_{\tilde{H}_0}$. This is a form of the Bogoliubov inequality, suitable for an application in conjunction with the method of canonical transformations [3, 4]. We make use of the splitting $\tilde{H} = \tilde{H}_0 + \tilde{V}$, of our transformed Hamiltonian, with

$$\tilde{H}_0 = \sum_{m,n} \langle \tilde{V}_{mn} \rangle_{\text{ph}} a_m^\dagger a_n + \sum_q \omega_q (b_q^\dagger b_q + \tfrac{1}{2}); \qquad \tilde{V} = \sum_{m,n} (\tilde{V}_{mn} - \langle \tilde{V}_{mn} \rangle_{\text{ph}}) a_m^\dagger a_n ,$$ (4)

where $\langle \ldots \rangle_{\text{ph}}$ denotes a thermal phonon average. The non-interacting Hamiltonian $\tilde{H}_0$ can be rewritten as $\tilde{H}_0 = \sum_k \epsilon(k) a_k^\dagger a_k + \sum_q \omega_q (b_q^\dagger b_q + 1/2)$, with $\epsilon(k) = \epsilon + \langle \tilde{V}_{kk} \rangle_{\text{ph}}$, where the Fourier transform of $\langle \tilde{V}_{mn} \rangle_{\text{ph}}$ is given by $\langle \tilde{V}_{kk} \rangle_{\text{ph}} = \sum_m \langle \tilde{V}_{mn} \rangle_{\text{ph}} e^{ik \cdot (R_m - R_n)}$. Since the canonical transformation employed leaves the total momentum invariant, the Bogoliubov inequality can be applied to the contributions to the overall free energy coming from different eigenspaces of the total momentum. The corresponding expressions for the bounds to the free energy read [4] (with Z_{ph} being the free-phonon partition function)

$$F_K = -\beta^{-1} \ln Z_{\text{ph}} - \beta^{-1} \ln \left(e^{-\beta V} \right)_{mm} ,$$ (5)

where $\tilde{V}$ is the matrix with the elements $(\tilde{V})_{mn} \equiv \langle \tilde{V}_{mn} \rangle_{\text{ph}}$, depending on $|m - n|$ only. The thermal phonon average of $\tilde{V}_{mn}$ can be calculated approximately, at the level of a second-order cumulant expansion. In the Einstein (dispersionless) phonon limit with one phonon mode of frequency ω, the result reads $\langle \tilde{V}_{mm} \rangle_{\text{ph}} = \varepsilon + (\lambda^2 - 2\lambda) \Delta$, for $m = n$, and for $m \neq n$ one arrives at the expression

$$\langle \tilde{V}_{mn} \rangle_{\text{ph}} = \varepsilon_{mn} e^{-\lambda^2 U_{mn}} + \left[\frac{e^{-\lambda^2 U_{mn}} - 1}{U_{mn}} + \sqrt{\frac{\pi}{U_{mn}}} (\lambda - 1) \, \text{erf} \left(\lambda \sqrt{U_{mn}} \right) \right] \Delta_{mn} ,$$ (6)

wherein the auxiliary quantities Δ, Δ_{mn} and U_{mn} are defined as: $\Delta_{mn} \equiv \omega \sum_q (g^q g^{-q})_{mn}$, $\Delta_{mm} \equiv \Delta$, $U_{mn} \equiv (1/2 + N_T) \Gamma_{mn}$, $\Gamma_{mn} \equiv G_{mm} + G_{nn} - g_{mn}^2$, $G_{mm} \equiv g_{mm}^2 + 1/2 \sum_{k \neq m} g_{mk}^2$, with $N_T = (e^{\beta \omega} - 1)^{-1}$

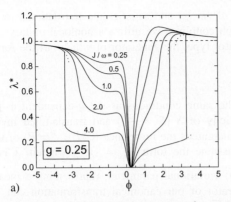
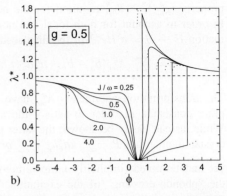

Fig. 1 Optimal dressing parameter λ^* as a function of the nonlocal coupling strength ϕ, at $T = 0K$, for different values of J/ω and local-coupling strengths a) $g = 0.25$ and b) $g = 0.5$. The dashed line indicates the limiting value $\lambda^* = 1$.

the thermally averaged phonon occupation number. In a one-dimensional nearest-neighbor model (with $\varepsilon_{m,m\pm1} \equiv J > 0$, $g_{m,m} \equiv g$, and $g_{m,m\pm1} \equiv \phi$), we define $\Gamma_{m,m\pm1} \equiv \Gamma_1$; $U_{m,m\pm1} \equiv U_1$; $\Delta_{m,m\pm1} \equiv \Delta_1$; $\tilde{V}_{m,m\pm1} \equiv \tilde{V}_1$; $R_m - R_{m\pm1} = \mp a$. It is straightforward to show that $\Gamma_1 = 2g^2 + \phi^2$, $\Delta = \omega(g^2 + \phi^2)$, and $\Delta_1 = \omega g\phi$. Upon minimizing the expression for the bound F_K to the free energy (involving the zeroth-order modified Bessel function $I_0(x)$)

$$F_K = \varepsilon + \left(\lambda^2 - 2\lambda\right)\Delta - \beta^{-1}\ln Z_{\mathrm{ph}} - \beta^{-1}\ln I_0(2\beta\langle\tilde{V}_1\rangle_{\mathrm{ph}}), \tag{7}$$

one obtains the self-consistency equation for the optimal value of the parameter λ at finite temperatures:

$$\lambda - 1 = \frac{1}{\Delta}\frac{I_1(2\beta\langle\tilde{V}_1\rangle_{\mathrm{ph}})}{I_0(2\beta\langle\tilde{V}_1\rangle_{\mathrm{ph}})}\frac{d}{d\lambda}(\langle\tilde{V}_1\rangle_{\mathrm{ph}}). \tag{8}$$

In the $\beta \to +\infty$ limit it goes over to the corresponding zero-temperature, bottom-of-the-band equation

$$\lambda - 1 = \sqrt{\frac{2\pi}{\Gamma_1}}\frac{\Delta_1}{\Delta}\,\mathrm{erf}\left(\sqrt{\frac{\Gamma_1}{2}}\lambda\right) - \left(\frac{J}{\Delta}\Gamma_1\lambda + 2\frac{\Delta_1}{\Delta}\right)e^{-\frac{1}{2}\Gamma_1\lambda^2}. \tag{9}$$

4 Results and discussion The Eq. (8) and (9) can be solved numerically for different choices of values of the parameters (J, ω, g, ϕ, T). It is judicious to expect that in the presence of both types of

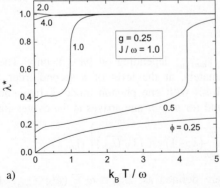
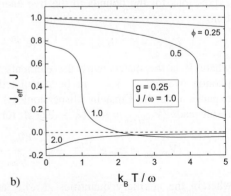

Fig. 2 Optimal values of λ and the effect on the renormalization of transfer integrals, for $J/\omega = 1.0$ and $g = 0.25$.

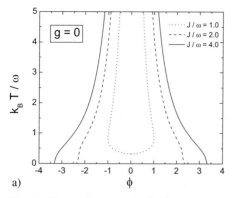 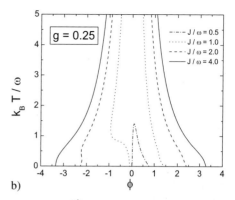

Fig. 3 Phase diagrams indicating regions of "strong dressing" ($\lambda^* > 0.5$) and "weak dressing" ($\lambda^* < 0.5$) in the $T - \phi$ plane, for different values of J/ω and local-coupling strengths a) $g = 0$ and b) $g = 0.25$.

couplings, the relative sign of the corresponding coupling constants may play a role. For definiteness, we fix the local coupling constant to a non-negative value ($g = 0, 0.25, 0.5$), and discuss the solutions of the two equations for both positive and negative signs of nonlocal coupling. By solving the zero-temperature Eq. (9), we clearly observe the non-monotonous dependence of the optimal value λ^* of the variational parameter on the nonlocal coupling strength, when both type of couplings are present (Figs. 1a, b). The striking feature of our observations is the existence of a "dressing minimum" at $\phi = g$. While the value of the relative coupling strength at which it appears, and the absolute value of the minimum, might be at least partly artifactual, the very existence of this minimum seems to be a robust characteristic, that can be ascribed to the ambivalent character of nonlocal coupling. The results obtained for the optimal value of λ and the corresponding effect on the renormalization $J_{\text{eff}}/J = \langle \tilde{V}_1 \rangle_{\text{ph}}/J$ of transfer integrals at finite temperatures are presented in Fig. 2a, b, for $g = 0.25$ and $J/\omega = 1.0$, respectively. It can be seen in Fig. 2b that the presence of strong nonlocal coupling can alter the sign of unrenormalized transfer integrals.

In Fig. 3a, b we depict "phase diagrams" in the $T - \phi$ plane, separating regions of "weak" ($\lambda^* < 0.5$) and "strong" ($\lambda^* > 0.5$) dressing, for $g = 0$ and $g = 0.25$, respectively. For any particular pair of phase boundary curves (i.e. for any fixed value of J/ω and the two possible signs of nonlocal-coupling strength) the region between them represents "weak dressing", and the outer region "strong dressing".

References

[1] M. Pope and C. E. Swenberg, Electronic Processes in Organic Crystals and Polymers, Second Edition, (Oxford University Press, New York-Oxford, 1999) and the references quoted therein.
[2] W. P. Su, J. R. Schrieffer, and A. J. Heeger, Phys. Rev. Lett. **42**, 1698 (1979).
[3] M. Wagner, Unitary Transformations in Solid State Physics (North-Holland, Amsterdam, 1986).
[4] D. Yarkony and R. Silbey, J. Chem. Phys. **65**, 1042 (1976).

Author Index

Information for conference organizers and guest editors

The third journal section *physica status solidi (c) – conferences and critical reviews* is devoted to the publication of proceedings, ranging from large international meetings to specialized workshops, as well as collections of topical reviews on various areas of current solid state physics research. The new series has been launched in December 2002 with volume **0** (2002/03). It is available both as a regular journal both online and in hardcover print volumes, to be delivered to subscribers, conference contributors and participants (upon arrangement with the organizers). Single copies of pss (c) may be ordered as a book using its ISBN number. Regular subscriptions to pss (c) are offered in combination with pss (a) and/or pss (b) .

Essential details concerning layout and organization of the new journal series are:

- pss (c) is published as a full hardcover-bound series, carrying a standard green-coloured cover design, individually adapted according to the organizers' request which includes conference designation, logo, names of Guest Editors etc.

- Proceedings issues contain all conference contributions which have been peer-reviewed and accepted by the Guest Editors. Upon special agreement between the pss journal editors and the Guest Editors, part of the conference papers may also be published simultaneously in an issue of pss (a) or (b). For all papers, strict criteria for journal publications, i.e. positive peer-review by independent referees, are obligatory. All papers are unambiguously citable as phys. stat. sol. (a), (b), or (c) journal articles and will be covered by standard reference databases.

- All articles are published online in PDF format at Wiley InterScience. Access for registered users (e. g. conference participants with special password) may be installed. The online version contains colour figures at no additional cost, regardless of their colour or black/white representation in print.

- The Editorial Office provides document templates and style files for Word and LaTeX, respectively, to be used by all authors, allowing an easy manuscript preparation and length estimate of their paper with respect to the page limits given by the organizers.

- The issue is completed by a table of contents in topical order, an author index, a preface, listings of conference committee members, organizers and sponsors, and any additional material, if desired.

- The usual service of the Editorial Office is available and includes support in the refereeing process, acceptance messages, PDF proofs (for typesetted papers), free PDF reprints (hardcopy reprints may be ordered) as well as individual communication with authors and organizers. The use of a Web-based software system for online submission and refereeing of papers is offered to Guest Editors.

- The editors of pss (c) aim at a timely, professional, and high-quality print and online publication of proceedings, typically within only four to six months after a conference.

- Various service packages for production are available, including either full typesetting of papers using electronic manuscript data or publication-ready delivery of manuscript files (prepared using the template/style files) by the organizers.

For further details as well as an individual offer for the publication of the proceedings of your forthcoming conference or of a special issue containing topical reviews, please contact the Editorial Office at pss@wiley-vch.de (for other contact information see the title page).